TOPICS IN CARBOCYCLIC CHEMISTRY

VOLUME 1

Topics in Carbocyclic Chemistry

Volume one

Edited by
Douglas Lloyd
University of St Andrews

Logos Press

Published by Logos Press Limited
in association with Elek Books Limited
2 All Saints Street London N.1

Library of Congress Catalog Card Number 74-80937
SBN 236.31132.8

Printed in Great Britain by
Spottiswoode, Ballantyne & Co. Ltd., London and Colchester

FOREWORD

Professor Wilson Baker, F.R.S.

Organic compounds are classified as aliphatic, carbocyclic, or heterocyclic, though in the very many cases where two or more such characteristic groupings are present, the classification chosen will depend on the relative chemical importance of these groupings to the particular investigation in hand, and perhaps even to the outlook of the investigator. Traditionally, however, ring compounds with attached aliphatic groups are referred to as cyclic, and any heterocyclic grouping serves to categorise a molecule as heterocyclic.

In these reviews it is the intention to deal, so far as possible, with carbocyclic compounds only, as borne out by this Volume I of the series with articles on the benzidine rearrangement, the bicyclo-[3,3,1]nonanes, Feist's acid, and the annulenes. The difficulty in keeping rigidly to carbocyclic substances is, however, apparent in the chapter on 'The Biosynthesis of Carbocyclic Compounds', where many heterocyclic compounds are encountered, as is inevitable in any reasonably comprehensive account of biosynthesis.

The justification for this new series is the need for reviews to enable both the specialist and the non-specialist to keep up with the accelerating expansion of original literature. This need is not wholly met by existing periodical publications, and moreover many areas of carbocyclic chemistry are ripe for review although they may not at present be under such extensive investigation as other branches of chemistry. It is the intention, as is evident from the five articles in this volume, to present the subjects with sufficient background to make them acceptable to all readers; the choice of the word 'Topics' rather than 'Advances' indicates this intention.

The standard of the articles in Volume I is uniformly very high and carries out the intention of the Editor and publishers with regard to coverage. To the individual worker the series will be of value because of the specific articles, rather than because the volumes deal essentially with carbocyclic chemistry, in the whole vast field of which he is scarcely likely to be interested. Nevertheless the series will fill a very real need and is to be warmly welcomed.

W. BAKER.

Department of Organic Chemistry
The University of Bristol
May 1969

PREFACE

The continuing expansion of original chemical literature makes it increasingly difficult for chemists to see their subject in perspective, and the accompanying spate of review publications represents an attempt to remedy this difficulty. This is indeed the justification for these secondary publications. In general it is convenient if review articles are grouped according to their subject matter, the limits of this being neither so narrow as to make them only of very specialist interest, nor so wide as to defeat the initial objective.

Various collections of reviews of the chemistry of heterocyclic compounds which have appeared in recent years have nicely exemplified this pattern, and the suggestion made to me, that a similar project dealing with the chemistry of carbocyclic compounds would be very useful, seemed a worthwhile one.

No volumes embracing carbocyclic chemistry as a whole are presently produced and it seemed to me, and to colleagues with whom I discussed the project, that this represented a very satisfactory area of chemistry to bring together. The older boundary between alicyclic chemistry and what was then called aromatic chemistry but is more precisely called the chemistry of benzene derivatives has been made obscure by the large amount of recent work on non-benzenoid aromatic compounds, and the present-day definitions of 'aromatic' make the older distinction in some ways outmoded. The present series of reviews will deal with any aspects of carbocyclic chemistry.

Review articles serve two especial purposes: to provide a non-specialist reader with a comprehensive survey of a subject for his general interest, or to provide an effective initial account of the most important information available on a particular topic as a prelude to a detailed search of the primary literature. For either of these purposes a rounded review by an expert in the field is more valuable than an article which deals only with recent advances. For this reason the present series has been entitled '*Topics* in Carbocyclic Chemistry' rather than 'Advances', and the authors have in general been asked to write their articles as overall reviews rather than as reviews of only recent work. As such it is hoped that the articles will be of much more general use and interest.

It is hoped to include articles dealing with carbocyclic chemistry from many standpoints, practical, theoretical, mechanistic and biochemical. It is also intended to include within any one volume articles of widely differing scope, from surveys of broad sections of the subject to more detailed accounts of more limited but interesting topics. All of these points are illustrated by the range of articles included in Volume I, and it is intended to maintain this breadth of approach in later volumes.

I should like to thank a number of people for their very real help to me, and especially my friends Professor Wilson Baker for writing an introduction to the series, and Professor John I. G. Cadogan for his constant helpful comments and suggestions. I am also most grateful for the co-operation and understanding of the publishers, and especially the untiring help of Mr. J. G. Mordue. Last and very far from least I must thank the authors for their kindness and hard work in contributing the articles, and their forbearance at all times with the editor.

I take no credit for the ideas put forward by the authors but I take full responsibility for any editorial errors which have crept in and apologise to authors and readers alike for them.

In conclusion may I say that I would appreciate any suggestions concerning topics which could usefully be considered for inclusion in later volumes.

St. Andrews, Fife — DOUGLAS LLOYD.
January 1969

CONTRIBUTORS

D. V. Banthorpe, Department of Chemistry, University College London, Gower Street, London, W.C.1.

D. H. G. Crout, Department of Chemistry, University of Exeter, Stocker Road, Exeter, England.

G. L. Buchanan, Chemistry Department, University of Glasgow, Glasgow, W.2, Scotland.

D. M. G. Lloyd, Department of Chemistry, Purdie Building, University of St. Andrews, St, Andrews, Fife, Scotland.

H. P. Figeys, Service de Chimie Organique, Faculté des Sciences, Université Libre de Bruxelles, Avenue F.-D. Roosevelt 50, Brussels, 5, Belgium.

CONTENTS

1

THE BENZIDINE REARRANGEMENT

D. V. Banthorpe

Chemistry Department, University College London

A. Nature of the Reaction

The acid-catalysed conversion of hydrazobenzene (I) into benzidine (II), equation (1), was discovered by Hofmann (1863); fifteen years later a minor product, diphenyline (III) was reported and the structures of both isomers were established (Schmidt and Schultz, 1878, 1881). This reaction was the prototype of a large class of aromatic rearrangements of outstanding theoretical interest which

$$\text{C}_6\text{H}_5\text{NHNHC}_6\text{H}_5 \text{ (I)} \xrightarrow{H^+} \text{NH}_2\text{C}_6\text{H}_4\text{–C}_6\text{H}_4\text{NH}_2 \text{ (II)} + \text{NH}_2\text{C}_6\text{H}_4\text{–C}_6\text{H}_4(\text{NH}_2) \text{ (III)} \quad (1)$$

give products that are of considerable importance in the dyestuffs industry. Semi-quantitative product-studies, especially those of Jacobson over the period 1892 to 1922, showed that ring and N-substituted hydrazo-arenes (of both the benzene and naphthalene series) could give o- and p-semidines (IV, V) with 2,N'- and 4,N'-linking of the aryl rings respectively, in addition to the 4,4′ and 4,2′-linked products of the original example, together with oxidation and disproportionation products (azo-compounds and fission amines). The whole range of products is summarised in scheme (2); many hydrazo compounds give essentially only one or two of these types and often

$$\text{A–NHNH–B} \xrightarrow{H^+} \begin{cases} \text{NH}_2\text{–A–B–NH}_2 + \text{NH}_2\text{–A–B(NH}_2) \\ \text{NH}_2\text{–A–NH–B (V)} + \text{A(NH}_2)\text{–NH–B (IV)} \\ \text{A–N=N–B} + \text{A–NH}_2 + \text{B–NH}_2 \end{cases} \quad (2)$$

ring substituents restrict the possibilities for coupling, but recent paper-chromatographic analyses of products from reaction of several substrates suggest that all the possible isomers are usually formed, albeit some in trace amounts (Vecera, Petranek and Gasparic, 1957a, b). The whole family of interconversions, the mechanism of which has remained obscure until recently, has become collectively known as the *Benzidine Rearrangement*.

Rearrangement usually occurs readily at 0° to 20° when a solution of the hydrazo-compound in an organic solvent is treated with dilute (< 2 N) mineral or organic acid or with hydrogen chloride gas, but electron-withdrawing ring-substituents stabilise the substrate and stronger acid may be required to effect reaction in a convenient time. Acidic ion-exchange resins promote rearrangement and are especially convenient as the basic products are absorbed on the resin and can be filtered off (Yamada, Chichata and Tsurvi, 1954, 1955). Most of the early rearrangements were carried out under ill-defined conditions by treatment of an azo-compound with an acidic reducing agent, especially stannous chloride in hydrochloric acid often under reflux, and the transiently formed hydrazo-compound was not isolated. Although this procedure is very convenient as a preparative method, much reductive fission to amines usually occurs in competition. Direct, but little used, routes from azo-compounds to rearrangement products follow from treatment with chlorine, bromine, or hydriodic acid (Zollinger, 1961), quinol sulphonates (Reeves and Andrews, 1967) or various acids in ethanol, acetone, or other organic media (Nesmeyanov and Golovnya, 1960). Abstraction of hydride ion from the solvent probably occurs in the last method, equation (3), and the yield (10 to 80 per cent) is very sensitive to the reaction conditions. Systematic studies of catalysis by Lewis acids have not been carried

$$C_6H_5N{=}NC_6H_5 \xrightarrow{H^+} C_6H_5\overset{+}{N}H{=}NC_6H_5 \xrightarrow{H^-} C_6H_5NHNHC_6H_5 \xrightarrow{H^+} \text{Products} \qquad (3)$$

out, but it is possible that under defined conditions co-ordination between metal ions (much as the stannous or titanous ion of reduction mixtures) and azo- or hydrazo-compounds could induce rearrangement. Of significance in this aspect is the rearrangement promoted by titanous chloride from azobenzene in near-neutral buffers where proton-catalysed isomerisation of any hydrazobenzene that is formed should be negligible (Veibel, 1954); but detailed kinetic investigations of the reduction of several azo-compounds with titanous chloride under more acidic conditions have shown that hydrazo-compounds accumulate during reaction to a considerable extent and are probably the rearranging species (Zollinger, 1961).

Rearrangement occurs when one or both of the hydrazo-nitrogens carry substituents such as methyl, acetoxy and phenyl groups

(Jacobson, 1922; Davies and Hammick, 1954; Holt and Hughes, 1954; Wittig *et al.*, 1957), and for *N,N′*-bridged species as shown in equation (4) (Wittig *et al.*, 1957; Wittig and Grolig, 1961). For the latter type of substrate, reductive scission is the sole reaction when $n = 2$ or 3, *o*-semidine and diphenylines accompany scission when $n = 4$ and 5, and *p*-benzidines are formed for $n = 10$. This progression illustrates the increasing flexibility of the structures. Heterocyclic compounds can also exhibit rearrangement: e.g. when one or both aromatic rings of the hydrazo-compound are sulphur-containing heterocycles (Beyer and Haase, 1957; Beyer and Kreutzberger, 1952; Federova and Mironova, 1962), phenylhydrazinoimidazoles (Pyl, Lahmer and Beyer, 1961) or other nitrogen-containing structures (Das-Gupta and Bose, 1929). However, 2,2′-hydrazo-pyridine and -quinoline can be recovered unchanged after refluxing for several hours with concentrated hydrochloric acid, presumably because protonation of the pyridine rings inhibits the protonation of the hydrazo-nitrogen which is necessary (see later) for rearrangement (Colonna and Risaliti, 1959; Beyer, Haase and Wildgrubbe, 1958). Phenylhydrazino-pyridines and -picolines rearrange under similar conditions in low (c. 25 per cent) yield accompanied by disproportionation and reductive chlorination (Beyer *et al.*, 1958; Colonna and Risaliti, 1956; Risaliti and Pentimalli, 1956). Reduction of azoferrocene under acid conditions gave no rearrangement product (Nesmeyanov, Perevalova and Nikitina, 1961).

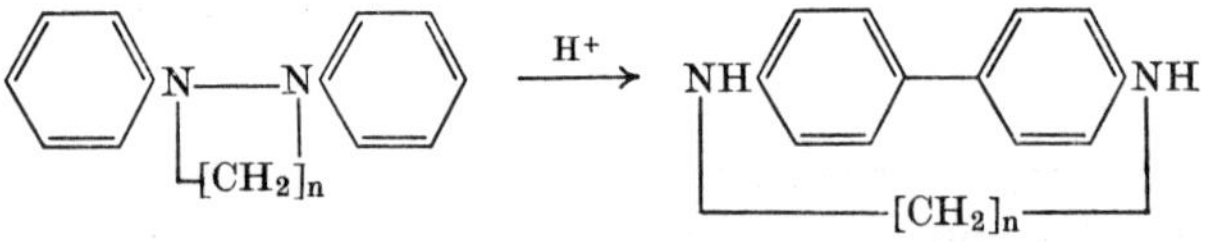

N—N $[CH_2]_n$ $\xrightarrow{H^+}$ NH NH $[CH_2]_n$ (4)

Several hundred papers have been published on the family of reactions, and three reviews are available (Jacobson, 1922; Vecera, 1958; Shine, 1967) in addition to a detailed mechanistic discussion (Banthorpe, Hughes and Ingold, 1964). This present review surveys the literature up to December 1967.

B. Proof of Intramolecularity

The benzidine rearrangement is one of a group of formally similar reactions in which an atom or group migrates from an amino nitrogen to the ring, equation (5), but in unique company with the rearrangement of *N*-nitroamines it is intramolecular rather than proceeding by fission and intermolecular recombination of the fragments. Early workers recognised the stereochemical restrictions to any intra-

$$C_6H_5NHR \xrightarrow{H^+} R\text{-}C_6H_4NH_2 \; (o \text{ and } p) \quad \{R = NHPh, NO_2, NO, CH_3, OH, \text{etc.} \qquad (5)$$

molecular mechanism that necessitated bond-breaking and -making at sites spanning the length of the aromatic ring and apparently involved turning the molecule inside out, and they suggested intermolecular pathways. One such route (Tichwinsky, 1903), although put forward in pre-electronic concepts, is equivalent to homolytic fission of the protonated substrate to give free radicals whose radical centres become transferred to *o* or *p* positions, the precise mode of incursion of protons being an adjustable detail. Two weighty arguments were soon adduced against this view (Jacobson, 1922). First, tetraarylhydrazines, $Ph'_2NNPh''_2$, although capable of undergoing acid-catalysed rearrangement gave no such products under conditions where homolysis to radicals was detectable (Wieland, 1912; 1915). Second, a survey of the previously reported rearrangements of some 66 hydrazo-compounds carrying different substituents in each ring, i.e. of the form A—B, indicated no symmetrical products A′—A′ and B′—B′, such as would be derived from the corresponding symmetrically substituted substrates A—A and B—B, which certainly should have arisen if kinetically-free homolysis products had been formed during reaction. The power of the second argument is lessened by the impossibility, some 40 years ago, of achieving a complete material balance in products; typically less than 70 per cent of the products were accounted for. However, modern techniques have vindicated the conclusion. Paper-chromatographic analysis of the products from acid-catalysed rearrangement of 1,2′-hydrazonaphthalene shows less than 0·5 per cent, if any, cross-products (Banthorpe, 1962b) and attempts to trap such products have proved fruitless. For example, less than 0·03 per cent of the ^{14}C-containing symmetrical product (VII, equation (6)) could be isolated when the ^{14}C-labelled substrate VI was rearranged in acid and a large amount of inactive VII was added, and was reisolated for radioactive assay

(Wheland and Schwartz, 1949). The same point was made by the observation that no benzidine (II) could be isolated from rearrangement of *m*-carboxyhydrazobenzene. Similar rearrangements of the

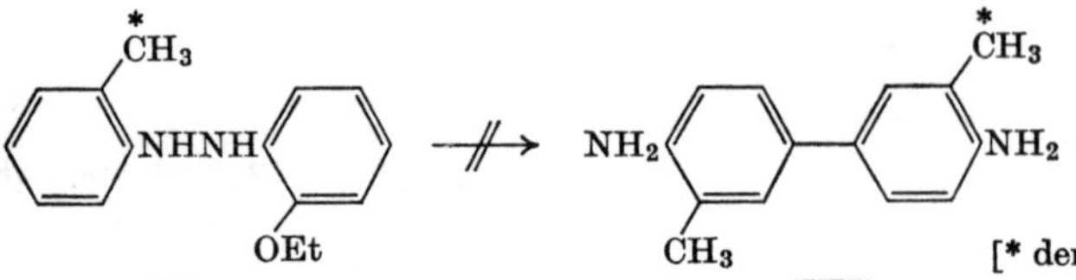

$\overset{*}{C}H_3$ NHNH OEt (VI) $\not\longrightarrow$ NH_2 NH_2 $\overset{*}{C}H_3$ CH_3 (VII) (6)

[* denotes C14 tracer]

o and *p* isomers were complicated by decarboxylation, but when allowance was made for this side reaction no intermolecular component could be detected (Bloink and Pausacker, 1950).

Another pre-electronic theory (Stieglitz, 1903), restated in modern terms, envisaged heterolysis and intermolecular recombination (equation (7)); the disposition of protons is again adjustable. Here the

$$\text{Ph}'\text{NHNHPh}'' \xrightarrow{H^+} [\text{Ph}'\overset{+}{\text{N}}\text{H}, \text{Ph}''\overset{-}{\text{N}}\text{H}] \longrightarrow \text{Products} \qquad (7)$$

previously detailed arguments are not particularly damaging, as a substrate, A—B, containing differently substituted rings, would be predisposed to heterolyse in a particular direction, e.g. to A^+B^-, leading to exclusive recombination of counter-ions and lack of cross-over products. However, on the basis of this mechanism concurrent reaction of two symmetrical hydrazo-compounds, A—A and B—B, which individually rearrange at comparable rates, would lead to unsymmetrical cross-products of the type A′B′ in addition to a calculable proportion of non-crossed-over material. The test was first applied to the co-rearrangement of 2,2′-dimethoxy- and 2,2′-diethoxy-hydrazobenzenes which differed in rate of rearrangement under the chosen conditions by a factor of six; melting-point curves of the product indicated that less than 6 per cent, if any, cross-product was produced (Ingold and Kidd, 1933). More recently, isotopic-dilution analysis of the product of mixed rearrangement of 2,2′-dimethyl and 2-^{14}C-methylhydrazobenzene gave less than 0·03 per cent cross-product (Smith, Wheland and Schwartz, 1952) and paper chromatography of the products from the three hydrazonaphthalene isomers (Banthorpe, 1962b) and substituted hydrazobenzenes (Vecera *et al.*, 1957) have made the same point with less, but adequate, precision. All the pairs of compounds studied differed in reactivity by less than a factor of 10, and exceptional precision is not necessary in analysis, as a few tenths of one per cent of cross-over under these conditions could always be attributed to an insignificant intermolecular pathway accompanying an overwhelmingly predominant intramolecular route.

No products resulting from attack of dissociated fragments on the solvent have ever been detected using precision chromatographic techniques (Banthorpe, 1962b), although such products would be characteristic of intermolecular mechanisms. As is now known, these examples which were monitored for intramolecularity include substrates rearranging by different mechanisms, and there is now no serious doubt that under the controlled experimental conditions used, i.e. rearrangement of the hydrazo-compounds at low temperatures (0–20°), the benzidine rearrangement is a completely intramolecular process, although no decision can be made for the more forcing acid-reductions of azo-compounds. As the *N*-nitroamine rearrangement is the only similar reaction rigorously proved to possess this property (Banthorpe, Thomas and Williams, 1965), the question arises as to the special properties of these two classes of compounds that permit and indeed require such a pathway.

C. Kinetic Studies

1. *Kinetic Form*

No progress could be made in elucidating the mechanism until kinetically-controlled studies on pure hydrazo-compounds were carried out; these were hindered by the absence of a convenient analytical method. The first such study on hydrazobenzene was made using a tedious gravimetric procedure, but although the results were semi-quantitative and the reaction medium was heterogeneous a first-order dependence on substrate and second-order dependence on acid was clearly indicated (Van Loon, 1904). This pioneer work was completely ignored by subsequent workers, probably because it predated *Chemical Abstracts*. The major breakthrough was the use of the redox dye, Bindschedler's Green, to assay unreacted substrate in a kinetic run (Dewar, 1946), and this analytical technique has been almost universally adopted. Recently, electroanalytical methods have been developed (Oglesby, Johnson and Reilley, 1966; Schwartz and Shain, 1965) following an earlier unsatisfactory and time-consuming procedure (Biilmann and Blom, 1924); ultra-violet spectrophotometry has also been used to measure both rates of disappearance of substrate and of appearance of products (Carlin and Odioso, 1954a, b); Carlin, Nelb and Odioso, 1951).

Despite Van Loon's results, most workers assumed a first-order dependence on acid (cf. Dewar, 1946) until a reinvestigation to modern standards of the hydrochloric acid-catalysed rearrangement of hydrazobenzene in 75 per cent ethanol–water confirmed the earlier conclusion (Hammond and Shine, 1950), and a similar kinetic form was discovered for the 3,3′- and 4,4′-dimethyl homologues under similar conditions (Carlin and Odioso, 1954; Carlin and Wich, 1958). The second-order law held for the 3,3′-compound when the chloride ion concentration was buffered at a series of different acidities and thus verified a genuine quadratic dependence on proton concentration, rather than the kinetically-equivalent situation of unit-order dependence on both proton and chloride concentration. The benzidine–diphenyline ratio in the products from hydrazobenzene was also shown to be independent of acidity, temperature, and ionic strength over the rather narrow ranges studied, suggesting that formation of both isomers follows the same kinetic law and the same fundamental mechanism (Carlin *et al.*, 1951; Croce and Gettler, 1953).

The recognition that two protons were accommodated in the transition state of the rate-determining step of the rearrangement led to two proposed mechanisms (Cohen and Hammond, 1953); cf. equation (8).

Mechanism 1

$$B + H^+ \underset{}{\overset{f}{\rightleftharpoons}} BH^+$$

$$BH^+ + H^+ \overset{f}{\rightleftharpoons} BH_2^{++} \overset{s}{\longrightarrow} \text{Products}$$

(8)

Mechanism 2

$$B + H^+ \overset{f}{\rightleftharpoons} BH^+$$

$$BH^+ + H^+ \overset{s}{\longrightarrow} \text{Products}$$

[B = Hydrazo-compound; f = fast; s = slow]

The common first step could hardly be other than a rapid equilibrium proton-transfer. The pK_b of a hydrazoarene has never been measured. At low acidities (c. 0·1 *N*) protonation as indicated by a change in ultra-violet absorption is undetectable, and the infra-red measurement of hydrogen-bonding capacity that has been successfully correlated with the pK_b of a variety of feeble bases gives unreliable results (Cooper, 1966); at higher acidities rearrangement intrudes. But the pK_b of hydrazobenzene can reasonably be estimated theoretically to be c. 14 (Dewar, 1946) and protonation of such a nitrogen-base should be almost unactivated. A recent claim (Lukashevich and Krolik, 1959) to have measured the pK_b by a method based on analysis of the precipitated acid-salts from competition of the hydrazo-compound with a substituted aniline of known basicity for a limited quantity of acid in an aprotic solvent is fallacious. The composition of the precipitate depends on solubilities and rates of precipitation, quite aside from complications due to concurrent rearrangement. Indeed, using this technique a difference in pK_b for aniline and *N*-methylaniline of about 5 can be estimated: the accepted value is about 0·6! (Banthorpe and Williams, unpublished). The status of the second protonation is more difficult to assess. Either a second equilibrium leads to a transiently-stable diprotonated species BH_2^{++}—formed in concentration about 10^{-12} times that of BH^+ (Sterba and Vecera, 1966)—which decomposes to products (equation (8), mechanism 1); or BH^+ breaks up in the rate-determining step on the approach, but not complete transfer, of the second catalytic proton (equation (8), mechanism 2). Mechanisms 1 and 2 correspond to specific and general acid-catalysis respectively. Three types of criteria have been applied to determine the favoured route and these will be discussed in section C.2.

Although protonation of BH^+ must be hindered by the repulsion of the first-bonded proton, it is unlikely that formation of BH_2^{++} could be rate-determining. Inability of the second proton-transfer to reach equilibrium would imply that the rate of breakdown of BH_2^{++} to products is faster than the rate of back-transfer of a proton to solvent. BH_2^{++} is a very strong acid and, as such a transfer would require essentially no activation energy, the rearrangement step could hardly be faster.

All rearrangements were originally believed to be second-order in acid and the practice arose of recording kinetics at single acidities and reporting third-order constants which were used to calculate heats and entropies of activation for use in mechanistic discussions (Croce and Gettler, 1953; Vecera and Petranek, 1960). Several compounds thus studied, e.g. 2-methoxy- and 2,2′-dimethoxy-hydrazobenzenes, are now known not to follow the quadratic acid law under the reaction-conditions chosen, and the demonstration of an order in acid of 1·6 for the reaction of 2,2′-dimethylhydrazobenzene in 95:5 ethanol: water over the range 0·03 to 0·1 *M* hydrochloric acid shattered this preconception (Carlin and Odioso, 1954a). Such non-integral order cannot remain constant with variation of acid concentration, but the measurements were made over an insufficient range to determine a variation in order. Similar non-integral (1·5 to 1·6) orders in acid for two 4,4′-disubstituted hydrazobenzenes were subsequently recorded, but with a complete absence of experimental details (Dewar and McNicol, 1959). These results were attributed either to appreciable storage of the substrate in mono- or di-protonated forms (Carlin and Odioso, 1954a, b) or to the incidence of a π-complex mechanism (Dewar, 1959) such as will be discussed in detail in Section G.1. Each theory gave a rate-law of the form of equation (9) which reduces to quadratic acid-dependence at low acidities, $b(H^+) < 1$, changing to unit dependence at higher acidities, $b(H^+) > 1$, with an intermediate non-integral range.

An alternative interpretation is that the rate law is as shown in equation (10) and corresponds to two independent concurrent mechanisms (Blackadder and Hinshelwood, 1957a); one first- and the other second-order in acid.

$$-\frac{dB}{dt} = \frac{a(\mathrm{B})\,(\mathrm{H}^+)^2}{1 + b(\mathrm{H}^+)} \tag{9}$$

[B = substrate; a and b are constants]

$$-\frac{dB}{dt} = k_2(\mathrm{H}^+)\,(\mathrm{B}) + k_3(\mathrm{H}^+)^2\,(\mathrm{B}) \tag{10}$$

[k_2 and k_3 are second- and third-order rate-constants]

The fundamentally different prediction of the second theory from the first is that the order in acid changes from one to two as the acidity is increased.

A reinvestigation of 2,2′-dimethylhydrazobenzene using 60 per cent dioxan–water as solvent demonstrated equation (10) to be quantitatively obeyed in that the order in acid changed from 1·4 to 2·0 as the concentration of perchloric acid varied from 0·01 to 0·5 *M*. Moreover, quantitative agreement was shown by the strict linearity of the plot of $[-(dB/dt)/(\mathrm{B})(\mathrm{H}^+)]$ versus (H^+), and the values of k_2 and k_3 could be dissected out (Banthorpe, Ingold, Roy and Somerville, 1962). Several substrates which show the extremes of either pure first- or second-order dependence on acid in the range of kinetic investigations, and seven examples of transitional kinetics have also been subsequently reported. The latter all obey equation (10) quantitatively.

All the available kinetic data are summarised in Tables 1 and 2. Most of the rearrangements were carried out at 0° with perchloric acid

TABLE 1

Kinetic order in (H^+) *for hydrazobenzenes*

Substituents in Ph groups	Range (H^+)	Solvent	Order in (H^+)	Reference
—, —	0·05–1·0 *M*	A	2	(a, b)
2-Me, 2′-Me	0·01–0·5 *M*	A	1·3–2·0	(c)
3-Me, 3′-Me	0·05–0·1 *M*	A	2	(d)
4-Me, 4′-Me	0·005–0·07 *M*	A	2	(a, e, l)
2-MeO, —	0·002–0·3 *M*	A	1·1–2·0	(f)
2-MeO, 2′-MeO	0·0001–0·05 *M*	A	1	(f)
4-MeO, —	7×10^{-6}–5×10^{-3} *M*	A	1	(f)
3-NH_2, 3′-NH_2	0·01–0·04 *M*	B	1	(g)
4-NHAc, —	0·007–0·1 *M*	A	1	(f)
2-F, 2′-F	0·1–0·8 *M*	A	2	(f)
2-Cl, 2′-Cl	0·8–2·8 *M*	A	2	(f)
4-Cl, 4′-Cl	0·1–1·0 *M*	A	2	(f)
4-Cl, —	0·07–1·0 *M*	A	2	(f)
2-Br, 2′-Br	0·2–2·0 *M*	A	1·2–1·9	(f)
4-Br, 4′-Br	0·1–0·5 *M*	A	2	(f)
2-I, 2′-I	0·7–1·6 *M*	A	1	(f)
4-I, 4-I′	0·05–0·5 *M*	A	2	(f)
2-Ph, 2′-Ph	0·9–1·6 *M*	A	2	(f)
4-Ph, —	0·004–0·6 *M*	A	2	(f)
4-Ph, 4′-Ph	0·006–0·02 *M*	B	1·8	(h, i)
4-NO_2, —	2·0–4·0 *M*	A	2	(f)
4-Vinyl, 4′-Vinyl	0·001–0·05 *M*	B	1	(j)
4-tBu, 4′-tBu	0·01–0·05 *M*	B	2	(k)
4-tBu, 4′-Cl	0·01–0·10 *M*	B	1·9	(k)

Solvent A = 60:40, Dioxan:water; Solvent B = 95:5, Ethanol:water.

References: (a) Bunton, Ingold and Mhala (1957). (b) Banthorpe, Hughes, Ingold and Roy (1962). (c) Banthorpe *et al.* (1962c). (d) Carlin and Odioso (1954b). (e) Carlin and Wich (1958). (f) Banthorpe, Cooper and Ingold (1967). (g) Clovis and Hammond (1963). (h) Shine and Stanley (1967). (i) Shine and Stanley (1965). (j) Shine and Chamness (1963). (k) Shine and Chamness (1967). (l) Hammond and Clovis (1963).

TABLE 2

Kinetic order in (H+) *for other hydrazoarenes*

RNHNHR′	Range (H+)	Solvent	Order in (H+)	Reference
1-$C_{10}H_7$, 1′-$C_{10}H_7$	5×10^{-6}–0·03 *M*	A	1	(a)
1-$C_{10}H_7$, 2-$C_{10}H_7$	0·01–0·15 *M*	A	1	(b)
2-$C_{10}H_7$, 2-$C_{10}H_7$	0·01–0·2 *M*	A	1·0–1·2	(c)
	0·001–0·02 *M*	C	1·0–2·0	(d)
1-$C_{10}H_7$, Ph	1×10^{-4}–0·4 *M*	A	1·0–2·0	(e)
2-$C_{10}H_7$, Ph	8×10^{-4}–0·6 *M*	A	1·1–2·0	(f)
N-Me-hydrazobenzene	3×10^{-4}–0·06 *M*	D	1·1–1·9	(g)

Solvents A and B as in Table 1; Solvent C = 75:25, acetone:water; Solvent D 25:75, methanol:water.

References: (a) Banthorpe, Hughes, and Ingold (1962a). (b) Banthorpe and Hughes (1962a). (c) Banthorpe (1962a). (d) Shine and Chamness (1963). (e) Banthorpe, Hughes and Ingold (1962b). (f) Banthorpe (1962c). (g) White and Preisman (1961).

as catalyst, and when ranges of acidity rising above 0·3 *M* were used the (H^+) was replaced by h_0 in the reckoning of kinetic orders (cf. Section C.2 for amplification). Table 3 gives a selection of values of k_2 and k_3 calculated for standard conditions from what is known about the dependence of rate on acidity and ionic strength. The generally-occurring enhancing and retarding effects on rate of electron-releasing and electron-attracting *o*- and *p*-substituents and the correlation of fast reactions with the one-proton mechanism are clearly seen. The observed rate-constants (cf. equation (8)) are products of equilibrium constants and rate constants for the step involving *N,N* scission, and if approximate allowance is made for the contribution from the former a reasonably good Hammett relationship (ρ, −6·1) can be demonstrated for the effect of *p*-substituents on the bond-scission step (Cooper, 1966).

Heats and entropies of activation for these compounds are also composite functions of equilibrium and rate-constants and cannot easily be correlated with structure, but two-proton mechanisms are reported to have $\Delta H^{\ddagger}$ in the range 19 to 22 kg. cal. mole^{-1} and ΔS^* −10 to +3 e.u. (Carlin and Odioso, 1954a, b; O'Sullivan, 1966; Sterba and Vecera, 1966) whereas $\Delta H^{\ddagger}$ for the one-proton process is, in the

TABLE 3

Comparison of k_2 and k_3 for RNHNHR′

R	R′	k_2	k_3	Order (H^+)
$2\text{-}ClC_6H_4$	$2\text{-}ClC_6H_4$	—	2×10^{-5}	2
$2\text{-}FC_6H_4$	$2\text{-}FC_6H_4$	—	$1{\cdot}6 \times 10^{-3}$	2
$4\text{-}ClC_6H_4$	$4\text{-}ClC_6H_4$	—	$7{\cdot}4 \times 10^{-3}$	2
H	H	—	1·0	2
$2\text{-}CH_3C_6H_4$	$2\text{-}CH_3C_6H_4$	0·01	5	Transitional
$3\text{-}CH_3C_6H_4$	$3\text{-}CH_3C_6H_4$	—	10	2
$4\text{-}Ph\text{-}C_6H_4$	—	—	400	2
$4\text{-}CH_3C_6H_4$	$4\text{-}CH_3C_6H_4$	—	800	2
$2\text{-}C_{10}H_7$	C_6H_5	0·3	3	Transitional
$1\text{-}C_{10}H_7$	C_6H_5	10	100	Transitional
$2\text{-}C_{10}H_7$	$2\text{-}C_{10}H_7$	250	—	1
$1\text{-}C_{10}H_7$	$2\text{-}C_{10}H_7$	600	—	1
$1\text{-}C_{10}H_7$	$1\text{-}C_{10}H_7$	1080	—	1
$4\text{-}MeOC_6H_4$	—	c. 5000	—	1

Rates calculated in 60% dioxan–water at 0° with $HClO_4$ as catalyst and ionic strength = 0·05. Under these conditions k_3 for hydrazobenzene is $1{\cdot}7 \times 10^{-3}$ sec^{-1} mol^{-2} l^2; k_2 is in sec^{-1} mol^{-1}l; k_3 is in sec^{-1} mol^{-2} l^2.

few examples studied, some 5 kg. cal. mole^{-1} lower (O'Sullivan, 1966).

The rates of a few rearrangements at pressures up to 5000 atmospheres (Osugi, Sasaki and Onishi, 1966) and under catalysis by polystyrene–sulphonic acid (Arcus, Howard and South, 1964) have been briefly reported. The reaction induced by the ion-exchange resin for hydrazobenzene was 120-fold faster than that catalysed by mineral acids at formally equivalent pH, perhaps owing to the possibility of near simultaneous transfer of two protons from the resin matrix to an absorbed hydrazo-molecule. A variety of dibasic organic acids did not promote a similarly enhanced reaction for hydrazobenzene (Cooper, 1966), but it would be worthwhile to investigate a substrate which reacts by the one-proton mechanism under catalysis by the ion-exchange resin.

2. *General and Specific Acid-Catalysis*

Mechanisms 1 and 2, equation (8), with the second proton transferred in either an equilibrium or a rate-determining step, can be kinetically distinguished by obeying specific hydrogen-ion and general acid-catalysis respectively. In particular, observation of specific acid-catalysis demonstrates that the proton is passed over from its acid in the transition state practically completely; the criterion is of degree rather than of rate of transfer (Banthorpe, Hughes, Ingold and Roy, 1962). Three types of approach have been made to characterise the mechanism of acid catalysis in representative hydrazoarenes.

(i) In the most direct approach, the rate of reaction of hydrazobenzene in 50 per cent aqueous ethanol has been measured in buffer solutions containing varying amounts of organic buffer-acid, but at constant buffer ratio, and thus at constant pH, and also with different organic acids added to give a fixed pH-meter reading (Cohen and Hammond, 1953). The results were considered to indicate catalysis by undissociated acid, but a Brönsted relationship could not be demonstrated, and a weakness of the procedure was that the quantity of acid necessary to effect reaction significantly altered the reaction medium from run to run. The sensitivity of rate to polarity of medium was ignored. Such results are thus inconclusive. A good candidate for general acid-catalysis is 4-methoxyhydrazobenzene, the extreme lability of which (Table 1) might result in rearrangement being induced by the approach, rather than by the transfer, of a proton; but studies using buffer solutions of various acids give no indication of this phenomenon (Cooper, 1966).

The alleged demonstration of general acid catalysis led to the study of the kinetics of rearrangement of deuterated hydrazobenzene. Because of the existence of the equilibrium (11),

$$\mathrm{HA} \rightleftharpoons \mathrm{H^+} + \mathrm{A^-} \tag{11}$$

the participation of either HA or of H^+ and A^- separately, i.e. the possibility of either slow transfer of a proton, or of equilibrium proton-transfer followed by the intervention of A^- acting as a base to remove a proton from some other site in the rearranging molecule, cannot be distinguished kinetically. But if proton-removal by base is rate-determining, a kinetic isotope-effect for deuterium-containing as compared with protium-containing substrates should be demonstrable. 4,4′-Di-deuterio-hydrazobenzene was originally reported to rearrange about 10 per cent slower than its protium analogue (Hammond and Grundemeier, 1955), but this small effect has been shown to be an artefact due to nonidentical conditions of acidity in the two series of runs that were compared, and no kinetic isotope-effect could be detected for either hydrazobenzene (two-proton mechanism) or 1,1′-hydrazonaphthalene (one-proton mechanism) in which either the *o*- or the *p*-positions were deuterated (Banthorpe, Hughes, Ingold and Humberlin, 1962; Banthorpe and Hughes, 1962b), the site of deuterium in the rings being proved by spectroscopic studies (Banthorpe and Hughes, 1962c). The rate-determining step is thus situated on the reaction co-ordinate well before the loss of aromatic hydrogen and there is no evidence for general acid-catalysis of the type originally proposed.

(ii) The second approach is based on correlation of rate with Hammett's acidity function, h_0. Over the range 0·05 to 0·25 *M* perchloric acid and in 60 per cent dioxan–water the rate of rearrangement of hydrazobenzene is strictly second-order in acid concentration, but at higher acidities the rate follows an increasingly higher power. An excellent linear correlation of rate with h_0 is obtained from 0·2 to 1·0 *M* acid (h_0 being definable in this solvent system) and an order in h_0 of 2·1 or 2·7 is obtained depending on whether the ionic strength is kept constant or allowed to vary over the series of runs (Bunton, Ingold and Mhala, 1957; Banthorpe, Hughes, Ingold and Roy, 1962). The higher order in media of unbuffered ionic strength reflects the large salt effect on both the rate-determining step and on h_0. Similar orders in h_0 of 2·0 to 2·3 have been demonstrated for *N*-1-naphthyl-*N*′-phenylhydrazine (Banthorpe, Hughes and Ingold, 1962b), *N*-2-naphthyl-*N*′-phenylhydrazine (Banthorpe, 1962c) and 2,2′-dimethylhydrazobenzene (Banthorpe, Ingold, Roy and Somerville, 1962). The acidities used were too low to allow the Bunnett treatment (1961) of acidity-function data to be meaningfully applied.

If all proton-transfers were at equilibrium, the rate should follow a scale of acidity designed to measure equilibrium proton-transferring power, i.e. h_0. If the transfers were kinetically-controlled, then the rate should correlate better with a scale of acidity more relevant to encounter probabilities, i.e. (H^+). Thus the above results, obtained in regions of acid concentration where h_0 and (H^+) diverge, clearly support Mechanism 1. The difference from the theoretical order in h_0 of 2·0 can be reasonably accommodated by the following circumstances: (a) A dependence on h_0h_+ rather than h_0^2 should be found; although h_+ has not been measured in the appropriate acid media, it is believed to be an approximately parallel function of h_0 in the range concerned; (b) the temperature of the kinetic measurements (0°)

differed from that (20°) of measurement of the h_0 scale; and (c) the protonation of a hydrazo-compound would not exactly parallel that of the indicator bases used for the h_0-determinations.

No h_0-correlations are available for the one-proton mechanism as such reactions proceed inconveniently rapidly before the region of divergence of h_0 and (H^+) is reached, but equilibrium-transfer of the single catalytic proton can hardly be doubted.

The variation of rearrangement-rate for hydrazobenzene in ethanol:water mixtures of various composition at constant pH is claimed to correlate better with $h_0.(H^+)$ than with h_0^2, and so to indicate rate-determining transfer of the second proton (Vecera, Synek and Sterba, 1960), but these results are inconclusive, as neither h_0 nor h_+ are definable, reproducible functions of acidity over this range of media.

(iii) The third approach is based on the observation of a solvent isotope-effect (k_{D_2O}/k_{H_2O}) of 4·8 when the aqueous fraction of the dioxan–water solvent was replaced by deuterium oxide in the re-arrangement of hydrazobenzene (Bunton *et al.*, 1957; Banthorpe *et al.*, 1962a); similar effects of 2·1 to 3·8 were observed for one-proton and transitional examples. (cf. Table 4). Current theory teaches that a reaction with pre-equilibrium proton-transfer shows a solvent isotope-effect of 1·5 to 2·0, whereas a rate-determining proton-

TABLE 4

Solvent isotope effects on hydrozoarenes

Substrate	(H^+) or (D^+)	Order in (H^+)	k_{D_2O}/k_{H_2O}	Reference
$1\text{-}C_{10}H_7NHNHC_{10}H_7\text{-}1$	0·01	1·0	2·3	(a)
$2\text{-}C_{10}H_7NHNHPh$	0·020	1·2	2·6	(b)
	0·31	1·8	3·8	
$2\text{-}CH_3C_6H_4NHNHC_6H_4CH_3\text{-}2$	0·01	1·3	2·1	(c)
	0·29	1·9	3·5	
PhNHNHPh	0·19	2·0	4·8	(a)

Effect on rates catalysed by $HClO_4$ at 0° of changing the solvent from 60 vol. dioxan + 40 vol. water to dioxan mixed with 40 vol. of deuterium oxide.

References: (a) Banthorpe, Hughes and Ingold (1962a). (b) Banthorpe (1962c). (c) Banthorpe, Ingold, Roy and Somerville (1962).

transfer should give a value less than 1·0. The uniform pattern from Table 4 indicates that the single proton of the one-proton mechanism and both protons of the two-proton mechanism are transferred in equilibrium steps. An objection (Dewar, 1963) that the basicity of the hydrazo-compound is changed on moving from protium to deuterium-containing solvents, in that nitrogen-bonded hydrogen is exchanged for deuterium in the latter case, seems invalid, as a secondary isotope effect could not account for the observed large rate-changes.

The available data thus support Mechanism 1, and no general acid-catalysis is detectable: the second proton must be, to all intents, completely transferred to nitrogen, and the transition state of rearrangement is indistinguishable from that derived from the di-protonated substrate. As the ratio of concentrations of BH^+ and BH_2^{++} is likely to be about $10^{12}:1$, the transfer of the second proton must lower the free energy of activation by about 18 kg. cal. mole^{-1} in order to allow the one- and two-proton mechanisms to coexist for the same substrate (Sterba and Vecera, 1966).

3. *Salt and Solvent Effects*

Most investigators have reported a strong acceleration of rearrangement by neutral salts (Banthorpe *et al.*, 1962a; Croce and Gettler, 1953; Hammond and Shine, 1950), and both one- and two-proton mechanisms are highly sensitive, the latter being slightly more so. The Debye–Hückel theory was applied to the variation of rate with ionic strength for two-proton catalysis of hydrazobenzene and was claimed to be consistent with a rate-determining step involving two positive ions, i.e. slow transfer of the second proton (Croce and Gettler, 1953). However, the high salt concentrations and the mixed solvent systems used invalidate application of the theory, and the very similar dependence of rate on ionic strength for 1,1′-hydrazonaphthalene where only one catalytic proton is employed (Banthorpe *et al.*, 1962a) refutes such naive conclusions.

The salt effect for the two-proton mechanism could be qualitatively explained by the concentration of two positive charges into the transition state of reaction from an overall more diffusively-charged initial state, but such an explanation is less satisfactory for the similar effect in the one-proton route. A more likely reason is that the salt effect reflects the creation of new charges as the reaction passes into the transition state (Banthorpe *et al.*, 1964). Recently, it has been pointed out that the observed salt effect on rate includes a contribution from the salt effect on h_0 or pH (and hence on the equilibrium formation of the protonated intermediates) as well as the direct effect on the breakdown of the intermediates (Sterba and Vecera, 1966), but this effect must be small as charge is neither created, destroyed, nor greatly dispersed in the protonation equilibrium.

A similar conclusion concerning the polarity of the transition state follows from the study of solvent effects. If the water-content of an aqueous-organic reaction medium is below a certain threshold, rearrangement is accelerated by a further reduction in the water content. This is illustrated for the two-proton mechanism of hydrazobenzene by the gradual rate increase over a total 5-fold range as the water content of aqueous dioxan or ethanol fell from 54 to 29 per cent and from 22 to 10 per cent respectively at constant stoichiometric pH (Sterba and Vecera, 1966). This is undoubtedly due to the increase in activity of the catalytic protons (an increase that cannot be quantitatively measured by a h_0 value, for Hammett scales of acidity cannot be set up in these media), as the water, which causes deactivation by solvation, is reduced in content (Banthorpe *et al.*, 1964). Under these conditions of low water concentration a correlation of high rate with solvent polarity can be demonstrated in a series of media of composition 95:5 in organic solvent: water where the major component varied in polarity over a range from acetonitrile to dioxan (Banthorpe *et al.*, 1962a).

Above the previously mentioned threshold of water concentration it is found that the rate of rearrangement rises rapidly with increase in water concentration. Thus the rate for 2,2′-hydrazonaphthalene (one-proton-mechanism) increases some 80-fold when the water in dioxan–water is increased from 30 to 52 per cent at constant stoichiometric hydrogen-ion concentration or pH-meter reading, a range in which the dielectric constant of the medium barely doubles (Banthorpe 1962a). A similar, although somewhat less spectacular increase is reported (Croce and Gettler, 1953) for hydrazobenzene in a series of ethanol–water mixtures. In the former example the rate is increasing almost asymptotically with increase in water concentration at the more aqueous limit of the range studied, and its dependence on dielectric constant of the reaction medium is qualitatively inconsistent with electrostatic theory as at present known, at which its application to pure and mixed solvents would be identical. However the dependence can be qualitatively understood if it is assumed that the electrostatic fields about the transition state are strong enough to separate the components of the solvent system, the more polar water preferentially solvating the charge centres (Banthorpe *et al.*, 1964). Again a highly polar transition state is strongly implied for both one- and two-proton mechanisms.

The overall picture of solvent effects appears to be that the rate passes though a sharp minimum in the region of 20–50 per cent water as the composition of the reaction medium is varied from pure organic component to pure water, at constant pH; the position of the minimum depending on the substrate and the organic component of the solvent.

D. Isolation of Intermediates

Early speculations that semidines were intermediates *en route* to benzidines and other rearrangement products were disproved by the demonstration of their complete stability under typical reaction conditions (Robinson and Robinson, 1918), but more recent work concerning the identity and status of intermediates is confused.

Claims have been made for the isolation of mono- and di-protonated salts following passage of dry hydrogen chloride into ethereal solutions of hydrazobenzene at –20° (Wieland, 1912a; Orelkin, Ryskaltschuk and Aizikovitsch, 1931) or by treatment of similar solutions with methyl iodide at 0° to 20° (Pongratz and Scholtis, 1942; Cohen and Hammond, 1953). The latter reaction was believed to proceed as in equation (12) to give a dihydriodide IX that was stable in dry ether

$$\begin{aligned} \mathrm{PhNHNHPh + MeI} &\longrightarrow \mathrm{(PhNHMe)_2^+ I_2^-} && \mathrm{(VIII)} \\ \mathrm{(PhNHMe)_2^+(I^-)_2} &\longrightarrow \mathrm{(PhNMe)_2 + 2HI} && \qquad (12) \\ \mathrm{PhNHNHPh + 2HI} &\longrightarrow \mathrm{(PhNH_2)_2^+(I^-)_2} && \mathrm{(IX)} \end{aligned}$$

but decomposed rapidly to rearrangement products on treatment with acid or water. A dihydrobromide was claimed to be formed in a similar fashion, and a detailed thermochemical investigation was carried through (Pongratz, Böhmert-Süss and Scholtis, 1944). Under closely controlled conditions methyl iodide promoted rearrangement of hydrazobenzene with the formation of tetramethylbenzidine (Pongratz and Scholtis, 1942), the rearranging species being presumably either VIII or IX; methylation of the initially-formed product follows.

Part of this work was subsequently refuted (Carlin *et al.*, 1951) and all has been severely criticised (Krolik and Lukashevich, 1952, 1953). The ether-insoluble precipitate of the 'dihydrochloride' salt is now claimed to be a mixture of fission and disproportionation products and the methyl iodide-promoted rearrangement is supposed to follow a route, equation (13), leading to a 2:1 complex of X and XI.

$$\begin{aligned} &2PhNHNHPh \longrightarrow PhN{=}NPh + 2PhNH_2 \\ &PhNH_2 + MeI \longrightarrow \underset{(X)}{(PhNMe_3)^+I^-} + \underset{(XI)}{(PhNH_3)^+I^-} \end{aligned} \qquad (13)$$

However, more rigorous analyses and characterisations of products will have to be carried out before the newer views can be accepted. This alleged failure to prepare di-hydrohalides of hydrazobenzene has been claimed to rule out the occurrence of a diprotonated substrate as a reaction intermediate (Lukashevich, 1964, 1967), but such experiments only demonstrate the non-existence or lability of the species under the aprotic conditions chosen and have no bearing on the situation in typical rearrangement conditions.

Methods have been developed to isolate certain mono-protonated hydrazobenzenes as acid salts, e.g., addition of hydrogen chloride in ether to a molar excess of hydrazobenzene or 3,3′-dimethylhydrazobenzene gave the monohydrohalide in good yield (Krolik and Lukashevich, 1953; Cooper, 1966). Both these compounds rearrange via the two-proton mechanism under kinetically-controlled conditions, but similar attempts to isolate mono-acid salts of 4-methoxyhydrazobenzene or 1,1′-hydrazonaphthalene which react via the one-proton mechanism give only rearrangement products (Banthorpe *et al.*, 1962a) although a claim is on record to have detected the formation of the hydrochloride of the latter compound in solution, by oxidising it to azonaphthalene (Lukashevich, 1964).

The only other evidence concerning reaction intermediates is the formation of oxidising agents during the later, rearrangement, steps. This is discussed on section K.2.

E. Product Studies

1. *General*

Tables 5 and 6 record all the available data on products obtained by accurate chromatographic or gravimetric analysis of rearrangements carried out under kinetic control. In one example (Clovis and Hammond, 1963) isotope-dilution analysis was used. The figures are mostly accurate to c. ±1 per cent actual value, although in all the examples so examined, trace products could be detected by paper chromatography; e.g. 2,2′, 2–*N*′ and 4–*N*′ linked products were qualitatively detected to the total extent of about 1 per cent in the products from hydrazobenzene (Vecera, Gasparic and Petranek, 1956, 1957; Hashimoto and Sunamoto, 1964; Hashimoto, Shinkai and Sunamoto, 1966; Cooper, 1966).

Table 7 includes some valuable results for 3,3′, 5,5′-tetrasubstituted compounds, unfortunately referring to heterogeneous, non-kinetically-controlled conditions and with non-quantitative product recovery. Except for a few scattered, unsystematic observations (cf. Clemo and Dawson, 1939; Clemo and Lee, 1954; Davies and Hammick, 1954; Ward and Pearson, 1959; Vecera and Petranek, 1960; Rakusan and Allan, 1967), the remainder of the literature is over forty years old and has been collected, correlated, and exhaustively discussed by Jacobson (1922). In his review, actual or surmised products from 87 hydrazobenzenes carrying CH_3, C_2H_5, Cl, Br, I, OH, OCH_3, OC_2H_5, CH_3COO, COOH, SO_3H, NH_2, $N(CH_3)_2$, and $NHCOCH_3$ substituents are recorded but the data are at best semi-quantitative and almost all refer to reactions conducted by treatment of azo-compounds with acid reducing-agents. Although a start has been made on putting this mass of information on a more quantitative footing, much remains to be done.

A survey of the information in Tables 5, 6 and 7, taken in association with the kinetics in Tables 1 and 2, leads to four general conclusions.

(i) Reactions forming diaminobiaryls, semidines or disproportionation products from a particular hydrazo-compound all have the same kinetic form and presumably all utilise a common rate-determining step. A good example is 4-chlorohydrazobenzene which shows strict unit order in hydrazo-compound and quadratic order in acid but gives 2,4′-, 2,*N*′- and *N*,4′-linked and disproportionation products in comparable quantities.

(ii) 2,2′-linkage only occurs for *m*-substituted compounds, for compounds with at least one naphthyl residue, and perhaps for *p*,*p*′-divinylhydrazobenzene. (In the last example the structure of the products was not rigorously proved.)

(iii) Semidine formation only occurs from 4-substituted compounds, and although unsymmetrical substitution of the rings allows the possibility of two *o*- or *p*-semidines (or diphenylines), e.g. via 2,*N*′- or *N*,2′-linkage, in fact only one isomer is formed.

TABLE 5

Rearrangement products of hydrazobenzenes

Compound substituents in Ph		Products %				
		4,4′-linked	2,4′-linked	2,N′-linked	N,4′-linked	Dispr.
—	—	73	27	—	—	—
2-CH_3	2′-CH_3	100	—	—	—	—
3-CH_3	3′-CH_3	100	—	—	—	—
4-CH_3	4′-CH_3	—	—	40	—	60
2-CH_3O	—	100	—	—	—	—
2-CH_3O	2′-CH_3O	95	—	—	—	5
4-CH_3O	—	—	—	55	24	20
4-$NHCOCH_3$	—	—	+[a]	+++	++[a]	70
2-F	2′-F	86	—	—	—	14
2-Cl	2′-Cl	94	—	—	—	6
4-Cl	4′-Cl	—	—	22	—	75
4-Cl	—	—	19	30	20	31
2-Br	2′-Br	95	—	—	—	5
4-Br	4′-Br	—	—	30	—	70
2-I	2′-I	100	—	—	—	—
4-I	4′-I	—	—	—	—	100
2-Ph	2′-Ph	90	—	—	—	10
4-Ph	—	—	+++	+++[b]	—	38
4-Ph	4′-Ph	—	—	25	—	75
4-NO_2	—	—	+++	+++[b]	20	40
4-Vinyl	4′-Vinyl	(100% 2,2′-linked)				
4-Bu^t	4′-Bu^t	—	—	47	—	53
4-Bu^t	4′-Cl	—	—	54	—	46

Conditions and references as in Table 1.
+, ++, +++ give qualitative magnitude.
[a] Products detected only at acidities well above kinetic range.
[b] Whether 2,N′ or N,2′ was not decided.

(iv) 4-Substituted SO_3H, COOH and possibly $COOCH_3$ groups may be ejected without their bonding electrons during the formation of 4,4′-linked products. Jacobson (1922) reports that Cl and Br can be likewise expelled (now with their bonding electrons)—but this may be a consequence of the reaction conditions (acid—stannous chloride) for such dismutation is not detected in kinetically-controlled reactions (Cooper, 1966; O'Sullivan, 1966).

Recently a detailed rationalisation of substituent effects has been

TABLE 6

Rearrangement products of other hydrazoarenes

RNHNHR′		Products % 4,4′-linked	2,2′-linked
1-$C_{10}H_7$	1-$C_{10}H_7$	61	39
1-$C_{10}H_7$	2-$C_{10}H_7$	—	100
2-$C_{10}H_7$	2-$C_{10}H_7$	—	100
1-$C_{10}H_7$	Ph	44	56
2-$C_{10}H_7$	Ph	—	100

Conditions and references as in Table 2.

TABLE 7

Rearrangement products of m-substituted hydrazobenzenes

3,3′-diR, 5,5′-diR′-Compound		Products %			
R	R′	4,4′-linked	2,4′-linked	2,2′-linked	Dispr.
CH_3	CH_3	27	30	11	8
Br	Br	19	26	8	18
Cl	Cl	36	21	16	16
F	F	69	10	2	7
H	H	79	11	0	5
CH_3	Br	49	20	2	11
Cl	Cl	60	14	+	+
NH_2, H	NH_2, H	—	—	40	—

References: First 6 compounds, Carlin and Foltz (1956); penultimate compound, Vecera and Petranek (1960); last compound, Clovis and Hammond (1963).

presented (Banthorpe *et al.*, 1964). This may be summarised (neglecting disproportion for the present) under six headings, (v) to (x).

(v) Effect of 4-substituents: Products are either 2,4′-linked or *o*-semidines. When 4-R is strongly electron-releasing (R = OC_2H_5, OCH_3, CH_3; c. 12 examples) the product is predominantly *o*-semidine, but when 4-R is either feebly polar or electron-attracting only

($R = H$, $COOCH_3$, $[(CH_3)_2NH]^+$, NO_2, c. 12 examples) the alternative pathway to 2,4′-linkage is brought into prominence. When 4-R shows both strong electromeric electron-release, and strong inductive electron-attraction ($R = Cl$, Br, I, Ph), both *o*-semidines and 2,4′-diphenylines are found, the latter usually predominating. In addition, for four examples ($R = OCH_3$; $NHCOCH_3$; Cl; NO_2) of totally different polar properties, 4′,*N* linkage intrudes.

(vi) Orientation of *o*-semidine linkage for 4-substituents: Either 2,*N*′- or *N*,2′-linkage is possible, but the observations indicate that the former is exclusively found when $R = OC_2H_5$, OCH_3, CH_3, Cl, Br, I, and $NHCOCH_3$. Of these, the three halogeno-hydrazo-compounds react via the two-proton mechanism and the methoxy- and acetylamino-compounds via the one-proton route, and so in this case both mechanism and polarity are different between these two groups of compounds. It would be worthwhile to examine 4-methylhydrazobenzene which would almost certainly follow the two-proton pathway, but would show opposite polarity to the halogens. All these 4-groups are capable of electromeric electron-release; unfortunately the direction of linkage of the semidine from the 4-nitro derivative, which is unique in only exhibiting electron-attraction, has not been characterised.

(vii) Effect of 4,4′-disubstituents: Provided that the substituent is not expelled, these groups jointly promote *o*-semidine formation, and if non-equivalent the direction of linkage is directed by a 4-substituent towards the 2-position of its own ring in the order of orientating strength of scheme (14). Some 7 examples illustrate this order, with no exceptions, which is of decreasing electron-release.

$$C_2H_5O > CH_3 > CH_3COO, I, [(CH_3)_2NH]^+ \qquad (14)$$

(viii) Effect of 2- and 2,2′-substitution: A single 2-substituent exhibiting strong electron-release favours 4,4′-linked products if the 4- and 4′- position of the other ring are unsubstituted, and for 2,2′-disubstitution each group directs in the same way (some 25 examples include $R = OC_2H_5$, OCH_3, CH_3, Cl). If the other ring is blocked at the 4-position a 4,2′-diphenyline is formed as the major product together with *p*-semidine (4,*N*′-linkage); some 5 examples are recorded.

(ix) Relative orientating strength of 2- and 2,2′-substituents: These cannot be compared as directly as when the substituents are at 4-positions because substituents which, when present separately in the 2-position, would orientate biaryl linkage to the 4-position, when present together in the 2- and 2′-positions can only lead to 4,4′-linking, a result which gives no information as to their relative orientating strengths. However, a 2- and a 4′-substituent are in competition if present together, as the former orientates towards the 4-position and the latter towards the 2′-position. Now the products differ, orientation of the former kind leading to 4,2′-linkage, whereas that of the second type leads to *N*,2′-linkage. The known six examples are mutually consistent and give the orientational order of control as in scheme (15). Again electron-release is indicated as the basis of orientation.

$$C_2H_5O > CH_3 > CH_3CONH \qquad (15)$$

(x) Effect of substituents at *m*-positions: Substitution of hydrazobenzene in the 3,3′,5,5′-positions by similar substituents promotes 2,2′-linkage in products. If the common *m*-group is fluorine, such orientation is accompanied by 4,4′- rather than 2,4′-linked products, and if it is chlorine, bromine or methyl with an increase in 2,4′- at the expense of 4,4′-bonding. These results point to a mixture of polar, as above, and steric effects.

The most important orientational effects may be summarised in a single statement (Banthorpe *et al.*, 1964): ‘A 4-substituent, if strongly electron-donating, leads to 2,*N*′-linking; if not to 2,4′-linking; and if both strongly donating (electromeric) and attracting (inductive), to both types of linking. An electron-donating substituent, if at 4, orientates linking towards 2; and if at 2, towards 4, with a strength paralleling that of its electron-donation.’

This statement has no exceptions in the presently available literature.

2. *Variation of Products with Acidity*

For a few individual hydrazo-compounds product analyses have been made at acidities corresponding to kinetically-controlled one-proton, transitional, and two-proton mechanisms. Even outside the province of kinetic control the direction of change of mechanism with increasing acidity is so clearly defined (cf. section C.1), that product analyses at high acidities can be confidently assigned, in certain examples, to two-proton mechanisms.

2,2′-Dimethyl- and 2-methoxy-hydrazobenzenes both give 100 per cent 4,4′-linked products by the defined one-proton mechanism, in the range of transitional kinetics, and in a higher, presumed two-proton, region (references in Table 5). But in contrast, *N*-1-naphthyl-*N'*-phenylhydrazine gives a yield of 4,4'-linked product increasing from 45 to 65 per cent, at the expense of 2,2'-linkage, as the acidity increases from the one- to the two-proton region of mechanism. The latter result can be rationalised by the electrostatic influence of the second proton in the transition state pushing the top of the molecule apart and lengthening *N*-*N* and 2-2'-relative to 4-4' distances. This is consistent with the former observation that when 4,4'-linkage predominates in the one-proton mechanism, a second proton in the transition state does not change the products.

Electrostatic influences may cause a product-change within the range of a particular mechanism. The product-proportions vary with acidity for the one-proton mechanism of 3,3'-diaminohydrazobenzene (Hammond and Clovis, 1963), and also possibly (in this case kinetic control was not applied) for 4-acetoaminohydrazobenzene (Banthorpe, Cooper and Ingold, 1967). In both of these examples protonation of the ring amino-groups of intermediates formed after the rate-determining step could have a decisive and sensitively-controlled influence on products.

Another type of effect of acidity on products is in the formation of carbazoles from hydrazo-compounds with one or two naphthyl residues. The 2,2'-linked species from these compounds in the range of the one-proton mechanism (cf. Table 6) include up to 20 per cent carbazoles (equation (16)), which are primary products and are not derived from the 2,2'-diaminobiphenyls under the reaction conditions (Banthorpe *et al.*, 1962a, 1964). At higher acidities when the two-proton route is proved or presumed to compete with or supersede the

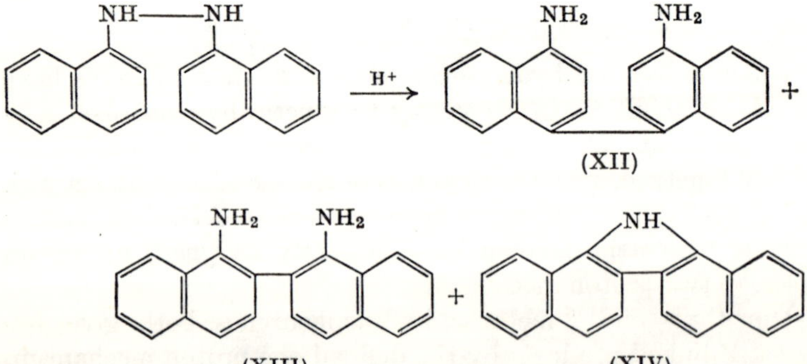

NH—NH $\xrightarrow{H^+}$ NH$_2$ NH$_2$ (XII) +

NH$_2$ NH$_2$ (XIII) + NH (XIV) (16)

one-proton mechanism, the proportion of carbazoles falls with a concurrent increase in yield of the 2,2′-linked diaminobiphenyl compound. For the reaction of equation (16) the carbazole yield falls from 17 per cent to 3 per cent as the acid concentration is increased from 1·3 *M* to 4·5 *M* perchloric acid. Probably a carbazole can only be produced at low acidities, for in order to secure elimination of ammonia from a presumed *o*-quinonoid intermediate (cf. section F.1)—one form of which is shown in equation (17)—one nitrogen must be nucleophilic and the other in a cationic form. At higher acidities a second proton can destroy the nucleophilic centre. Jacobson's (1922)

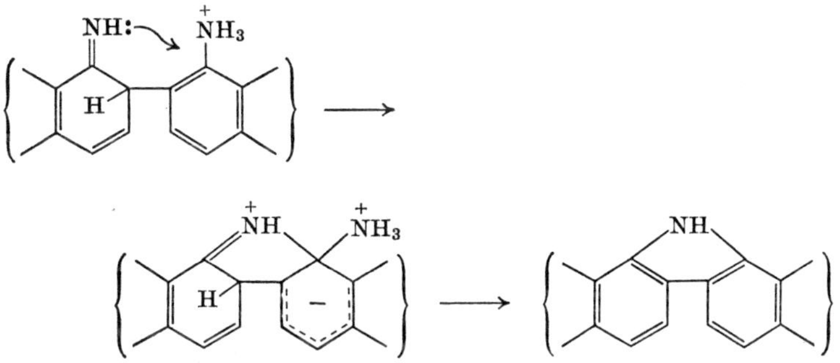

(17)

list of products does not include carbazoles: presumably the high acidities used in the stannous chloride–acid method gave negligible or small yields of these products that were not detected with the crude analytical methods then available.

Products from reaction of hydrazobenzene over the range 0·5 *M* to 17 *M* catalysing acid have been reported (Allan and Chmatal, 1964; Allan and Rakusan, 1966). The proportions vary considerably at high acidities from those obtained under the more usual reaction conditions, but this is not surprising in view of the high dielectric constant of the reaction medium and the low activity of water molecules (which act as proton-acceptors to complete the rearrangement) in these experiments.

3. *Products from Deuterated Substrates*

2,2′- and 4,4′-dideuterio-derivatives of 1,1′-hydrazonaphthalene have been rearranged under conditions of the one-proton mechanism in both 60 per cent dioxan–water and 95 per cent ethanol–water and the products compared with those from the isotopically-normal compound (Banthorpe, Hughes, Ingold and Humberlin, 1962). In both solvents, neither 2- nor 4-deuteration affected the ratio of 4,4′- to total 2,2′-ring-coupled products; 2-deuteration but not 4-deuteration did change the internal ratio of the two 2,2′-linked products XIII and XIV (equation (16)). Analyses are in Table 8. Similar studies with

TABLE 8

Products from deuterated and isotopically-normal 1,1′-*hydrazonaphthalene*
(0·05 *M* $HClO_4$; 60% dioxan–water; 0°)

	Products %		
	4,4′-linked (XII)	2,2′-linked (XIII)	Carbazole (XIV)
Protium compound	63·6	17·0	16·7
85% 2,2′-Dideuterio-compound	63·1	6·5	29·5
90% 4,4′-Dideuterio-compound	62·5	18·1	18·5

2- and 4-deuterated hydrazobenzenes gave no isotope effect on either the 4,4′- or 2,4′-linked products (Banthorpe and Hughes, 1962b). In terms of reasonable routes that can be written for reaction, with intermediates I_1 and I_2 losing protons to give 2,2′- (or 2,4′-) and 4,4′-linked products (schemes (18) and (19)), these results indicate the former pathway, as the alternative would show a product isotope effect owing to the competition for I_1. However the details of the

$$BH^+ \text{ or } BH_2^{++} \begin{cases} \nearrow I_1 \longrightarrow \text{2,2' or 2,4'-linked products} \\ \searrow I_2 \longrightarrow \text{4,4'-linked products} \end{cases} \quad (18)$$

$$BH^+ \text{ or } BH_2^{++} \longrightarrow I_1 \begin{cases} \nearrow I_2 \longrightarrow \text{4,4'-linked products} \\ \searrow \text{2,2' or 2,4'-linked products} \end{cases} \quad (19)$$

formation of the two 2,2′-linked species (cf. equation (17)) do allow a large isotope effect and imply activated proton-loss.

F. The Polar-Transition-State Mechanism

1. *General*

The transition state for most organic reactions is not far removed in electronic and geometric structure from either the initial or final states and so can be readily visualised in outline if not in detail. The benzidine rearrangement is novel in that its transition state is so unlike these datum points as to be obscure, although the intramolecularity suggests that the aromatic rings become parallel in this configuration. As a result, several distinct mechanisms have been proposed, although it was not until recently that an attempt was made to define the actual route by the study of kinetics and products.

The mass of data summarised in the previous two sections has been correlated on the basis of a so-called 'polar-transition-state mechanism' (Banthorpe *et al.*, 1964) which is a development of more restricted schemes (Hughes and Ingold, 1941; 1950) that were framed in then-popular resonance terminology. The modern theory can accommodate all the recent information.

The mono- or di-protonated hydrazo-molecule is considered, on this basis, to heterolyse to an incipient mono- or di-cation. For a di-protonated species one charge is anchored on nitrogen and the other is carried by the ring, predominantly residing, to minimise electrostatic repulsion, on the 4-position (cf. equation (20)). For the monocation, as formed from 1,1′-hydrazonaphthalene, the charge is delocalised as in equation (21). The positive charge represents an electron-deficiency and resides on the 2- and 4-positions. For each mechanism the incipient

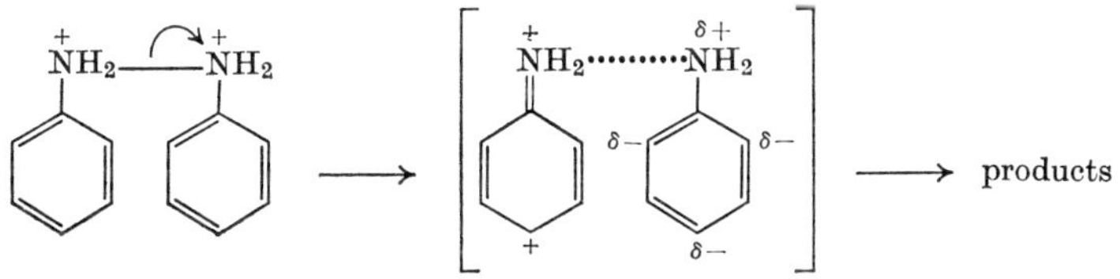

$\overset{+}{N}H_2$—$\overset{+}{N}H_2$ ⟶ [$\overset{+}{N}H_2$⋯⋯$\overset{\delta+}{N}H_2$ ($\delta-$, $\delta-$, $\delta-$; $+$)] ⟶ products (20)

Transition state

cation is formed adjacent to an incipient neutral species capable of undergoing activated electrophilic attack at the 2- and especially 4-ring positions. Thus a strong electrostatic interaction between potential bonding sites, together with the polar character of the *N*,*N* bond in the transition state allowing a low bending-constant of the *C*,*N* bond, permits energetically-cheap changes of shape to meet the stereochemical requirements for rearrangement, and a concerted

NH—$\overset{+}{N}H_2$ → [Transition state: NH, $\overset{\delta+}{N}H_2$, $\delta+$, $\delta-$, $\delta+$, $\delta-$] (21)

Transition state

↓

Products

movement of the electronically distinct *quasi*-fragments into coplanarity is possible. This may be assisted by electrokinetic interactions between the π-orbitals of the rings (such as occur in graphite), and also by the low effective dielectric constant of the medium at the atomic distances involved. The weakly directed electrovalencies develop into well-directed covalencies and internal substitution occurs to lead to rearrangement products. An important factor permitting approach of the potential bonding sites is the above-mentioned relaxation of bending force-constants (Hammick and Mason, 1946) which allows a 4-4′ distance of about 1·50 Å to be achieved; in the absence of this relaxation the distance is about 4·0 Å but even this latter distance would permit fairly strong electrovalencies to develop. Energetically, rearrangement is favourable (Hammick and Mason, 1946; 1949): the energy released by the establishment of the biaryl covalency (c. 59 kg. cal. mole^{-1}) is quite adequate to allow for *N,N* fission (c. 24 kg. cal) and the transformation of a benzenoid to a quinonoid ring (32 kg. cal).

When a family of isomeric products is formed from a particular hydrazo-compound the transition-states can be regarded as a family of cols, all within a few kT of one another but on separated reaction coordinates, the energy-relationships of which can be delineated by a study of the variation of isomer ratio with temperature. It is unlikely that the transition states collapse directly to the finally isolated products. The differing sensitivities to medium of rate and products (cf. section E) indicate intermediates at a later stage than the transition state, and the lack of a kinetic isotope-effect shows that the future biaryl bond is established before either of the aromatic hydrogens that have to be displaced are lost. A family of σ-bonded intermediates of quinonoid structure, analogous to those occurring in aromatic S_E2 reactions, are likely; they lose protons to give benzenoid products in subsequent fast steps (Banthorpe *et al.*, 1964).

This description of the rearrangement process has much in common with the anodic coupling of *N*,*N′*-dimethylaniline and *N*-methylaniline to give the corresponding methylated benzidines (Galus and Adams, 1962; Galus, White Rowland and Adams, 1962; Galus and Adams, 1963; Mizoguchi and Adams, 1962). In this case it was proved by a variety of electrochemical and other techniques that 2-electron oxidation occurred to give the dication (XV) which coupled with an amine molecule to give the products. Electrolytic or cerous salt

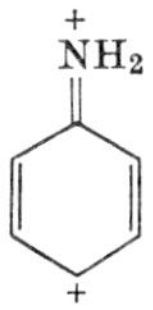

(XV) 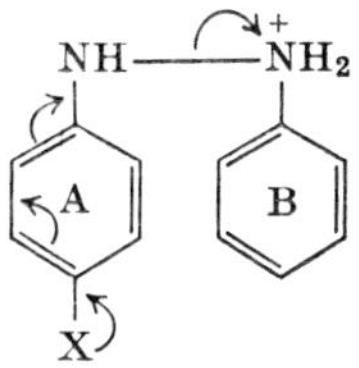 (XVI)

oxidation of aniline however involved 1-electron transfer to form anilino-radicals that coupled to form benzidine with considerably less efficiency than the previous route (Mohlinger, Adams and Argersinger, 1962; Sterba, Sagner and Matrika, 1965).

2. *Application to Kinetics*

Data from Tables 1 and 5 show that under a particular set of conditions, the rearrangements, oxidations and disproportionations of a hydrazo-compound have the same kinetic form, and suggest a common basic mechanism as is embodied in the above theory.

The polar-transition-state model readily accommodates the two acid-catalysed routes. The first-added proton prepares for *N*,*N* heterolysis by producing an electronic dissymmetry. For hydrazo-molecules with sufficiently electron releasing-groups (X) in ring A (XVI), protonation on the nitrogen of ring B will induce rearrangement with electron-transfers as shown. If X does not possess sufficient electron-release the second catalytic proton has to be called into play at the other nitrogen to destabilise the molecule: this second proton destroys the favoured dissymmetry, in fact it tends to favour homolysis, but it also weakens the *N*,*N* link by adjacent-change repulsion and allows the electron-transfers outlined in the previous sub-section to take place.

The clue to the factors which control the change from the one- to the two-proton mechanisms came from application of the theory to the first example of transitional kinetics, 2,2′-dimethylhydrazobenzene (Banthorpe *et al.*, 1964). The 2-methyl group donates electrons to its ring and favours heterolysis of the *N*,*N* bond, but it also reduces the basicity of its hydrazo-nitrogen relative to the situation in aniline. This base-weakening effect, probably due to steric inhibition by the 2-group of the development of the solvation shell accompanying protonation of the amino group, has the effect of increasing the electron-affinity, which is built-up on protonation, to an extent which is beyond the capacity of the solvation shell to satisfy. This affinity has to be satisfied by the taking of electrons from the *N*,*N* bond, so causing extensive polarisation, and the development of an electrostatic character that is transferred to the other bonds during rearrangement. This polarisation is aided, in the example considered, by electron-release from the 2-methyl group in the other ring. In principle, either factor, the release of electrons from substituents or the base-weakening effect, could raise the one-proton mechanism to predominance, but for many substituents the two factors act oppositely, e.g. for a 4-nitro group electron-attraction accompanies base-weakening, and for a 4-methyl group electron-release accompanies base-strengthening; reliable predictions cannot therefore be made. Only in the cases such as 2,2′-dimethyl substitution do both factors collaborate and raise the one-proton route to importance.

All our present information, as summarised in Tables 1 and 2, can be rationalised along these lines. The benzo-substituents of 1- and 2-naphthyl residues, which are more electron-releasing and more base-weakening than 2-methyl substituents, do promote the one-proton route almost exclusively. Structurally, 4-4′-divinyl groups should show features analogous to 2-naphthyl residues and again the one-proton route is promoted. Further, 2,2′-dimethoxy-substitution should be similar to a 1-naphthyl residue as regards electron-supply and should show about the same base-weakening effect as 2,2′-dimethyl substitution; here the one-proton mechanism is found although the 2-methoxy-substituted compound shows transitional kinetics. A 4-methoxy group, although slightly more base strengthening, shows much greater electron-release and promotes the one-proton route exclusively.

If the change from the two-proton mechanism, the mode for unsubstituted hydrazobenzene, to the one-proton route is assisted by the electron-release factor, a rate-increase, relative to that for hydrazobenzene, is expected. On the other hand if the base-weakening factor controls the mechanistic change, the latter should be accompanied by a large drop in absolute rate. When both factors are important, marginal rate differences should occur from that of the parent. The data of Table 3 and other results not included there (Banthorpe, Cooper, Ingold and O'Sullivan, 1967) can be well interpreted on this basis as the following few examples show.

The rate-constants for hydrazobenzene and its 2,2-diX-derivatives (X = F, Cl, Br) are in the order 100,000 : 160 : 2 : 1·8 for the two-proton mechanism and the corresponding rate for the iodo-compound is undetectably slow. This is the order expected from a combination of increasing base-weakening caused by the presence and bulkiness of the 2-groups and decreasing electron-release due to lessening overlap of the *p*-orbitals of the halogens with the aromatic π-system. The rates for the one-proton route for 2,2′-dimethyl, 2,2′ dibromo, and 2,2′-di-iodo-compounds lie in the approximate proportions 3000 : 1·5 : 1—the last being the most extreme case known of the elevation to prominence of the one-proton mechanism due to repression by deactivating substituents of the two-proton route. 4,4′-Dibromo-, -dichloro-, and -di-iodo-hydrazobenzenes do not possess suitably positioned substituents to show the base-weakening properties that can promote the one-proton route and all follow the two-proton route with only mild (c. 300-fold) deactivation compared with hydrazobenzene.

The very polar transition state required on the basis of our accepted mechanism is also consistent with salt and solvent effects on rate. Such effects on both types of acid-catalysed mechanisms (cf. section C.3) indicate that the dominant polarity is due to charge created in the transition state rather than that carried over by the catalytic proton(s).

3. *Application to Products*

The pattern of products from unsubstituted hydrazonaphthalenes and hydrazobenzene under the conditions usually employed, with the importance of 2,2′- relative to 4,4′-linkage in the former but the complete absence of 2,2′-linkage in the latter, at first sight suggests electronic control in intermediate or transition states of quinonoid character (Hammick and Mason, 1946). For it is well known that the *p*-quinonoid bond-arrangement is greatly favoured for benzene compounds, whereas the *o*-quinonoid alternative plays an important part in the reactions of naphthalenes. However, such theories, based on reactivities in independent aromatic ring-systems, cannot account for the entire picture. In particular the approximate 70:30 proportion of 4,4′- and 2,4′-linkage from hydrazobenzene, accompanied by *no* 2,2′-linked compound, and the comparable quantities of 4,4′- and 2,2′-linked products from both 1,1′-hydrazonaphthalene and *N*-1-naphthyl-*N*′-phenylhydrazine accompanied by *no* 2,4′-products, show that the true explanation must be found in some more detailed mutual relationship between the rings. The polar-transition-state theory, using considerations of shape-changes and the distribution of reactivity in the rearranging structures, can rationalise all the current product-data: several examples follow.

(a) Hydrazobenzene

The charge-distribution in the transition-state is shown in scheme (20). Under the usual rearrangement conditions the rings fold into parallel planes and 4,4′-linkage predominates, but also the positive charge at 4 can be attracted to the centroid of the negative charge in the counter-fragment and the flexibility of the polarised *N*,*N*-bond allows relative displacement of the rings to permit appreciable 2,4′-linkage. Any change in reaction conditions, such as the use of solvents of low dielectric constant (in which electrostatic interactions are magnified) or of high temperatures (when thermal motion is increased), which favour ring-displacement would be expected to favour 2,4′- and even 4,*N*′- or 2,*N*′-products. Various sets of results support this prediction (Vecera *et al.*, 1957a, b; Vecera and Petranek, 1958; Allan and Rakusan, 1966; Vecera, Synek and Sterba, 1960). *N*,4′- and *N*,2′-linkage (to give identical products to those from the previous linkings in this particular example) would not occur, as electron-deficiency necessary for bonding is not present in an ammonium centre.

(b) 1,1′*Hydrazonaphthalene*

Here a transition state as in scheme (21) is formed, in which one aromatic fragment has nearly one positive charge residing on the 2- and 4-positions rather than on the nitrogen which has the larger effective nuclear charge, whereas the other fragment is neutral but dipolar. The two aryl residues come together congruently without relative longitudinal displacement to give 4,4′- and 2,2′-, but no 2,4′-linked products. A similar situation holds for the one-proton rearrangement of *N*-1-naphthyl-*N*′-phenylhydrazine: here the phenylamine moiety is the more basic and the naphthyl group is the better donor of the two aryl groups and so the quasi-cationic fragment resembles that from hydrazonaphthalene and the quasi-neutral fragment resembles that from hydrazobenzene. Again congruence without displacement leads to 2,2′ and 4,4′-linked products.

(*c*) *Substituted Hydrazobenzenes*

Analogous reasoning rationalises the products. For compounds with electron-releasing 4-substituents, the transition state has a charge distribution as in scheme (22) (for the two-proton mechanism), and 2,*N*′-linkage occurs. Again electrophilic attack by either the ′onium

NH_2—$\overset{+}{N}H_2$ (X) ⟶ [$\overset{+}{N}H_2$⋯⋯$\overset{\delta+}{N}H_2$; X^+ ; δ− δ− δ−] ⟶ Products (22)

Transition state

positive charge of the nitrogen or group X cannot occur, but the positive charge on the other nitrogen, caused by electron deficiency, can be so utilised. 2,*N*′-Linkage occurs for $X = OCH_3$. With $X = CH_3$, hyperconjugative electron-release in the transition state occurs and the quinonoid bond-distribution of scheme (22) is less fully developed, whereas with X = Hal, electron-release is weaker still, as conjugation is restricted by the electronegativity and size of substituent, and now only part of the product is 2,*N*′-semidine (cf. Table 5). When 4-X is insufficiently electron-donating to produce these effects the configuration of the transition state will revert towards that of hydrazobenzene itself, and as 4,4′-linkage is excluded by the presence of the substituent, 2,4′-linkage will arise. This last type forms part of the products for all the 4-halogenated compounds and all the product from the parent hydrazobenzene, if we exclude the major 4,4′-product. A similar detailed analysis has been presented for 2-substituents

(Banthorpe *et al.*, 1964): *p*-semidine formation will be considered in section H.

(*d*) A recent report of a so-called meta-rearrangement (Rakusan and Allan, 1966), equation (23), is worthy of comment, although the study failed to meet modern standards in that no attempt at kinetic control (or interpretation) was offered, and the out-dated technique of conducting the rearrangement by acid-reduction of the azo-compound was used. This observation can be rationalised as follows (Ingold,

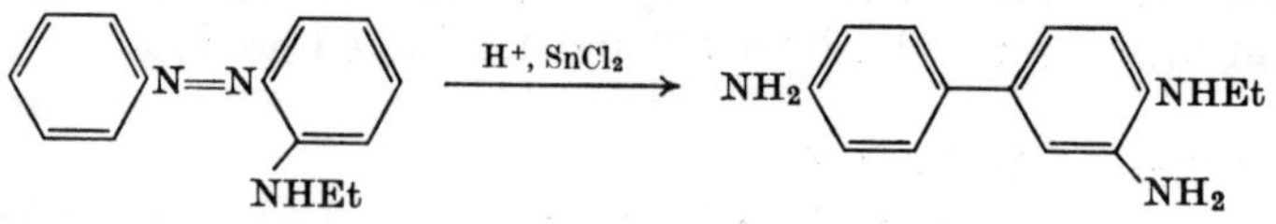

N=N NHEt $\xrightarrow{H^+, SnCl_2}$ NH_2 NHEt NH_2 (23)

personal communication): If kinetic studies of the pure hydrazo-compound were made, 4,4′-coupling at low acidities in the region of the one-proton mechanism would probably be found but if the acidity were increased into the region of the two-proton mechanism, the 4,5′-coupled product would occur, for here an unprecedented situation arises in the cationic fragment. Previously studied 2-substituents, although supplying electrons in the transition state, are also predisposed to regain them as products are formed, but here the substituent can retain its charge and allow the hydrazonitrogen to recover the electrons. The situation is outlined in scheme (24).

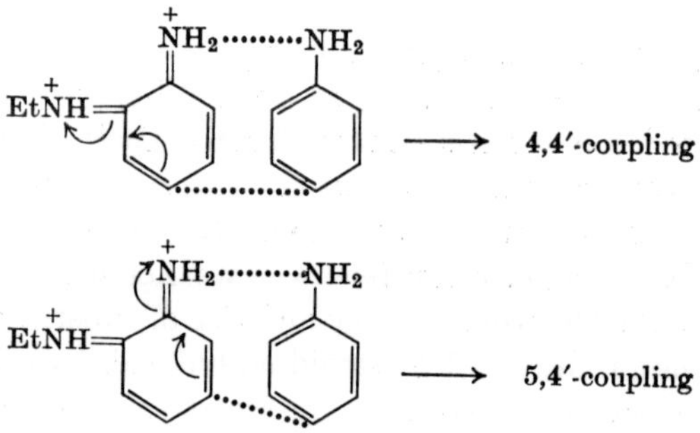

$\overset{+}{N}H_2 \cdots\cdots NH_2$ $Et\overset{+}{N}H=$ → 4,4′-coupling

(24)

$\overset{+}{N}H_2 \cdots\cdots NH_2$ $Et\overset{+}{N}H=$ → 5,4′-coupling

G. Other Proposed Mechanisms

The recent accumulation of experimental data has allowed the polar-transition-state theory to be put on a firm footing, but previously speculation was rife and several other mechanisms were proposed.

Of these, two are deserving of mention as seriously attempting to account for the intramolecularity and for stereochemical problems.

1. *The π-Complex Mechanism*

This is an application of a general theory that π-complexes are intermediates in a variety of organic reactions (Dewar, 1949). Although known to be in error on many specific points, this theory in its general application has never been formally repudiated and so it has often come to be accepted by default. For the benzidine case, the *N*,*N* bond is supposed to split and be replaced by a delocalised covalent π-bond between the rings which holds them in parallel planes with the possibility of mutual rotation. Products follow from replacement of the delocalised bond by a localised interatomic σ-bond. This model is superficially attractive and has been vigorously (but somewhat uncritically) argued on several occasions (Dewar, 1949, 1959, 1963; Dewar and Marchand, 1965). Although theoretical arguments for the existence and stability of such π-complexes are tenuous (Coulson and Dewar, 1947), a claim to justify the concept by molecular-orbital calculations has appeared (Snyder, 1962). However in its predictions the theory completely fails to meet the test of the new experimental facts.

Dewar originally assumed one-proton catalysis. On this basis *N*,*N′* fission led to an incipient electron-deficient species $(C_6H_5NH)^+$ the vacant orbital of which overlapped with the π-orbital of the quasi-amine (in much the same way as complexes between, say, silver ion and aromatics are formed) and anchored the two fragments. 4,4′-, 4,2′-, 2,*N′*-Linking, etc., followed via rotation of the components of the complex. The discovery of the two-proton mechanism and of transitional kinetics was an embarrassment for this theory, as diprotonated π-complexes were considered unlikely, and could only be accommodated on a scheme (25).

$$\phi NHNH\phi \overset{H^+}{\rightleftharpoons} \phi\overset{+}{N}H_2NH\phi \rightleftharpoons \pi\text{-complex} \xrightarrow{H^+} \text{Products} \quad (25)$$

Application of the steady-state approximation to the π-complex gave an equation of the form of (9) (section C.1) allowing the extremes of one- or two-proton mechanisms or intermediate kinetics. But the two last kinetic forms require rate-determining proton transfer, for which we have seen (section C.2) there is no evidence;

and also equation (9) cannot accommodate the observed form of the variation of order with acidity.

The π-complex theory also falls down on three important specific kinetic observations. First, when the theory was adjusted as above to account for the second catalytic proton it was pointed out that whereas formation of 4,4′-linked benzidine was second-order in acid, semidine formation should be linear (Dewar, 1951) because non-proton-assisted rotation of the initially formed π-complex would have to be faster than proton-assisted breakdown of the same complex. Contrary to this prediction, semidine formation is second-order in acid in several examples, and indeed always has the same kinetic form as the reaction leading to 4,4′-linked or other biaryl products. An analogous discrepancy between theory and fact at this time was the identical kinetic form for benzidine and diphenyline-formation from hydrazobenzene; as diphenyline is supposed to originate from a rotated π-complex it should also follow the one-proton mechanism.

Second, any π-complex formed from 1,1′-hydrazonaphthalene should be more stable than the corresponding species from hydrazobenzene; this follows from theoretical considerations and from experimental data for metal-ion and halogeno-complexes. Consequently if rearrangement follows from attack by a proton on a π-complex as is postulated by Dewar to account for the two-proton kinetics of hydrazobenzene, the naphtho-compound should require the second catalytic proton *a fortiori*. In fact it reacts much more rapidly than hydrazobenzene with one-proton kinetics.

Third, it has been claimed that transitional kinetics or the one-proton mechanism arise for 4,4′-disubstituted hydrazobenzenes when bulky substituents interfere with the stability of the π-complex and cause its rotation into a more sterically-favoured situation, with consequent lowering of kinetic-order in acid from the modal two. Examples of this effect, but with no supporting experimental data, were claimed for 4-chloro-4′-t-butyl and 4-methyl-4′-chlorohydrazobenzenes (Dewar and McNicol, 1959); but the previously recorded quadratic dependence on acidity shown by 4,4′-dimethylhydrazobenzene (Bunton *et al.*, 1957; Carlin and Wich, 1958), which did not fit this theory, was ignored. More recently the 4-chloro-4′-t-butyl compound has been reinvestigated and shown to follow approximately two-proton kinetics (order $\sim 1{\cdot}9$), and the corresponding di-t-butyl compound, which should be an outstanding example of transitional or one-proton kinetics if the π-complex route were involved, was shown to follow strictly the two-proton mechanism over the range studied (Shine and Chamness, 1967).

The transition states of formation and decomposition of the π-complexes have not been defined and so modifications could still be built into the theory to account for the observed salt and solvent effects, and no prediction is possible.

π-Complex theory is no more successful in accounting for products. A survey of the literature led to the proposal of a set of 'simple rules' (Dewar, 1949, 1959) for the prediction of products from substituted hydrazobenzenes, and these rules were claimed to be consistent with this theory; similar but more restricted rules had been put forward earlier (Ritter and Ritter, 1931). A reappraisal of this claim has shown that there is no observational basis for these rules (Banthorpe *et al.*, 1964).

The π-complex theory cannot account for the products from hydrazo-compounds with one benzene and one naphthalene residue. The explanation proposed for the predominance of 2,2′- and 4,4′-linkage on hydrazonaphthalenes required the locking effect of the extended π-orbitals preventing relative rotation of the rings (unlike the situation in hydrazobenzenes when overlap between only two aromatic rings can take place). Thus when one naphthalene ring is replaced by a benzene residue the possibility of free rotation should be introduced; however the pattern of products as found in the hydrazonaphthalenes still persists.

Finally, one of the firmest props of the theory was the alleged lack of rearrangement products from 4,4′-diphenylhydrazobenzene (Bell, Kenyon and Robinson, 1926). This was attributed to steric factors and π-cloud repulsions preventing π-complexing and hence rearrangement (Dewar 1959), and was contrasted to the situation for 2,2′-diphenylhydrazobenzene where such factors were absent and rearrangement was normal. Reinvestigation of the 4,4′-compound has shown that although 80 per cent oxidation and disproportionation occurs, the balance *is* rearrangement (Shine and Stanley, 1967): the latter being not appreciably less than is found with other 4,4′-disubstituted substrates (Banthorpe *et al.*, 1967).

Thus the π-complex theory incorrectly predicts and cannot account for the situation with respect to both kinetics and products. A specific test was devised by isolating unreacted 3,3′-diaminohydrazobenzene initially labelled with ^{15}N at the hydrazo-group after 50 per cent reaction (Hammond and Clovis, 1963): rotation within a π-complex would have led to tracer scrambling between the hydrazo and amino groups, but none was detected.

Recently the theory has been adapted in an attempt to incorporate two-proton kinetics without introducing general acid-catalysis (Dewar and Marchand, 1965): a diprotonated π-complex is now introduced despite previous theoretical objections to such a species (Snyder, 1962; Dewar, 1963) and various new steps are proposed. This adaption merely reduces the theory to a tenuous statement, of no predicative value, that the potentially dissociable fragments are held together during reaction. Nevertheless the basic idea dies hard and π-complexes have been postulated, with no accompanying evidence, as being possible intermediates following the polar-transition state on the routes to rearrangement or disproportionation (Hammond and Clovis, 1962, 1963; Shine and Chamness, 1967).

The intermediary of a novel 'proton-sandwich' whereby a catalytic proton is situated between, and holds together, the rings of a rearranging hydrazo-compound (Ferstandig, 1963) has been very soundly criticised (Dewar and Marchand, 1965), and is quite unable to account for the kinetics and products of rearrangements in general. A charge-transfer complex, related to a diprotonated π-complex, has been proposed (Murrell, 1962) but there is no evidence that complexes of the required structure are stable above about $-80°$, and such species would be molecular complexes, held by Van der Waals interactions rather than covalently bonded π-complexes.

2. *Radical-Pair Mechanism*

Tetra-arylhydrazines form radicals under conditions where no rearrangement is detectable (Wieland, 1908; Schlenck and Bergmann, 1927), i.e. dimerisation is the preferred mode of reaction; but it was implied that protonation of such radicals prevented *N*,*N*-coupling and favoured 4,4′-linkage in order to minimise electrostatic interactions (Wieland 1913). This appears to be a valid view, e.g. no 'rearrangement' products are formed from intermolecular coupling of the neutral amino radicals produced on oxidation by various techniques of primary arylamines (Edward, 1954; Fox and Waters, 1965; Morgan and Aubert, 1962; Orr, Sims and Manson, 1956) but 4,4′-coupling predominates on oxidation of 1-naphthylamine in acid conditions (Sah and Yuin, 1939).

If ion-radicals are intermediates in rearrangement, and calculations suggest that they are considerably more stable than the corresponding neutral radicals (Snyder, 1962), they must be generated within a solvent cage in which repeated collision leads to recombination, in order to preserve intramolecularity (Kenner, 1932). Such a mechanism was proposed to account for the variation of products from hydrazobenzene with changing solvent (Vecera, Synek and Sterba, 1960); although it is difficult to see how the conclusion follows from the reported results.

A major criticism of this radical-pair mechanism is the difficulty of explaining why the fragments stay caged with no detectable leakage or attack on the solvent shell. The 100 per cent primary recombination required is rarely approached for neutral radicals in other reactions and seems unlikely for ion-radicals of the same charge.

Another objection is that the radical-pair mechanism cannot accommodate the one-proton mechanism. In the two-proton mechanism homolysis could be reasonably stimulated by adjacent-charge repulsion, but a transition state containing one proton would show little such tendency and the observed easy concurrence of one- and two-proton mechanisms would be unlikely. Nor can the problem be overcome by proposing a heterolytic one-proton mechanism accompanying a homolytic two-proton mechanism. When we start from a symmetrical hydrazo-compound and effect first a mono-substitution and then a symmetrical disubstitution with the same group X, we observe two successive shifts of mechanism in the same direction; on the basis of the proposal these shifts should be in opposite directions. A dichotomy of homolytic and heterolytic mechanisms also cannot account for the very similar salt and solvent effects shown by the one- and two-proton mechanisms. These should be very different for the two types of fission and in particular formation of radicals could not allow sufficient polarity to develop in the transition state to display the large observed effects. Homolysis should be favoured due to the influence of magnetic catalysis if iodine is either substituted in a hydrazo-compound near to the *N*,*N*-bond or is present in the solvent (Leffler, Faulkner and Petropoulos, 1958), but 2,2′-di-iodohydrazobenzene rearranges with the slowest rate on record and furthermore it uses the one-proton mechanism which can only be heterolytic. Nor is rearrangement of hydrazobenzene accelerated when conducted in media containing dissolved iodine or organo-iodine compounds (Banthorpe, Cooper, Ingold and O'Sullivan, 1967; O'Sullivan, 1966).

The products of rearrangements cannot be rationalised by the radical-pair hypothesis. The orientation of biaryl and semidine linking is a polar phenomenon in which each ring influences the other, and it is inconsistent with dissociated fragments. Independent radicals would possess an internal distribution of reactivity for a particular structure that is independent of its origin, and this is certainly not observed (Dewar, 1959). For example, protonated amino radicals can be shown theoretically to carry about one seventh of the free electron density on the nitrogen atom, and so contrary to experience such intermediates would be expected to lead to appreciable yields of semidines irrespective of the substituents in the parent hydrazo-compound.

Most experiments to detect the formation of radicals during rearrangement have failed. No electron spin resonance (e.s.r.) signal could be detected during typical reactions proceeding by either one- or two-proton mechanisms and polymerisation of added monomers could not be induced (Banthorpe, Bramley and Thomas, 1964). E.S.R. signals have been recorded during rearrangement of certain *N*,*N*-bridged hydrazobenzenes of the type in equation (4), section A, but were assigned to side-oxidations of products (Snyder, 1962; Wittig, Borzel, Neumann and Klau, 1966; Shine and Stanley, 1967). A bridged diradical was trapped by nitric oxide in one of these reactions (Wittig *et al.*, 1966), but such trapping does not prove that the diradical was an intermediate in rearrangement. The formation of polymeric products on attempted rearrangement of tetraarylphenylhydrazines was attributed to the attack of radical-ions on rearrangement products (Seidel and Hammond, 1963; Hammond, Seidel and Pincock, 1963). These substrates are expected to show a great tendency to homolysis, and only one example (from 4,4′-di-vinylhydrazobenzene) of such products from hydrazoarenes is on record (Shine and Chamness, 1963). Radical intermediates have also been suggested in the rearrangements of 3,3′-diamino-hydrazobenzene and 4,4′-dimethylhydrazobenzene (Clovis and Hammond, 1963; Hammond and Clovis, 1963), but again this is sheer speculation based on the observation that for the latter compound an oxidising agent was generated during reaction which oxidised Würster base to Würster blue. A similar oxidation could not be demonstrated for 4,4′-dichlorohydrazobenzene (Banthorpe and Williams, unpublished work) and the identification of the oxidising agent as a radical is arbitrary.

Radical-pair formation has recently lost favour with one group of its protagonists (Sterba and Vecera, 1966) but, as with π-complexes, it has been suggested as occurring in a kinetically insignificant step following steps involving polar-transition states (Shine and Stanley, 1967). There is no evidence to support this viewpoint.

3. *Miscellaneous Theories*

An early heterolytic mechanism (Robinson, 1941) was based on a so-called theory of small electronic oscillations which has little theoretical justification and which explicitly brushed aside problems of intramolecularity and stereochemistry. An even more unphysical and obscure approach postulated 'resonating semi-polar bonds' in the transition state (Shamin-Ahmed and Hasan, 1952).

A most attractive proposal (Brownstein, Bunton and Hughes, 1956) invisaged 4,4′-linkage as having progressed via a 2,2′-linked intermediate, scheme (26), with obvious adaptions to other modes of biaryl and semidine linking. The presence of the branching point

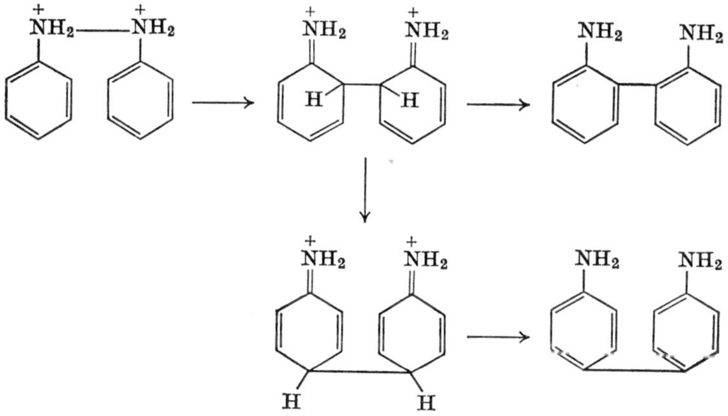

(26)

would allow a product isotope effect caused by 2,2′-dideuteration in a substrate such as 1,1′-hydrazonaphthalene forming comparable quantities of 2,2′ and 4,4′-linked products, but this is not found. Also it has proved difficult to develop a rationalisation of the pattern of products from substituted hydrazobenzene using this mechanism.

An ill-defined mechanism has been repeatedly advocated (Lukashevich and Krolik, 1959; Lukashevich, 1964, 1967) which denies, on the strength of the preparative experiments outlined in section D, the existence of diprotonated substrate on the reaction path, and questions the use of kinetics for elucidating the composition of the transition state. Reaction is supposed to be promoted by formation of a complex between mono-protonated substrate and an undissociated molecule of acid, but the absence of general acid catalysis is ignored, and the whole outlook seems completely erroneous (Shine, 1967).

H. Rearrangements Forming *p*-Semidines

p-Semidines were reported as the major rearrangement products of 4-methoxy-, 4-ethoxy-, 4-amino- and 4-acetoamino-hydrazobenzenes and as minor products from 4-chloro- and 4-bromo-compounds (Jacobson, 1922), although these compounds are usually difficult to isolate in good yield owing to air-oxidation. The acid-stannous-chloride reduction of azo-compounds was used for all save the 4-chloro-compound, and here the azo-benzene was reduced by ammonium sulphide before treatment with hydrochloric acid (Jacobson, 1909).

Subsequently 4-ethoxy and various other hydrazobenzenes were reduced and rearranged by hydrogen–Raney nickel in acetic acid and no *p*-semidine was isolated (Hammick and Munro, 1950). Calculations of bonding-distances from a model with reduced bending force-constants in the transition state showed that the distance for *p*-semidine formation was approximately 4·9 Å compared with 2·5 Å for *o*-semidine formation, and it was concluded—ignoring the special case of 4-chlorohydrazobenzene—that the presence of heavy metal ions was necessary for *p*-semidine formation, the favourable stereochemistry for this route apparently only being achieved through metal-complexing. This general conclusion was derived from a single set of experiments in which no attempt was made to achieve a product balance, and the interpretation has been doubted (Vecera and Petranek, 1960; Shine and Stanley, 1967).

More recent work (cf. Table 5) shows that 4-methoxy-, 4-acetamido-, 4-chloro- and 4-nitro-hydrazobenzenes all give up to 24 per cent *p*-semidine when rearranged in the absence of heavy metal ions, and a somewhat smaller value is obtained on reinvestigation of 4-ethoxyhydrazobenzene under kinetically-controlled conditions, also in the absence of potential chelating catalysts (Shine, Baldwin and Harris, 1968). Comparative studies on 4-amino- and 4-acetoamino-hydrazobenzene under a variety of conditions, e.g. (*a*) 60 per cent dioxan–water, perchloric acid, (*b*) heterogeneous conditions with ethereal hydrochloric acid and (*c*) reduction of the azo-compound with acid-stannous chloride or Raney nickel in acetic acid, all showed qualitatively similar products containing about 20 per cent *p*-semidine, and this type of compound could be detected as a few per cent of the products from hydrazobenzene itself (Cooper, 1966). Of particular interest was the formation of *p*-semidine at acidities above 1·5 *N*; this is presumably due to protonation changing the +E effect of the group to a –I effect which changes the properties towards that shown by chlorine, a substituent known to give *p*-semidine. There is thus no observational basis for Hammick and Munro's theories, and the polar transition state theory can be shown by calculation to accommodate the bonding distances involved when the two rings are relatively displaced, for here the *N*,4′-distance can be reduced to c. 3 Å which would allow an electrostatic bond of quite sufficient strength to be formed (Banthorpe, Cooper and Ingold, 1967). It is possible that *p*-semidine formation may result when heterolysis occurs in the opposite direction from normal. Such a function would be harder to achieve but would be efficient in leading to these products.

Both *o*- and *p*-semidine formation should be favoured when the rings can be displaced parallel to each other in the transition state as is favoured at high temperatures and when the reaction is carried out in media of low dielectric constant. In agreement, semidines account for appreciable proportions of products when hydrazobenzene is treated with a solution of hydrogen chloride in benzene (Dziurzynski, 1908; Lukashevich and Krolik, 1948, 1959). Heterogeneous reaction conditions also appear to favour these products (Vecera and Petranek, 1958; Krolik and Lukashevich, 1960).

J. Thermal Rearrangement

2,2′- and 4,4′-linked rearrangement products are formed when solutions of 1,1′- and 2,2′-hydrazonaphthalene in benzene or ethanol are heated at 95° (Krolik and Lukashevich, 1949). Subsequently a long-standing report (Meisenheimer and Witte, 1903) that 2,2′-hydrazonaphthalene gave 2,2′-diamino-1,1′-dinaphthyl on refluxing in alcoholic alkali was proved to be a thermal, rather than a base-catalysed, process (Shine, 1956). Several other reactions reported to proceed at elevated temperatures under the influence of sometimes exotic catalysts, e.g. the rearrangement of hydrazobenzene promoted by a solution of phosgene in benzene under reflux (Klauke and Bayer, 1963), are almost certainly similar thermal processes.

Thermal, also called no-proton or non-catalytic, rearrangement has been shown to be general for hydrazoarenes. If precautions are taken to conduct the reaction in the absence of oxygen, very little disproportionation or oxidation accompanies rearrangement of hydrazonaphthalenes (Shine and Trisler, 1960; Shine, Huang and Snell, 1961; Banthorpe, 1964) although this side-reaction is always more pronounced for hydrazobenzenes (Vecera, Gasparic and Petranek, 1957a; Banthorpe, 1964). Qualitatively the products of thermal and acid-promoted rearrangements are the same (Vecera *et al.*, 1957a; Shine *et al.*, 1961; Banthorpe, 1964; Banthorpe and Hughes, 1964a) and the differences in proportions of products may largely reflect the difference in reaction conditions. Thus semidines are favoured in acid-catalysed rearrangement when non-polar media and high temperatures are employed, and the same products commonly occur on thermal rearrangement in dry organic solvents, either under reflux or in sealed, heated tubes, or in the melt.

Early mechanistic speculations concerned either an unspecific but direct electronic reorganisation (Krolik and Lukashevich, 1949) or a radical mechanism (Vecera *et al.*, 1957a). Rearrangements of certain hydrazobenzenes, hydrazonaphthalenes and mixed benzene-naphthalene compounds have been proved strictly intramolecular by the lack of cross-over products from either the single rearrangement of unsymmetrical substrates A—B or from concurrent rearrangements of compounds A—A and B—B (Vecera *et al.*, 1957a; Shine *et al.*, 1961; Banthorpe 1964; Banthorpe and Hughes, 1964b). Recent kinetic studies on the reactions of 1,1′- and 2,2′-hydrazonaphthalenes in protic (methanol, ethanol) and dipolar aprotic (methyl cyanide, acetone etc.) solvents (Shine, 1956; Shine and Trisler, 1960; Banthorpe and Hughes, 1964a) and a large amount of qualitative data on the rates of reaction of 1,1′-hydrazonaphthalene in some 22 solvents of all classes (Lukashevich and Krolik, 1962) have enabled the mechanism to be delineated. The main kinetic conclusions are:

(i) The reaction is first-order in substrate.

(ii) The rate in hydroxylic solvents is much greater than in non-hydroxylic. For the former class the rate falls with polarity:

$$H_2O > MeOH > EtOH > i\text{-}PrOH > t\text{-}BuOH.$$

For non-hydroxylic media the rate still follows polarity:

$$CH_3CN > (CH_3)_2CO > \text{Dioxan} > \text{tetrahydrofuran} > \text{hydrocarbons}.$$

(iii) The rate order is:

$$1,1'\text{-Hydrazonaphthalene} > 2,2'\text{-Hydrazonaphthalene} \gg \text{Hydrazobenzene}.$$

The last inequality is derived from qualitative observations, as rates of decomposition of hydrazobenzenes have not been measured.

The solvent effects strongly suggest that although the initial state is neutral and only weakly polar, considerable charge separation is formed in the transition state (in polar media at least) and the thermal rearrangement develops largely electrostatic bonds similar to those proposed by the polar-transition-state theory for proton catalysis; i.e. despite the differences in net charge the no-, one- and two-proton mechanisms follow the same pattern. The thermal route is not activated by the uptake of protons and so requires temperatures about 100° greater than those typically used in acid-catalysis. This rules out the possibility of observing transitional kinetics and overlap between the no- and one-proton mechanisms. Hydroxylic solvents probably exhibit pseudo-acid catalysis by hydrogen-bonding to the hydrazo-nitrogen (Shine, 1956).

The transition state for the thermal process of hydrazobenzene probably consists of a quasi-cationic portion, like that described for one-proton catalysis, linked to a quasi-anionic fragment, figure XVII

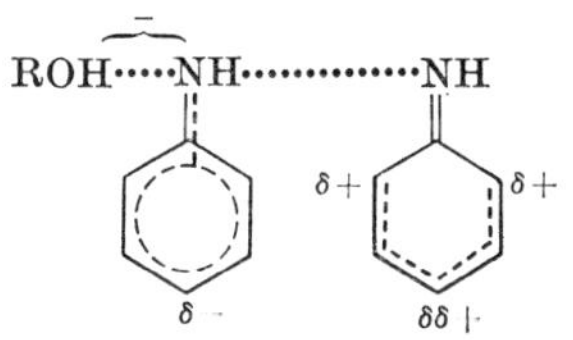

(XVII) 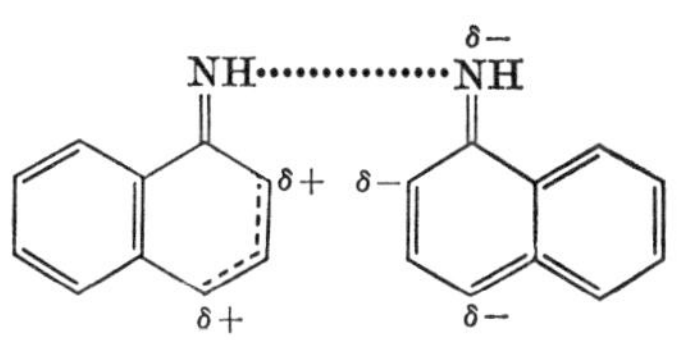 (XVIII)

(Banthorpe *et al.*, 1964). The quasi-cation will have its positive charge distributed between the 2- and 4-positions, but with most on the former because the *exo*-double bond is essentially static and so electron-delocalisation is concentrated on the pentadiene system the ends of which capture most of the charge. The anionic fragment has a large fraction of the negative charge residing on nitrogen, the element of highest effective atomic number, although in a protic solvent this charge will be neutralised and dispersed by hydrogen-bonding as shown. Nearly all the rest of the negative charge will reside on the most remote 4-ring position, where in a substituted anilide residue it will be carried also partly by any electron-absorbing 4-substituent. For hydrazobenzene there is no close correlation between this electronic configuration and that of the two-proton state. Transition state XVII favours 2,4-linkage, a preference accentuated in a solvent of low polarity where all electrostatic interactions are increased.

Experiment indeed shows that in contrast to the predominant 4,4′-linking characteristic of acid-catalysis, the thermal rearrangement predominantly leads to 2,4′-linked products in ethanol and methylcyanide, but in the latter solvent it is partially, and in the less-polar acetone largely, replaced by 2,*N*′-semidine.

For 1,1′-hydrazonaphthalene the transition state should have a configuration XVIII which bears a much closer resemblance to the one-proton mechanism, and the rings will approach congruently to give similar products. This view is verified by the experimental results and similar products are found from the other hydrazonaphthalenes (Banthorpe, 1964) although a solvent-dependence difference in ratio of products is also superimposed. Insomuch as the activation energy of any reaction is used mainly in advancing the leading bond-fission, which here is assumed to be *N*,*N*-heterolysis as in the acid-promoted mechanisms, the parallelism of rate of the no- and one-proton routes is reasonable.

An important feature of the thermal rearrangement is the effect of 2- and 4-deuteration (Banthorpe and Hughes, 1964a). Kinetic-isotope-effects (kH/kD) of 2 to 6, values quite normal for reactions whose rate-determining step involves C—H fission, are found for the reaction of 1,1′-hydrazonaphthalene in ethanol, acetone and methylcyanide. This indicates that at least one of the two protons that have to be lost from the ring to effect rearrangement is pushed forward to an early portion of the reaction co-ordinate into the region of the rate-determining step. On the basis of the polar-transition-state mechanism this is expected to apply to the loss of the first proton (Banthorpe *et al.*, 1964), and can be rationalised as follows. The electrostatic interactions in the transition state are cation : dipole for the one-proton mechanism, but cation : anion for the thermal route. The electrons forming the new biaryl bond must come from the less positive aromatic ring; i.e. from the dipolar ring in the one-proton mechanism and from the quasi-anionic ring in the thermal transition state. Either electron-donating ring will then receive a substitute pair of electrons from its hydrogen-atom that is expelled as a proton. This proton, always from the less positive of the two aromatic rings, will be the first to be expelled; the second will be subsequently lost from the pseudo-cationic moiety. But that portion of the structure is the same for the two transition states. Hence isotopic differences between the mechanisms will be essentially only felt in the loss of the first aromatic hydrogen.

In comparison with the situation for the one-proton mechanism a considerably facilitated hydrogen-transfer from the ring in the quasi-anionic portion would be expected for the thermal mechanism, the reason being the tendency for this proton to move to the adjacent main centre of negative charge in the neighbouring ring, viz., the nitrogen atom (the corresponding nitrogen for the one-proton mechanism is the positive end of a dipole). If the recovery of electrons from the hydrogen of the aromatic ring were made so much easier for the thermal mechanism that they occurred concurrently with supply of electrons from the ring to form the new biaryl bond, then a kinetic isotope-effect would occur on deuteration. The tendency of the aromatic hydrogen to move to the nitrogen of the adjacent ring is reduced if the latter is hydrogen-bonded and its negative charge is dispersed. Hence the transfer of the aromatic proton would be less important in polar media and this is consistent with the observations that low isotope effects occur when ethanol is solvent.

In addition to the previously discussed pattern of products there are two other fields in which differences between the non-catalytic and acid-catalytic rearrangements for hydrazonaphthalenes appear systematic. One is that when acid conditions are replaced by neutral there is a shift away from 4,4′- towards 2,2′-linking, and the shift strengthens markedly as the solvent becomes less polar along the series ethanol, methylcyanide, acetone. This is consistent with the strong ion–ion interaction in the transition state of the thermal reaction which would lead to stronger *N*,*N*-attraction and hence a smaller shift from initial to final position of bonding than in the acid-promoted mechanism. This special feature would appear more markedly in generally less polar solvents. Secondly, both one- and no-proton mechanisms produce carbazoles from hydrazonaphthalenes, but the latter does not do so when an alcoholic solvent is used which is made strongly basic with the corresponding alkoxide ion (Lukashevich and Krolik, 1962). Here what would have been carbazole appears as an extra quantity of 2,2′-linked diamine. This is a strong argument that bond changes, alike in the one- and no-proton mechanisms, are essentially heterolytic, for the elimination of ammonia in carbazole-formation requires one nitrogen to be in basic and the other in cationic form, the latter being generated in the no-proton route from either complete hydrogen-transfer or via hydrogen-bonding from the solvent. Too much protonation will destroy the basic centres and too little the cationic; evidently the alkoxide ions deprotonate the reacting system.

Full tables of the product data on which these conclusions are based may be found in a recent discussion (Banthorpe *et al.*, 1964).

This proposed heterolytic mechanism only applies to the situation in protic and some dipolar-aprotic media. In benzene complicated kinetics are found for rearrangement of the hydrazonaphthalenes and tar-formation occurs to considerable extents in this solvent, in hydrocarbon solvents and in the melt. The reactions still appear intramolecular but nothing is known about mechanism. Caged-radical formation is possible although no e.s.r. signal could be obtained from rearranging solutions (Banthorpe, Bramley and Thomas, 1964). The π-complex theory appears conceptually inapplicable to these processes.

An interesting adaption of the thermal rearrangement is the formation of tetramethylbenzidine from treatment of azobenzene with methyl iodide at 100° (Pongratz *et al.*, 1944). Methylation may precede or follow rearrangement but the details are obscure.

K. Oxidation and Disproportionation

1. *Promoted by Heat and Ultra-Violet Radiation*

Early studies on the acid-stannous chloride reduction of azo-compounds often led to considerable fission to amines (Jacobson, 1922). It is now known that acids, bases, heat and ultra-violet radiation can all induce oxidation and disproportionation of hydrazo-compounds, and under certain conditions these can compete with or supersede rearrangement.

The initial studies on the decomposition of hydrazobenzenes in ethanol at 140° in the absence of oxygen reported formation of azo-compounds and fission amines but overlooked the later-discovered accompanying rearrangement. The reactions were first-order in substrate and a radical-dissociation mechanism to form the species PhNH was proposed (Stieglitz and Curme, 1913a, b). It was pointed out (Wieland, 1912a, b) that a fragmentation was not necessitated by the results and oxidation and reduction could occur with equation (27) as the rate-determining step. In support of this view was the

$$\phi\text{NHNH}\phi' \longrightarrow \phi\text{N{=}N}\phi' + 2\text{H} \qquad (27)$$

discovery (Wieland, 1915) that thermal decomposition of unsymmetrically-substituted hydrazo-compounds gave the two fission amines in equivalent amounts and only one (unsymmetrical) azo-compounds in each example. The radical-dissociation mechanism would lead to some 'cross-over' azo-compound accompanied by a not necessarily 1:1 correspondence of the amines. However, hydrogen could not be detected amongst the products of the related disproportionation-oxidation promoted by ultra-violet light of wavelength 230 to 270 mμ. Therefore a radical route was assumed in which the orientation of encounter of the fragments was governed by their polarity and so led to disproportionation rather than recombination to substrate, as shown in scheme (28) (Weiss, 1940).

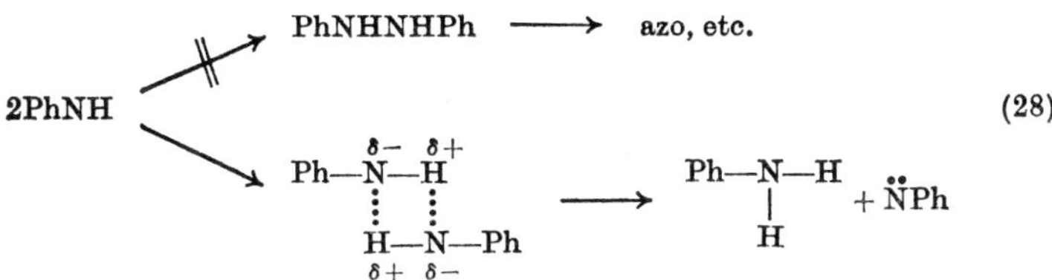

Nevertheless, for the thermal reaction at least the demonstration of only one azo-product from an unsymmetrical hydrazobenzene implies intramolecular oxidation, and this has been recently verified. Decomposition of ^{15}N-labelled hydrazobenzene in ethanol at 150° gave azobenzene with no randomisation of tracer such as would have occurred if kinetically independent radicals had been formed (Holt and Hughes, 1953) and a similar result was obtained for the UV-promoted decomposition (Holt and Hughes, 1955). A direct reduction of one hydrazo molecule by another in a four-centre transition state is ruled out by the first-order kinetics and a scheme (29) was proposed

$$\begin{aligned}
\mathrm{PhNHNHPh} &\xrightarrow{\text{slow}} \begin{cases}\mathrm{PhNH\dot{N}Ph + \dot{H}} \\ \text{or } \mathrm{2Ph\dot{N}H}\end{cases} \\
\mathrm{PhNH\dot{N}Ph} &\longrightarrow \mathrm{PhN{=}NPh + \dot{H}} \\
\mathrm{PhNHNHPh + \dot{H}} &\longrightarrow \mathrm{PhNH_2 + Ph\dot{N}H} \qquad (29) \\
\mathrm{PhNHNHPh + Ph\dot{N}H} &\longrightarrow \mathrm{PhNH_2 + PhNH\dot{N}Ph}
\end{aligned}$$

for both types of reaction with various chain-terminating steps leading to PhNHNHPh, $PhNH_2$ or PhN=NPh. Similar ^{15}N-tracer studies showed that azobenzene from the thermal decomposition of triphenylhydrazine (Ph_2NNHPh) involved randomisation of tracer (Holt and Hughes, 1955); a different but as yet unelucidated mechanism must obtain in this case. It is not clear if the photolytic reaction gave any rearrangement, but the thermal decomposition of hydrazobenzene is now known to do so.

2. *Promoted by Acids*

Disproportionation and oxidation accompanying the acid-catalysed rearrangement of a pure hydrazobenzene was first characterised in 1933 (Ingold and Kidd). Table 5 shows that this can account for from 20 to 100 per cent of the total disappearance of substrate for 4,4′-disubstituted hydrazobenzenes and the phenomenon is of special interest for the light it sheds on the intermediates of the rearrangement process.

The constancy of kinetic form during particular runs, and in different runs for substrates with varying proportions of rearrangement and disproportionation-oxidation, a constancy embracing both the one- and two-proton routes, shows that the oxidation-disproportionation shares transition states of the same stoichiometric composition as that for rearrangement. In particular the percentage of non-rearrangement products is independent of temperature, ionic strength and acidity over the range of kinetically-controlled conditions (Carlin and Wich, 1958) and is similarly also independent of the initial concentration of hydrazo-compound (Cooper, 1966; Shine and Stanley, 1967). As an example, the percentage of disproportionation products varied only within the range 20 to 23 per cent as the initial concentration of 4-phenylhydrazobenzene was varied 20-fold, the other reaction conditions being held constant (Cooper, 1966). The unique first-order dependence on substrate for the disproportionation-oxidation of all studied hydrazobenzenes, irrespective of the order in acid, invalidates earlier views (Dewar, 1949), that have been doubted and rejected (Vecera *et al.*, 1960; Shine, 1967) but recently revived (Lukashevich, 1967), that the rate-determining step is the reaction of a mono- or di-protonated species with a neutral hydrazo molecule. The kinetic form shows that the second hydrazo molecule that must enter the reaction to complete the stoichiometry of equation (30) must do so later than the rate-determining step and at a low energy on the reaction co-ordinate so as to make its intervention essentially irreversible.

$$2\phi\text{NHNH}\phi' \longrightarrow \phi\text{N}{=}\text{N}\phi' + \phi\text{NH}_2 + \phi'\text{NH}_2 \qquad (30)$$

The following data enable the reaction to be further delineated.

(i) Intermolecular reactions involving radical-like nitrenes or their protonated species do not occur. Such intermediates would attack the solvent (Bamberger, 1921) to give amino-phenols and other oxidation products which have never been observed during rigorous chromatographic screening of the products of several disproportionations (Cooper, 1966). Also, no retardation of rate is caused by the addition of radical inhibitors and induction of polymerisation of added reactive monomers cannot be achieved (Cooper, 1966); nor can e.s.r. signals be detected (Shine and Stanley, 1967).

(ii) Again, if intermolecular, the best trap for a singlet ϕN species is another ϕ'N species, and so cross-over should occur in azo-products. But substrates AB go to A′B′ only: in particular no detectable 4,4′-dichloroazobenzene resulted from reaction of 4-chlorohydrazobenzene (Cooper, 1966).

(iii) Crossed-oxidation can however be achieved. If a slow-reacting hydrazo-compound is added to a solution of a much faster-reacting compound under conditions such that the first reacts insignificantly but the latter almost completely during the reaction-time, it is found that the unreactive-compound is oxidised to azo-compound and an equivalent quantity of the reactive-compound is cleaved to fission amines. By application of this principle azobenzene was isolated from a reaction solution containing hydrazobenzene and its 4,4′-dimethyl analogue (Hammond and Clovis, 1963) but the experimental record was rather unsatisfactory in that autoxidation may have occurred during working-up of the products. The point has since been proved with greater rigour for this pair of compounds and for 4-acetamidohydrazobenzene and hydrazobenzene (Cooper, 1966; Banthorpe, Cooper, Ingold and O'Sullivan, 1967) and also for 4,4′-diphenyl- and 4,4′-dichloro-hydrazobenzenes (Shine and Stanley, 1967). In all these systems no azo-compound is formed from the more unreactive component unless the reactive component is present to accept reducing power. Rearranging solutions of 4,4′-dimethylhydrazobenzene can also oxidise Würster base to Würster blue (Hammond and Clovis, 1963). Such a reaction cannot be carried out by products or reactants but the base may be regarded as a vinylogue of hydrazobenzene, and so it is quite reasonable that an intermediate that can oxidise one should oxidise the other.

In summary, an oxidising intermediate is formed during disappearance of the hydrazo-compound that can manifest itself by oxidation of either unreacted substrate or of additives. There is one recorded example (4,4′-dimethylhydrazobenzene) where presumed oxidation of the aqueous alcoholic solvent leads to a lack of equivalence of formation of azo-compound (20 per cent) and fission-amine

(40 per cent), the inbalance corresponding to an equivalent amount of formation of acetaldehyde which, however, has not been isolated (Carlin and Wich, 1958; Hammond and Clovis, 1963). An obvious possible oxidising species is the dication that has fully separated from the heterolysing substrate without being captured *en route* to rearrangement, but in the example of 4,4′-dimethylhydrazobenzene, at least, this is ruled out by lack of exchange of tracer between the 4-methyl group and solvent that would be expected if an equilibrium (31) were set up (Hammond and Clovis, 1963).

$$\overset{+}{N}H_2{=}C_6H_4{-}\overset{+}{}CH_3 \rightleftharpoons \overset{+}{N}H_2{=}C_6H_4{=}CH_2 + H^+ \qquad (31)$$

The above observations can readily be interpreted on the polar transition state theory, despite an erroneous view (Shine and Stanley, 1967; Shine, 1967b) that this theory cannot accommodate the rate-determining formation of an intermediate prior to the product-forming step. The quinonoid intermediates, one of which is shown in scheme (32), can accept electrons as formally shown in a bimolecular reaction with a neutral hydrazo-compound, after the rate-determining

(Coupled process)

$$PhNHNHPh' \longrightarrow PhN{=}NPh' + 2e^- + 2H^+$$

$$2e^- + \overset{+}{N}H_2{=}C_6H_4(H){-}(H)C_6H_4{=}\overset{+}{N}H_2 \longrightarrow 2NH_2C_6H_5 \qquad (32)$$

(XX)

step, to give the observed products. In order to accommodate the constant proportion of non-rearranged products with variation of substrate concentration, the reasonable assumptions must be made that formation of the species XX is irreversible and that when 4- or 4,4′-substituents that cannot be ejected are present, *all* the species XX goes through to fission products. Thus the presence of additional neutral, potentially oxidisable species cannot increase the yield of these products, for XX is not, in these cases, partitioned between rearrangement and disproportionation. When no 2- or 4-substituent is present, direct competition between rearrangement and disproportionation is possible, and now the proportions of these two reactions should be dependent on the initial concentration of substrate, for the products (*although not the kinetics*) of disproportionation are dependent on the bimolecular reaction. The only relevant data (Vecera *et al.*, 1960) record a small increase in non-rearrangement products as the initial concentration of hydrazobenzene is increased. The effect is small, but it is in the expected direction and would be worth re-examination.

For most hydrazobenzenes, excluding those with 4- or 4,4′-substituents, rearrangement is at least some 20- to 100-fold faster than oxidation-disproportionation and the factor is even greater for hydrazonaphthalenes. On the above theory, the prediction could be made that the intermediate quinonoids (2,4′-, 4,4′-linked etc.) would be stabilised in media of low polarity and in aqueous solutions at high acidities—in both conditions when proton loss from the intermediates is different to achieve—and so the redox process leading to amines and azo-compounds might be here allowed to compete with rearrangement on more favourable terms. High temperatures, too, should favour the normally less important process. The available data do indicate that oxidation-disproportionation is favoured under these conditions (Vecera *et al.*, 1960; Hashimoto and Sunamoto, 1964; Allan and Chmatel, 1964). Particularly pertinent is the study of products from *m*-tetrasubstituted hydrazocompounds under conditions of high acidity (65 per cent sulphuric acid) in heterogeneous conditions at high temperatures (90°). Here some 7 to 18 per cent oxidation-disproportionation is found for substrates where direct competition between this mode of decomposition and rearrangement is possible, and for which oxidation-disproportionation is negligible under the kinetically-controlled conditions of low acidities in aqueous media at 0 to 20°.

Of practical interest is the reduction in azo-formation that is achieved, e.g. in the case of 4-chlorohydrazobenzene (Yamada *et al.*, 1954, 1955), when an ion-exchange resin is used as catalyst, compared with that observed under normal kinetically-controlled conditions. Here the approach of the second substrate molecule to the bonded quinonoid intermediate is presumably hindered.

The occurrence of oxidation and disproportionation has led to speculations concerning the intermediary of π-complexes or radicals after the rate-determining step (Hammond and Clovis, 1963; Shine and Stanley, 1967) but there is no evidence for such species. The polar-transition state theory and its reasonable intermediates can accommodate all the current data.

3. *Promoted by Bases*

Oxidation to azo-compounds has been reported after direct treatment of hydrazonaphthalenes with base (Shine, 1956), and when the use of buffers has depressed acid-catalysed routes but brought into prominence reactions catalysed by buffer-anions (Banthorpe *et al.*, 1962a). An attractive route is scheme (33), but difunctional catalysis as in scheme (34) by 2-hydroxypyridine has not been detected

$$\text{PhNHNHPh}' \xrightarrow{B^-} \text{Ph}\bar{\text{N}}\text{NHPh}' \longrightarrow \text{PhN} + \text{Ph}'\bar{\text{N}}\text{H} \longrightarrow \text{PhN{=}NPh} + \text{Ph}'\text{NH}_2 \qquad (33)$$

$$\text{[2-hydroxypyridine: N} \cdots \text{HNPh; OH} \cdots \text{NHPh}'] \longrightarrow \text{PhN} + \text{H}_2\text{Ph}' + \overset{+}{\text{N}}\text{H (2-oxidopyridinium, O}^-) \qquad (34)$$

(Cooper, 1966). Oxygen quantitatively oxidises hydrazobenzene to azobenzene and a dianionic intermediate has been proposed. The catalysis by metal ions, especially cupric salts, has been studied in detail (Blackadder and Hinshelwood, 1957b).

No details concerning intramolecularity or kinetics of any of these reactions are known.

L. Related Rearrangements

The closest relative of the benzidine rearrangement is the Fischer-Indole Synthesis. This involves an arylhydrazone of an aldehyde or ketone and is believed to precede via the *o*-benzidine-like rearrangement as in scheme (35). No *p*-rearrangement has been recorded: this would have a very unfavourable stereochemistry and if it occurred would lead to products that would form tars (as are detected) under the reaction conditions. The reaction has been exhaustively reviewed (Robinson, 1963). Of especial interest is the observation that phenols react as their keto-tautomers, and so direct analogies to the benzidine rearrangement can be achieved. An example is reaction (36). Very low yields of products are obtained and tracer-studies using ^{15}N show

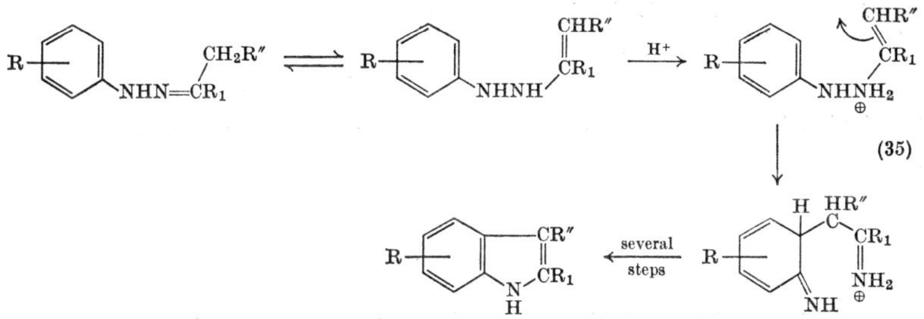

(35)

that the route shown occurs concurrently with a pathway via an *o*-semidine (Holt and McNae, 1964). The products are the same as those from the acid-catalysed rearrangement of *N*-2-naphthyl-*N'*-phenylhydrazine, but the conditions of the two reactions are so far removed that no rationalisation of the proportions is worth while.

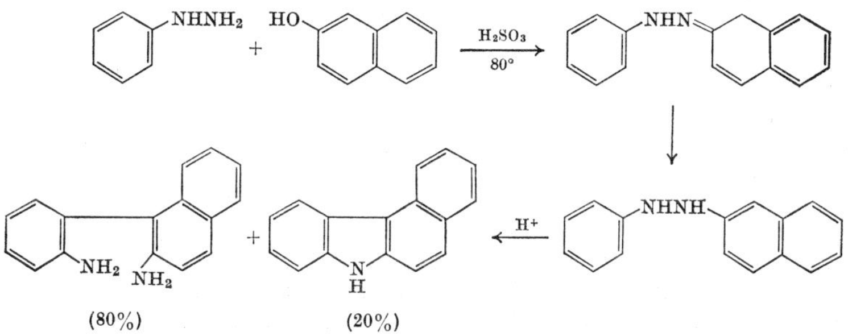

(36)

(80%) (20%)

Other analogues of the benzidine rearrangement are the heat and acid-promoted isomerisations of compounds of the general formula ϕ-XY-ϕ' ($\phi = \phi' =$ aryl) to products which usually are analogous to semidines. A brief list includes: X = S, Y = S (Shine and Bear, 1957); X = N, Y = O (Cox and Dunn, 1963); X = N, Y = S (Möller, 1958); and X = O, Y = NH (Parish and Whiting, 1964). Although speculations as to π-complex intermediates have been advanced, in no cases has the mechanism been defined or even characterised as intermolecular or intramolecular.

ACKNOWLEDGMENTS

I would like to express my deep gratitude to the late Professor E. D. Hughes and to Professor Sir Christopher Ingold who introduced me to the study of this rearrangement some 9 years ago, and with whom I have had numerous stimulating discussions and constant advice and help concerning the subject.

I also thank Sir Christopher Ingold, Dr. A. Cooper and Mrs. June Winter for reading the manuscript and making many helpful suggestions.

References

Allan, Z. J. and Chmatal, V. (1964) *Coll. Czech. Chem. Comm.* **29**, 531.
Allan, Z. J. and Rakusan, J. (1966) *Coll. Czech. Chem. Comm.* **31**, 3555.
Arcus, C. L., Howard, T. J. and South, D. C. (1964) *Chem. Ind.* (*London*) 1756.
Bamberger, E. (1921) *Liebigs. Ann.* **424**, 233.
Banthorpe, D. V., Hughes, E. D. and Ingold, C. K. (1962a) *J. Chem. Soc.* 2386.
Banthorpe, D. V. and Hughes, E. D. (1962a) *J. Chem. Soc.* 2402.
Banthorpe, D. V. (1962a) *J. Chem. Soc.* 2407.
Banthorpe, D. V. (1962b) *J. Chem. Soc.* 2413.
Banthorpe, D. V., Hughes, E. D. and Ingold, C. K. (1962b) *J. Chem. Soc.* 2418.
Banthorpe, D. V. (1962c) *J. Chem. Soc.* 2429.
Banthorpe, D. V., Ingold, C. K., Roy, J. and Somerville, S. M. (1962) *J. Chem. Soc.* 2436.
Banthorpe, D. V., Hughes, E. D., Ingold, C. K. and Roy, J. (1962) *J. Chem. Soc.* 3294.
Banthorpe, D. V., Hughes, E. D., Ingold, C. K. and Humberlin, R. (1962) *J. Chem. Soc.* 3299.
Banthorpe, D. V. and Hughes, E. D. (1962b) *J. Chem. Soc.* 3308.
Banthorpe, D. V. and Hughes, E. D. (1962c) *J. Chem. Soc.* 3314.
Banthorpe, D. V. and Hughes, E. D. (1964a) *J. Chem. Soc.* 2849.
Banthorpe, D. V. (1964) *J. Chem. Soc.* 2854.
Banthorpe, D. V. and Hughes, E. D. (1964b) *J. Chem. Soc.* 2860.
Banthorpe, D. V., Hughes, E. D. and Ingold, C. K. (1964) *J. Chem. Soc.* 2864.
Banthorpe, D. V., Bramley, R. and Thomas, J. A. (1964) *J. Chem. Soc.* 2890.
Banthorpe, D. V., Thomas, J. A. and Williams, D. L. H. (1965) *J. Chem. Soc.* 6135.
Banthorpe, D. V., Cooper, A. and Ingold, C. K. (1967) *Nature, Lond.* **216**, 232.
Banthorpe, D. V., Cooper, A., O'Sullivan, M. and Ingold, C. K. (1968) *J. Chem. Soc.* (*B*) 605 *et seq*.
Banthorpe, D. V. and Williams, M. R. (unpublished observations).
Barnes, C. S., Pausacker, K. H. and Badcock, W. E. (1951) *J. Chem. Soc.* 730.
Bell, F., Kenyon, J. and Robinson, P. H. (1926) *J. Chem. Soc.* 1239.
Beyer, H. and Kreutzberger, A. (1952) *Chem. Ber.* **85**, 333.
Beyer, H. and Haase, H. J. (1957) *Chem. Ber.* **90**, 66.
Beyer, H., Haase, H. J. and Wildgrube, W. (1958) *Chem. Ber.* **91**, 247.
Biilmann, E. and Blom, J. H. (1924) *J. Chem. Soc.* **125**, 1719.
Blackadder, D. A. and Hinshelwood, C. N. (1957a) *J. Chem. Soc.* 2904.
Blackadder, D. A. and Hinshelwood, C. N. (1957b) *J. Chem. Soc.* 2911.
Bloink, G. J. and Pausacker, K. H. (1950) *J. Chem. Soc.* 950.

Brownstein, S., Bunton, C. A. and Hughes, E. D. (1956) *Chem. Ind.* (*London*) 981.
Bunnett, J. F. (1961) *J. Am. Chem. Soc.* **83**, 4956 *et seq.*
Bunton, C. A., Ingold, C. K. and Mhala, M. (1957) *J. Chem. Soc.* 1906.
Carlin, R. B., Nelb, R. G. and Odioso, R. C. (1951) *J. Am. Chem. Soc.* **73**, 1002.
Carlin, R. B. and Odioso, R. C. (1954a) *J. Am. Chem. Soc.* **76**, 100.
Carlin, R. B. and Odioso, R. C. (1954b) *J. Am. Chem. Soc.* **76**, 2345.
Carlin, R. B. and Foltz, G. E. (1956) *J. Am. Chem. Soc.* **78**, 1992.
Carlin, R. B. and Wich, G. S. (1958) *J. Am. Chem. Soc.* **80**, 4023.
Clemo, G. R. and Dawson, E. C. (1939) *J. Chem. Soc.* 1114.
Clemo, G. R. and Lee, T. B. (1954) *J. Chem. Soc.* 2417.
Clovis, J. S. and Hammond, G. S. (1963) *J. Org. Chem.* **28**, 3290.
Cohen, M. D. and Hammond, G. S. (1953) *J. Am. Chem. Soc.* **75**, 880.
Colonna, M. and Risaliti, A. (1956) *Gazz. Chim. Ital.* **86**, 288.
Colonna, M. and Risaliti, A. (1959) *Gazz. Chim. Ital.* **89**, 2493.
Cooper, A. (1966) Mechanism of the benzidine rearrangement. *Ph.D. Thesis* (*London*).
Coulson, C. A. and Dewar, M. J. S. (1947) *Disc. Faraday. Soc.* **2**, 54.
Croce, L. J. and Gettler, J. D. (1953) *J. Am. Chem. Soc.* **75**, 874.
Cox, J. R. and Dunn, M. F. (1963) *Tetrahedron Letters* 985.
Das-Gupta, B. C. and Bose, P. K. (1929) *J. Ind. Chem. Soc.* **6**, 495.
Davies, D. W. and Hammick, D. Ll. (1954) *J. Chem. Soc.* 475.
Dewar, M. J. S. (1946a) *J. Chem. Soc.* 406.
Dewar, M. J. S. (1946b) *J. Chem. Soc.* 777.
Dewar, M. J. S. (1949) *Electronic Theory of Organic Chemistry*, Oxford Univ. Press, London, p. 233.
Dewar, M. J. S. (1951) *Ann. Reports. Prog. Chem.* (*London*) **48**, 126.
Dewar, M. J. S. (1959) *Theoretical Organic Chemistry*, Kekule Symposium, Butterworths, London, p. 195.
Dewar, M. J. S. and McNicol, H. (1959) *Tetrahedron Letters* No. 5, 22.
Dewar, M. J. S. (1963) *Molecular Rearrangements*, edited P. de. Mayo, Vol. 1, Interscience, New York, p. 323.
Dewar, M. J. S. and Marchard, A. P. (1965) *Ann. Rev. Phys. Chem.* **16**, 338.
Dziurzymski, M. (1908) *Bull. Acad. Sci. Cracovie*, 401 [*Chem. Abstr.* **2**, 2790].
Edward, J. T. (1954) *J. Chem. Soc.* 1464.
Fedorova, I. P. and Mironova, G. F. (1962) *Zh. Obshch. Khim.* **32**, 1893.
Ferstandig, L. L. (1963) *Tetrahedron Letters* 1235.
Fox, W. M. and Waters, W. A. (1965) *J. Chem. Soc.* 4628.
Galus, Z. and Adams, R. N. (1962) *J. Am. Chem. Soc.* **84**, 2061.
Galus, Z., White, R. M., Rowland, F. S. and Adams, R. N. (1962) *J. Am. Chem. Soc.* **84**, 2065.
Galus, Z. and Adams, R. N. (1963) *J. Phys. Chem.* **67**, 862.
Hammick, D. Ll. and Mason, S. F. (1946) *J. Chem. Soc.* 638.
Hammick, D. Ll. and Mason, S. F. (1949) *J. Chem. Soc.* 1939.
Hammick, D. Ll. and Munro, D. C. (1950) *J. Chem. Soc.* 2049.
Hammond, G. S. and Shine, H. J. (1950) *J. Am. Chem. Soc.* **72**, 220.
Hammond, G. S. and Grundemeier, W. (1955) *J. Am. Chem. Soc.* **77**, 2444.
Hammond, G. S., Seidel, B. and Pincock, R. G. (1963) *J. Org. Chem.* **28**, 3275.
Hammond, G. S. and Clovis, J. S. (1962) *Tetrahedron Letters* 945.

Hammond, G. S. and Clovis, J. S. (1963) *J. Org. Chem.* **28**, 3283.
Hashimoto, S. and Sunamato, J. (1964) *Kogyo. Kagaku. Zasshi.* **67**, 2090 [*Chem. Abstr.* **63**, 4119].
Hashimoto, S., Shimkai, I. and Sunamoto, J. (1966) *Kogyo. Kagaku. Zasshi.* **69**, 290 [*Chem. Abstr.* **67**, 137].
Hofmann, A. W. (1863) *Proc. Roy. Soc.* **12**, 576.
Holt, P. F. and Hughes, B. P. (1953) *J. Chem. Soc.* 1666.
Holt, P. F. and Hughes, B. P. (1954) *J. Chem. Soc.* 764.
Holt, P. F. and Hughes, B. P. (1955) *J. Chem. Soc.* 98.
Holt, P. F. and McNae, C. J. (1964) *J. Chem. Soc.* 1759.
Hughes, E. D. and Ingold, C. K. (1941) *J. Chem. Soc.* 608.
Hughes, E. D. and Ingold, C. K. (1950) *J. Chem. Soc.* 1638.
Ingold, C. K. and Kidd, H. V. (1933) *J. Chem. Soc.* 984.
Jacobson, P. (1909) *Liebigs. Ann.* **367**, 304.
Jacobson, P. (1922) *Liebigs. Ann.* **428**, 76.
Kenner, G. W. (1932) *J. Chem. Soc.* 711.
Klauke, E. and Bayer, O. (1963) *D.R.P.* 1154091 [*Chem. Abstr.* **60**, 457].
Krolik, L. G. and Lukashevich, V. O. (1949) *Dokl. Akad. Nauk. SSSR* **65**, 37.
Krolik, L. G. and Lukashevich, V. O. (1952) *Dokl. Akad. Nauk. SSSR* **87**, 229.
Krolik, L. G. and Lukashevich, V. O. (1953) *Dokl. Akad. Nauk. SSSR* **93**, 663.
Krolik, L. G. and Lukashevich, V. O. (1960) *Dokl. Akad. Nauk. SSSR* **135**, 1139.
Leffler, J. E., Faulkner, R. D. and Petropoulos, C. C. (1958) *J. Am. Chem. Soc.* **80**, 5435.
van Loon, J. P. (1904) *Rec. Trav. Chim.* **23**, 62.
Lukashevich, V. O. and Krolik, L. G. (1948) *Dokl. Akad. Nauk. SSSR* **63**, 543.
Lukashevich, V. O. and Krolik, L. G. (1959) *Dokl. Akad. Nauk. SSSR* **129**, 117.
Lukashevich, V. O. and Krolik, L. G. (1962) *Dokl. Akad. Nauk. SSSR* **147**, 1090.
Lukashevich, V. O. (1964) *Dokl. Akad. Nauk. SSSR* **159**, 1095.
Lukashevich, V. O. (1967) *Tetrahedron* **23**, 1317.
Meisenheimer, J. and Witte, K. (1903) *Ber.* **36**, 4153.
Miller, B. (1964) *J. Am. Chem. Soc.* **86**, 1127.
Mizoguchi, T. and Adams, R. N. (1962) *J. Am. Chem. Soc.* **84**, 2058.
Mohlinger, D. N., Adams, R. N. and Argersinger, W. J. (1962) *J. Am. Chem. Soc.* **84**, 3618.
Möller, F. (1958) *Methoden der Organischen Chemie*, Vol. 11(2), 4th Edition edited Houhen-Weyl, Thieme, Stuttgart, p. 876.
Morgan, L. R. and Aubert, C. C. (1962) *Proc. Chem. Soc.* **73**.
Murrell, J. N. (1962) in *Chem. Soc. Special. Publ.* No. 16, 118.
Nesmeyanov, A. N. and Golovoyya, R. V. (1960) *Dokl. Akad. Nauk. SSSR* **133**, 1337.
Nesmeyanov, A. N., Perevalova, E. G. and Nikitina, T. V. (1961) *Dokl. Akad. Nauk. SSSR* **138**, 1118.
Oglesby, D. M., Johnson, J. D. and Reilley, C. N. (1966) *Anal. Chem.* **38**, 385.
Orelkin, P. U., Ryskaltschuk, O. V. and Aizikovitsch, P. (1931) *J. Gen. Chem. USSR* **1**, 696.
Orr, S. F. D., Sims, P. and Manson, D. (1956) *J. Chem. Soc.* 1337.
Osugi, J., Sasaki, M. and Onishi, I. (1966) *Rev. Phys. Chem.* (*Japan*) **33**, 100.

O'Sullivan, M. (1966) The acid-catalysed rearrangements of hydrazobenzenes. M.Sc. Thesis, London.
Parish, J. H. and Whiting, M. C. (1964) *J. Chem. Soc.* 4713.
Pongratz, A. and Wüstner, H. (1940) *Ber.* **73**, 423.
Pongratz, A. and Scholtis, K. (1942) *Ber.* **75**, 138.
Pongratz, A., Böhmert-Süss, S. and Scholtis, K. (1944) *Ber.* **77**, 651.
Pyl, T., Lahmer, H. and Beyer, H. (1961) *Chem. Ber.* **94**, 3217.
Rakusan, J. and Allan, Z. J. (1966) *Tetrahedron Letters* 4955.
Rakusan, J. and Allan, Z. J. (1967) *Coll. Czech. Chem. Comm.* **32**, 2882.
Reeves, R. L. and Andrews, R. W. (1967) *J. Am. Chem. Soc.* **89**, 1715.
Risaliti, A. and Pentimalli, L. (1956) *Ann. Chim.* **46**, 1050.
Ritter, J. J. and Ritter, F. O. (1931) *J. Am. Chem. Soc.* **53**, 670.
Robinson, G. M. and Robinson, R. (1918) *J. Chem. Soc.* **113**, 639.
Robinson, R. (1941) *J. Chem. Soc.* 220.
Robinson, B. (1963) *Chem. Rev.* **63**, 373.
Sah, P. T. and Yuin, K-H. (1939) *Rec Trav. Chim.* **58**, 751.
Schlenck, W. and Bergmann, E. (1927) *Liebigs Ann.* **463**, 281.
Schmidt, H. and Schultz, G. (1878) *Ber.* **11**, 1754.
Schmidt, H. and Schultz, G. (1881) *Liebigs Ann.* **207**, 320, 348.
Schulte-Frohlinde, D. (1957) *Liebigs Ann.* **612**, 131.
Schuler, P. (1944) Dissertation, Berlin [*Chem. Abstr.* **40**, 6069].
Schwartz, W. M. and Shain, I. (1965) *J. Phys. Chem.* **69**, 30.
Seidel, B. and Hammond, G. S. (1963) *J. Org. Chem.* **28**, 3280.
Shamin-Ahmed, S. and Hasan, H. (1952) *J. Ind. Chem. Soc.* **29**, 955.
Shine, H. J. (1956) *J. Am. Chem. Soc.* **78**, 4807.
Shine, H. J. and Bear, J. L. (1957) *Chem. Ind.* (*London*) 565.
Shine, H. J. and Trisler, J. C. (1960) *J. Am. Chem. Soc.* **82**, 4054.
Shine, H. J., Huang, F. T. and Snell, R. L. (1961) *J. Org. Chem.* **26**, 380.
Shine, H. J. and Chamness, J. T. (1963) *J. Org. Chem.* **28**, 1232.
Shine, H. J. and Stanley, J. P. (1965) *Chem. Comm.* 294.
Shine, H. J. and Chamness, J. T. (1967) *J. Org. Chem.* **32**, 901.
Shine, H. J. and Stanley, J. P. (1967) *J. Org. Chem.* **32**, 905.
Shine, H. J. (1967a) *Tetrahedron Letters* 4043.
Shine, H. J. (1967b) *Aromatic Rearrangements*, Elsevier, Amsterdam, p. 126.
Shine, H. J., Baldwin, M. A. and Harris, C. (1968) *Tetrahedron Letters* 977.
Smith, J. M., Wheland, G. W. and Schwartz, W. M. (1952) *J. Am. Chem. Soc.* **74**, 2282.
Snyder, L. C. (1962) *J. Am. Chem. Soc.* **84**, 340.
Sterba, V., Sagner, Z. and Matrka, M. (1965) *Coll. Czech. Chem. Comm.* **30**, 2475, 2477.
Sterba, V. and Vecera, M. (1966) *Coll. Czech. Chem. Comm.* **31**, 3486.
Stieglitz, J. (1903) *Am. Chem. J.* **29**, 62.
Stieglitz, J. and Curme, G. O. (1913a) *Ber.* **46**, 911.
Stieglitz, J. and Curme, G. O. (1913b) *J. Am. Chem. Soc.* **35**, 1143.
Tichwinsky, M. (1903) *J. Russ. Phys. Chem. Soc.* **35**, 667.
Vecera, M., Gasparic, J. and Petranek, J. (1956) *Chem. Ind.* (*London*) 99.
Vecera, M., Gasparic, J. and Petranek, J. (1957a) *Chem. Ind.* (*London*) 299.

Vecera, M., Gasparic, J. and Petranek, J. (1957b) *Coll. Czech. Chem. Comm.* **22**, 1603.
Vecera, M. and Petranek, J. (1958) *Coll. Czech. Chem. Comm.* **23**, 249.
Vecera, M. (1958) *Chem. Listy* **52**, 1373.
Vecera, M., Synek, L. and Sterba, V. (1960) *Coll. Czech. Chem. Comm.* **25**, 1992.
Vecera, M. and Petranek, J. (1960) *Coll. Czech. Chem. Comm.* **25**, 2005.
Veibel, S. (1954) *Canad. J. Chem.* **32**, 638.
Ward, E. and Pearson, B. D. (1959) *J. Chem. Soc.* 3378.
Weiss, J. (1940) *Trans. Faraday Soc.* **36**, 856.
Wheland, G. W. and Schwartz, J. R. (1949) *J. Chem. Phys.* **17**, 425.
White, W. N. and Preisman, R. (1961) *Chem. Ind.* (*London*) 1752.
Wieland, H. (1908) *Ber.* **41**, 3498.
Wieland, H. (1912a) *Ber.* **45**, 492.
Wieland, H. (1912b) *Liebigs Ann.* **392**, 127.
Wieland, H. (1913) *Ber.* **46**, 3296.
Wieland, H. (1915) *Ber.* **48**, 1095, 1098.
Wittig, G., Joos, W. and Rathfelder, P. (1957) *Liebigs Ann.* **610**, 180.
Wittig, G. and Grolig, J. E. (1961) *Ber.* **94**, 2148.
Wittig, G., Borzel, P., Neumann, F. and Klau, G. (1966) *Liebigs Ann.* **691**, 109.
Yamada, S., Chichata, I. and Tsurui, R. (1954) *Pharm. Bull.* (*Japan*) **2**, 59 [*Chem. Abstr.* **50**, 214].
Yamada, S., Chichata, I. and Tsurui, R. (1955) *Jap. Patent* 2374 [*Chem. Abstr.* **51**, 14803].
Zollinger, H. (1961) *Azo and Diazo Chemistry*, Interscience, London, pp. 301, 302.

2

THE BIOSYNTHESIS OF CARBOCYCLIC COMPOUNDS

D. H. G. Crout

Department of Chemistry, University of Exeter
Exeter, Devon

A. Preface

The aim of this chapter has been to give a broad picture of the current status of our knowledge of some of the main pathways for the biosynthesis of carbocyclic compounds, covering as many aspects of this topic as possible. Many of the subjects included have been reviewed in detail, and where this is so an attempt has been made to avoid duplication by including only enough background material to make the main discussion intelligible to readers coming fresh to the field.

It can be taken as axiomatic that where biosynthetic pathways are concerned, speculation behaves in an ideally Parkinsonian manner, tending to fill just as much space as is made available to accommodate it. Speculation has therefore been kept to a minimum and emphasis has been laid on describing what is actually known about biosynthetic pathways from experimental investigations. Although this has meant the exclusion of a number of interesting and attractive hypotheses, it is hoped that this exercise might in fact be useful in highlighting the strengths (and weaknesses) of some current biosynthetic theories.

Some of the topics discussed are based on large numbers of original papers, and where this is so, reference is sometimes made to previous review articles and summaries. It is hoped that this will make for clarity without unduly reducing the accessibility of the original papers.

B. Introduction

The organic chemist has traditionally regarded the study of the biosynthesis of natural products as coming within his range of competence even though the means by which such compounds are produced—through the intermediacy of living cells—differ greatly at first sight from synthetic procedures used in the laboratory. The fundamental premise on which this assumption is based is that reactions taking place in living cells should be understandable in terms of the principles used to rationalise similar reactions studied under laboratory conditions. This premise is held to be valid in spite of the fact that chemical reactions in the cell are mediated by complex organic catalysts, the enzymes, and take place under vastly different conditions from those normally applied in the laboratory. This principle was stated in tentative form in 1907 by Collie and has been reaffirmed often since (Collie, 1907; Birch and Donovan, 1953). It has received ample support from enzymatic studies, which have given information on ways in which biosynthetic intermediates acquire the increased activation necessary for reactions to be able to take place under cellular conditions. For instance, in condensation reactions between carboxylic acid esters *in vivo*, extra activation is achieved by conversion into the thioester of an enzyme or co-enzyme:

$$\text{R.COOH} + \text{HS-enzyme} \longrightarrow \text{R—CO.S-enzyme}$$

It is well known that thioesters are much more reactive in Claisen-ester type condensations than oxygen esters. Baker and Reid found

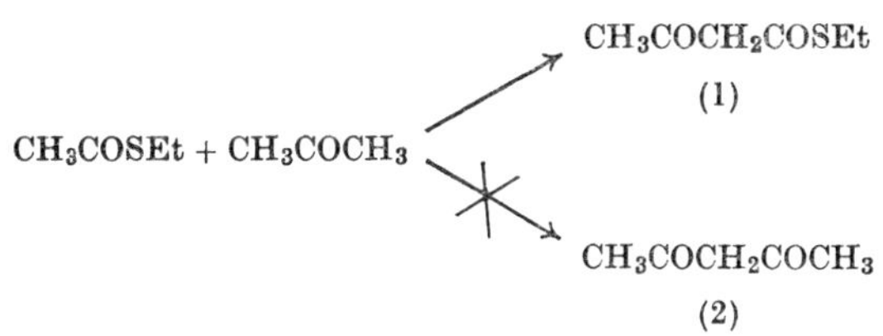

FIG. 1. The condensation of ethyl thioacetate with acetone.

that under Claisen condensation conditions, a mixture of ethyl thioacetate and ethyl acetate gave predominantly ethyl acetothioacetate (1). Even in the presence of acetone, ethyl thioacetate underwent preferential self condensation; in this reaction (Fig. 1) there was apparently no formation of acetylacetone (2) as might have been expected (Baker and Reid, 1929).

This is just one example of the way in which biological reactions can be made to take place under 'mild' conditions, although they can still be seen to resemble closely normal laboratory processes.

C. EXPERIMENTAL METHODS IN BIOSYNTHETIC INVESTIGATIONS

Before considering any detailed evidence for specific biosynthetic pathways, it will be helpful to review briefly the various ways in which such information can be obtained.

5

The methods by which biosynthetic processes are studied can be broadly divided into four categories. First, circumstantial evidence for biosynthetic pathways can be obtained by considering the structural relationships between natural products. Frequently sequences of related compounds can be found in which the progression from one compound to the next can be rationalised in terms of known *in vitro* reactions or established biosynthetic processes. In recent years, very complete analyses have been made of the metabolites produced by individual micro-organisms, plants, etc. and such sequences can provide very powerful evidence for postulated biosynthetic pathways. In the same way, common structural features in a series of compounds can suggest possible common precursors.

Second, reactions *in vitro* which lead to compounds or structural elements which also occur among natural products often suggest the operation of the same type of reaction on similar precursors *in vivo*. An early example of this type of argument was presented by Collie (*loc. cit.*) who noted that base-catalysed condensations of synthetic poly β-keto compounds led to phenolic products of a type frequently found in nature. This observation prompted him to suggest a similar biogenesis for the natural products (section H.1).

To take another example, it is now considered that a large variety of carbon–carbon and carbon–oxygen bond forming reactions in nature involve free radical intermediates generated by one-electron oxidation of phenols (Scott, 1965). Again, the possibility of such reactions occurring *in vivo* was largely suggested by the observation of closely related reactions in the laboratory. This type of oxidative coupling is discussed in more detail later (section H.8).

This approach has been extended to the study of the reactivity of model compounds under simulated biosynthetic conditions, by the synthesis of postulated intermediates or model compounds which can be treated with reagents analogous to those suspected of being involved in the biosynthetic process. Experiments of this kind can give valuable information on the relative extent to which biological reactions are governed on the one hand by the purely 'chemical' properties of the reactants and on the other by the specific operation of enzyme control.

Sometimes, this method has been refined to the point where experiments have been carried out under 'physiological conditions', i.e. in dilute aqueous solution at pH $\sim$7. The formation of compounds with structures identical with or closely related to those of natural metabolites is regarded as supporting evidence for the occurrence of similar reactions *in vivo*. Implicit in such conclusions, of course, is the premise mentioned at the beginning of the introduction.

Some such reactions proceed so extraordinarily readily under 'physiological' conditions that it has been suggested from time to time that these do not require the intermediacy of enzymes *in vivo*. However, it is extremely doubtful if any cellular reaction, no matter how facile it may seem in the laboratory, takes place completely independent of enzyme control.

An excellent example is provided by the flavanone–chalcone interconversion (3) $\leftrightarrows$ (4) (Fig. 2). This interconversion takes place

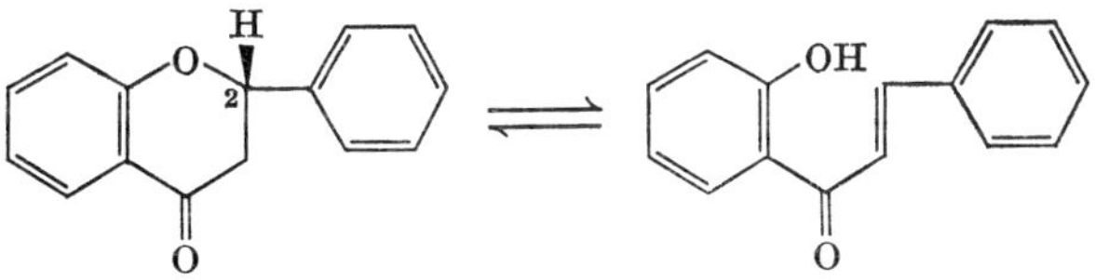

Flavanone (3) Chalcone (4)

Fig. 2. The flavanone–chalcone interconversion.

so readily that flavanone specimens are often tinged with yellow due to the presence of small amounts of the corresponding chalcone. In solution, an equilibrium exists between the two forms, emphasising the ease with which they can be interconverted. In spite of this it appears that the conversion of chalcones into flavanones *in vivo* comes under enzymatic control, since naturally occurring flavanones are optically active due to the generation of an asymmetric centre at C(2) (as in (3)) during ring closure. An enzyme from soybean which catalyses this specific reaction has recently been isolated and partially purified (Moustafa and Wong, 1967).

These first two approaches, however, can at best only provide circumstantial evidence for biosynthetic pathways. Their great value lies in the understanding they bring to biosynthetic reactions from investigations of similar reactions in the laboratory and in the stimulus they provide for experimentation that can lead to direct proof.

The third and most recent method, which does provide such direct evidence, involves the use of radioisotopes as 'tracers'. The incorporation of activity from a suitably labelled precursor provides direct evidence for the utilisation of the precursor in the biosynthesis of a metabolite. The recent rapid improvement in the availability of radioisotopes and labelled substrates, and of sensitive and accurate counting equipment, notably scintillation counters, has made this the method of choice for organic chemists, and to a large extent, biochemists also.

Non-radioactive isotopes such as ^{2}H, ^{18}O and ^{13}C have been used from time to time, the pattern of incorporation being determined by nuclear magnetic resonance or mass spectrometry. These methods, however, although very powerful and convenient in many instances, are considerably less sensitive than those employing radiotracers and for this reason will probably not become as widely used.

The fourth approach might be called the 'biological' method. This embraces a wide variety of experimental techniques ranging from the investigation of the properties of isolated enzyme systems to the use of microbial mutants which have genetic deficiencies resulting in the deletion or inactivation of enzymes responsible for specific steps in a biosynthetic sequence. Information can be gleaned from studies with such 'blocked' mutants in various ways. Occasionally, the mutant will accumulate large quantities of the intermediate involved in the blocked enzymatic step; the structure of the intermediate can then be determined by isolation and degradation in the usual way. Sometimes, valuable information on the nature of the true biosynthetic intermediates can be obtained by examining the effect of added substrates in compensating for the blocked step. The reappearance of a metabolite which is produced by the parent strain but not by the mutant, when an artificial substrate is added to the culture medium of the mutant, provides good (but not infallible) evidence that the substrate is a true biosynthetic intermediate.

The final proof for a biosynthetic pathway must lie in the isolation and purification of the enzymes involved and the direct demonstration of the sequential conversion of the postulated intermediates into the product. Unfortunately, experiments such as this represent a major undertaking and only limited progress has been made in elucidating the individual enzymatic steps in the biosynthesis of most complex metabolites. There are notable exceptions, however; the mechanistic features of the individual steps in the biosynthesis of cholesterol, for example, are known in remarkable detail, even down to the finest points of stereochemistry (Clayton, 1965).

D. Biosynthesis and Biogenesis

It will be useful here to clarify the distinction between biosynthesis and biogenesis, terms which are often used synonymously, although a real and useful distinction can be made between them. By biogenesis* we mean the biological *origin* of a metabolite in terms of the primary metabolites used in its synthesis. If, by direct experimentation, a simple precursor (e.g. acetate) is shown to be incorporated into a metabolite, this can be taken as evidence for the precursor being involved in its *biogenesis*. To describe the *biosynthesis* we need, in addition, to have information on the individual chemical reactions by means of which this precursor is transformed into the metabolite.

The difference between the terms biogenesis and biosynthesis reflects the normal chronological sequence in biosynthetic investigations. For example, the *biogenesis* of many benzenoid compounds has been shown to involve glucose, because the direct utilisation of this precursor, suitably labelled, in the formation of these metabolites has been demonstrated. However, the *biosynthesis* only became clarified by much further experimentation when the variety of intermediates, such as shikimic acid (section I), which lie between glucose and the final aromatic metabolites (e.g. phenylalanine) could be demonstrated and the individual chemical transformations explained.

E. Primary and Secondary Metabolites

Metabolites are usually classified as *primary* or *secondary*. Although no hard and fast line can be drawn between the two, it is useful to regard primary metabolites as compounds which appear to be necessary for the proper functioning of cells in general, and secondary metabolites as compounds which appear to be restricted to particular biological families, genera or even individual species, but which do not appear to be required by other organisms. To put it crudely, and hence to lose a little accuracy, primary metabolites may be described as 'known to be generally useful' and secondary metabolites as 'not known to be generally useful'. A particular metabolite, according to

* The term *biogenesis* was first used by T. H. Huxley in his Presidential Address to the British Association in 1870, where he defined it as the hypothesis 'that living matter always arises by the agency of pre-existing living matter' (Huxley, 1870). If we paraphrase this definition and say that 'chemical compounds characteristic of living matter (always) arise by the agency of pre-existing chemical compounds characteristic of living matter', we come close to the definition given above.

this definition, may, of course, be promoted from one category to another when further investigations reveal it to be more widespread than previously supposed. An example of this is seen in the history of the diterpenoid metabolites, the gibberellins (Grove, 1961). These were first isolated, in Japan, from the fungus *Gibberella fujikuroi* which infects rice (Bakanae Disease). The disease is characterised in its early stages by a dramatic elongation of the ears of the growing plant. The gibberellins, which are responsible for this phenomenon, were at first regarded only as biological curiosities by chemists (and as an agricultural nuisance by the Japanese). Since the initial discovery, however, gibberellins have been found in the embryos and immature seedlings of a number of widely different plant species, and it is apparent that they fulfil some more general function than at first realised (Grove, 1961).

Even a brief analysis of 'who does what' in the field of biosynthesis will reveal that, according to the definition just given, organic chemists tend to be more interested in secondary metabolites whereas biochemists are usually concerned with the more fundamental primary metabolites.

It is perhaps unfortunate that there should be this fragmentation of effort, which has resulted from time to time in some mild criticism of the methods and attitudes of one side by the other. The division of labour, however, is not fortuitous, but is a reflection of purely practical considerations. The biochemical techniques which employ blocked mutants and cell-free enzyme systems are more powerful where primary metabolites are involved. For example, the non-production of a primary metabolite usually has a very obvious effect on the development of an organism. A common result with micro-organisms which lack an intermediate enzyme essential for the production of a primary metabolite is that the organism refuses to grow unless the deficiency is compensated for by adding the missing intermediate. This technique is well exemplified in the investigations leading to the identification of shikimic acid as a key intermediate in aromatic amino acid biosynthesis (section I). With secondary metabolites, which are often genus-specific and sometimes species-specific, the choice of organism with which one can work is greatly reduced and the usefulness of mutants is also reduced, since a mutant strain which fails to produce a given secondary metabolite often displays no other abnormality and continues to grow and develop as well as the parent strain. The biosynthesis of secondary metabolites has been studied using blocked mutants in a few cases (for example, in the tetracycline series), but a programme of this kind is not lightly undertaken since usually many hundreds of mutant strains must be examined before one is found in which an informative block has appeared (Hassall, 1965). The appearance of an enzymatic block is marked by the production of new metabolites, some of which may be seen by inspection to be possible precursors of the metabolite under investigation. However, as secondary metabolites by definition are not necessary for the proper functioning of the organism, the precursor-product relationship can only be inferred. It is quite possible that the new metabolite may not be a real intermediate but simply the product of a branch in the main biosynthetic pathway. Definite proof can sometimes be obtained from incorporation experiments with the whole organism or cell-free extracts, whereby the presumed intermediate can be shown to be converted into the product. Although biosynthetic pathways deduced from studies with blocked mutants usually rest on an inferential basis, nevertheless, in the absence of alternative techniques, their great value cannot be denied.

As a consequence of the considerable effort involved in these biological approaches, the biosynthesis of secondary metabolites has mainly been pursued with the help of radiotracers. Experimentation in this field involves the synthesis of labelled precursors, the isolation and purification of natural products, and the devising of specific degradation procedures to permit the location of radioactivity in labelled metabolites. Since these are all activities familiar to the organic chemist, it has naturally fallen to him to undertake a major role in the study of secondary metabolism.

F. Interpretation of Results

The great success achieved by the application of radiotracer methods to biosynthetic problems has occasionally led to the uncritical assumption that the incorporation of radioactivity into a metabolite from a labelled compound demonstrates that the latter is a true biosynthetic intermediate. This is not necessarily true since utilisation of a precursor may be due to prior breakdown into smaller molecules which enter the general metabolic pool of small biosynthetic intermediates to be incorporated *de novo*. The wide variety of biosynthetic pathways open to simple compounds such as acetic and pyruvic acids means that some incorporation of these into practically any metabolite will be observed, although possibly at a very low level. High incorporations naturally have greater significance than low ones, but on the other hand, failure of a labelled substrate to be incorporated does not necessarily mean that it is not a true intermediate. Factors such as membrane permeability, transport phenomena, competing metabolic pathways and diurnal variation in metabolic activity may all affect incorporation rates, sometimes drastically.

Striking examples of the influence of such factors have recently been described by Goodwin (Goodwin, 1966). When etiolated (light deprived) maize seedlings were exposed to light, active biosynthesis of a number of compounds began in specialised cellular bodies, the chloroplasts. Many of these compounds are known to be wholly or in part derived from acetate via mevalonic acid (page 157). However, when ^{14}C-labelled mevalonate was administered to such seedlings, some of these compounds became labelled but not others. The sterols, pentacyclic terpenes and ubiquinone became strongly labelled whereas little or no activity was incorporated into β-carotene or the terpenoid side chains of chlorophyll, plastoquinone, tocopheryl quinone or vitamin K_1. However, when $^{14}CO_2$ was administered, an exactly opposite pattern of incorporation was observed. These results have been attributed to the effects of enzyme segregation and specific membrane permeability in the developing chloroplast. Some anomalous results from experiments on terpene biosynthesis may be explained by the operation of similar controls on the incorporation of labelled substrates (Banthorpe and Turnbull, 1966; Ruddat, Heftmann and Lang, 1965).

It is also important, when assessing the significance of overall incorporations, to take into account the actual amount of precursor administered. If a plant, for example, is producing x moles of a metabolite in a given period, and an amount X moles of a labelled precursor is taken up and metabolised in the same period, then the maximum possible incorporation, if X is greater than x, is $x/X.100$ per cent (assuming that biosynthesis of one molecule of the metabolite requires only one molecule of the precursor and that the metabolism is not disturbed by the administration of the precursor). The limitation on incorporation may be quite severe when milligramme amounts of labelled compound are administered. This is particularly important with complex precursors, since laboratory synthesis of these almost invariably involves dilution of the radioactivity and the purified precursor will thus have a relatively low specific activity. This means that rather large amounts of precursor have to be administered in order to obtain useful incorporations.

Under these conditions the precursor itself may disturb the metabolism of the organism, even though it is a normal metabolite. In an investigation of the incorporation of tiglic acid into the tigloyl ester tropane alkaloids of some *Datura* species, it was observed that administration of more than 20 mg sodium tiglate to a mature plant resulted in the appearance of definite toxic symptoms; this limited the amount of precursor that could be fed (Woolley, 1968).

The general problem of attaching significance to absolute incorporation rates can be largely overcome by the use of specifically labelled precursors, preferably labelled at more than one position or with more than one isotope. If the same ratio of activities is found at the labelled positions in the metabolite as in the precursor, then incorporation of the intact precursor can be safely assumed (although there are exceptional circumstances where even this may not be true) (Austin and Meyers, 1966).

Conclusions drawn from absolute incorporation rates can also be made more reliable by using internal standards for comparison. The effectiveness of two ^{14}C-labelled precursors can be compared if both are administered with a tritium-labelled substrate which is known to be incorporated into the metabolite (or vice versa). Comparison of the $^{14}C:^{3}H$ ratios in the metabolite then gives a better insight into the relative efficiencies of the precursors than direct comparison of the ^{14}C incorporations themselves. The justification for this approach is that qualitative and quantitative differences in, for example, individual plants used in a biosynthetic investigation, may produce large differences in absolute incorporation rates, even for the same precursor. However, the *ratio* of the incorporation rates for two differently labelled precursors is less likely to be affected and may be expected to remain fairly constant.

Grisebach has recently used this method to provide convincing evidence that dihydroflavonols are not involved in isoflavone biosynthesis. Tritium-labelled kaempferol (5) and dihydrokaempferol (6) (Fig. 3) were fed, together with [^{14}C]-phenylalanine (known to be a precursor of flavones and isoflavones), to buckwheat seedlings. The relative dilutions of tritium and ^{14}C were then determined in the isolated cyanidin (7) and quercetin (8)

$$\left(\text{dilution} = \frac{\text{specific activity of precursor}}{\text{specific activity of product}}\right).$$

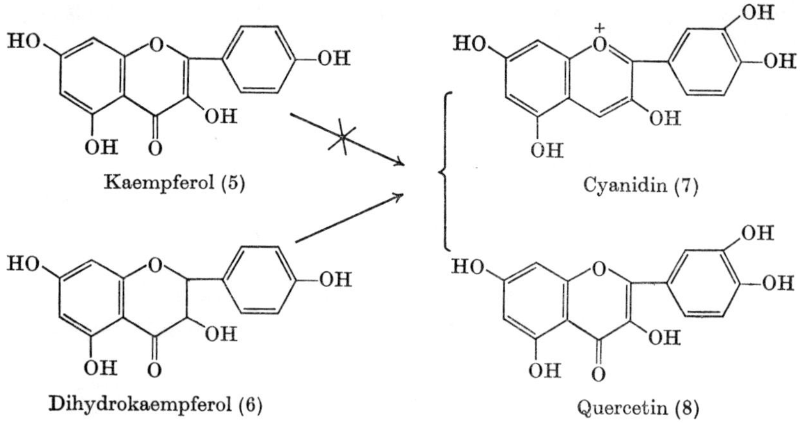

Kaempferol (5)

Dihydrokaempferol (6)

Cyanidin (7)

Quercetin (8)

FIG. 3. The role of dihydroflavonols in flavonoid and isoflavonoid biosynthesis.

In the experiment with [^{3}H]-dihydrokaempferol (6) and [^{14}C]-phenylalanine, the ^{3}H : ^{14}C ratio in the mixture fed was 9·5 : 1, whereas the relative activity of ^{3}H and ^{14}C in the isolated cyanidin (7) and quercetin (8) was 8·3 : 1, and 12·9 : 1 respectively, showing that dihydrokaempferol and phenylalanine were incorporated into these compounds to approximately the same extent. In the corresponding experiment with [^{3}H]-kaempferol (5), the initial ^{3}H : ^{14}C ratio dropped from 7·0 : 1 to 0·7 : 1 in cyanidin (7) and 1·8 : 1 in quercetin (8), indicating that relative to dihydrokaempferol, kaempferol was a less efficient precursor (Fig. 3).

A much more dramatic result was obtained when the incorporations

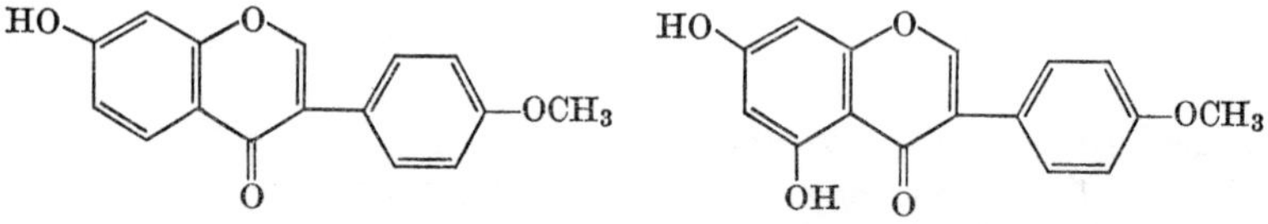

Formononetin (9)

Biochanin-A (10)

into the isoflavones formononetin (9) and biochanin-A (10) were investigated. When [^{3}H]-dihydrokaempferol and [^{14}C]-phenylalanine (^{3}H : ^{14}C ratio 2·6 : 1) were administered, the isolated formononetin (9) had a ^{3}H : ^{14}C activity ratio of $7{\cdot}4 . 10^{-3}$: 1, the corresponding value for biochanin-A being $1{\cdot}5 . 10^{-2}$: 1. A similar result was obtained in feeding experiments with [^{3}H]-kaempferol. These figures provide convincing evidence that dihydroflavonols such as (6) do not serve as precursors of isoflavones *in vivo*, contrary to previous suggestions (Barz, Patschke and Grisebach, 1965).

Swain has recently given an admirable account of the pitfalls to be negotiated in the interpretation of results from radiotracer experiments. He has also discussed the conditions that must be satisfied before a biosynthetic pathway can be said to be definitely established (Swain, 1965).

G. Major Pathways in the Biosynthesis of Carbocyclic Compounds

At present, three major biosynthetic routes to carbocyclic compounds are recognised; the acetate–malonate pathway leading to compounds known generically as polyketides,* the biogenetically related acetate–mevalonate pathway to the terpenes and steroids, and the glucose–shikimic acid pathway to phenylalanine and related aromatic compounds.

These three routes appear to act as main production lines of biosynthetic activity. A vast variety of natural products is derived from them by variation in the main pathways and by secondary transformations of the primary products. Frequently more than one of these major pathways can be seen to operate in the production of a metabolite. This is particularly evident in the flavonoids (as in (11)),

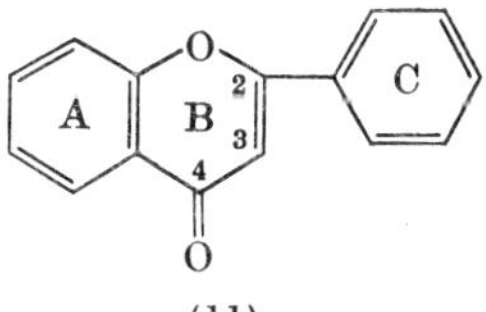

(11)

where ring C and C(2–4) of ring B represent a C_6–C_3 unit derived from the glucose–shikimic acid–phenylalanine pathway, whereas ring A is built on by the addition of three C_2 units derived from acetate, in a manner characteristic of the acetate–malonate pathway. Natural products derived from both these pathways are common; terpenes and

* The term *acetogenin* has recently been coined to encompass all metabolites derived wholly or in part by this route (Richards and Hendrickson, 1964). We prefer to use here the term *polyketide*, first used by Collie and later adopted by Birch, who was instrumental both in developing the theory and testing it experimentally.

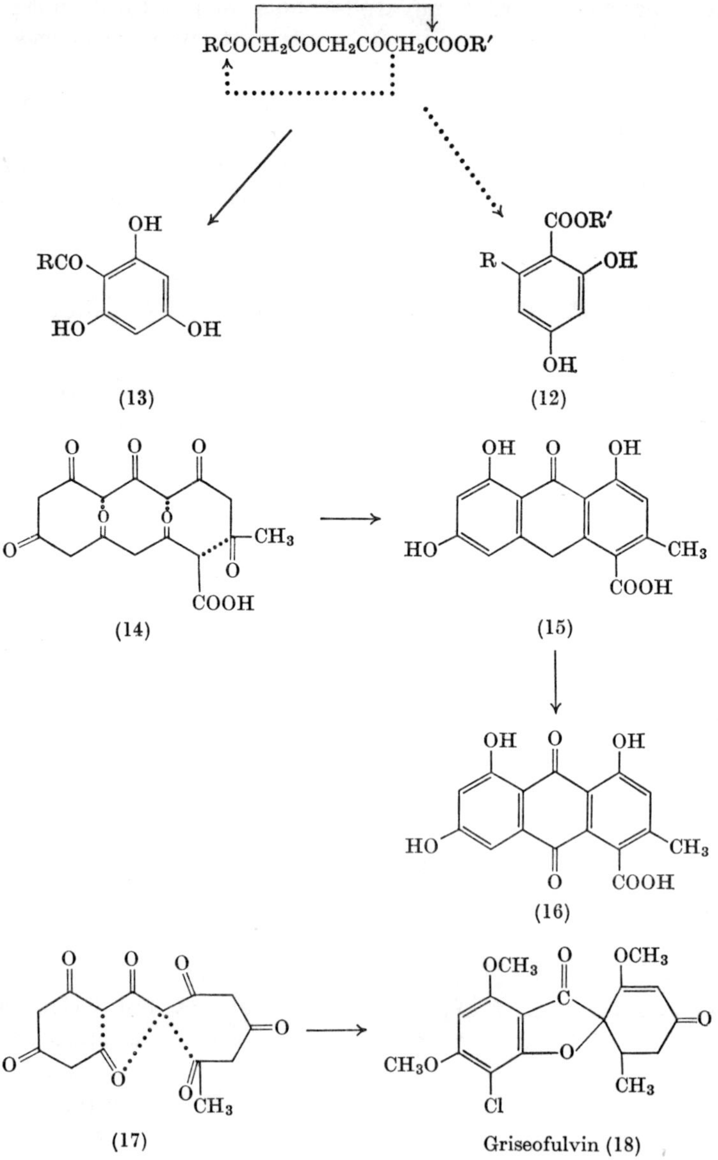

$RCOCH_2COCH_2COCH_2COOR'$

(13) (12)

(14) (15)

(16)

(17) Griseofulvin (18)

FIG. 4. Some theoretical predictions for polyphenol biosynthesis according to the acetate hypothesis.

steroids derived via the acetate–mevalonate pathway, however, are not usually elaborated by transformations associated with the other two routes, but instead seem largely to undergo oxidative modification as a means of producing variations of the basic structures.

The biosynthesis of the terpenes and steroids has been discussed in detail in several recent reviews (Richards and Hendrickson, 1964; Clayton, 1965; Pridham, 1967). For this reason and for reasons of space, the present chapter will be concerned with the remaining two major pathways and with other minor pathways which have not been recently discussed.

H. Polyketides

1. *The Acetate Hypothesis*

This was originally developed by Birch in order to rationalise the biogenesis of a large number of naturally occurring phenolic compounds (Birch and Donovan, 1953; Birch, 1957).

The central idea in this hypothesis was that linear poly-β-keto compounds, produced from acetate by multiple Claisen-ester type condensations, could cyclise in various ways to give polyphenolic products. An aldol-type condensation of such an intermediate (Fig. 4) could lead to orsellinic acid derivatives (12) whereas Claisen-type condensation would give rise to acylphloroglucinol derivatives (13).

The hypothesis was extended to include more complex metabolites such as the anthraquinone (16) which could be produced from the poly-β-keto compound (14) via the anthrone (15). By postulating further modification of the initial cyclised product, through oxidation, reduction, *O*- and *C*-methylation, further cyclisations, etc., the biosynthesis of metabolites such as griseofulvin (18) (from the precursor (17)), could be explained.

Although the elaboration and experimental verification of this hypothesis with ^{14}C-labelled substrates were primarily due to Birch, the central idea had been proposed by Collie nearly half a century before (Collie, 1907). Collie had observed that poly-β-keto compounds such as diacetylacetone ((19), Fig. 5) condensed under basic conditions to give products (e.g. orcinol (20) and the naphthalene (21)), resembling naturally occurring phenolic compounds. In his own words, Collie 'wished to call attention to the manner in which the group $.CH_2.CO.$ ...can be made to yield by means of the simplest reactions a very large number of interesting compounds; the chief point of interest

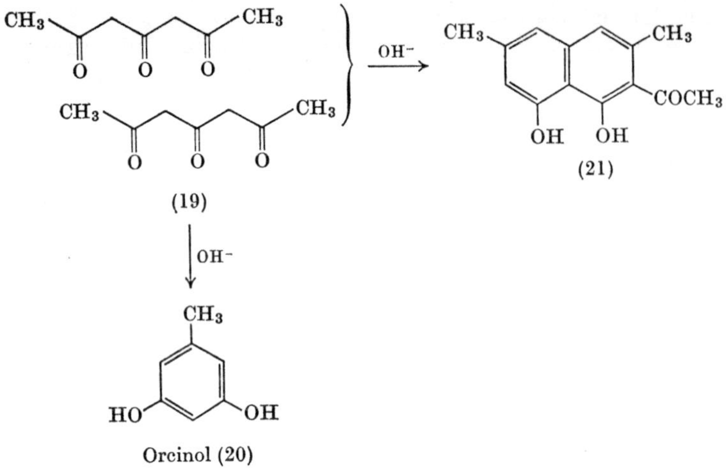

$CH_3COCH_2COCH_2COCH_3$ (19) $\xrightarrow{OH^-}$ (21)

(19) $\xrightarrow{OH^-}$ Orcinol (20)

Fig. 5. Base-catalysed condensation reactions of diacetylacetone.

being that these compounds belong to groups largely represented in plants'.

Although we can now see that Collie attempted to push his hypothesis too far by including carbohydrates and terpenes within its scope, his paper was remarkably prophetic. It was also noteworthy for containing an early statement of the rationale behind laboratory investigations of natural processes: 'although the reactions taking place in the plant are different from those in the laboratory, still much may be done by studying, in the laboratory, reactions which might conceivably occur in plants, such as those involving hydration, dehydration, oxidation, reduction, polymerisation and condensation at or as near the ordinary temperature as possible. Moreover, the effect on these reactions of faintly acid or alkaline solutions is worth while investigating'.

In 1953, when Birch and Donovan presented their hypothesis, there was a precedent for postulating the biogenesis of naturally occurring phenolic compounds from acetate, since evidence had already been obtained that both steroids and fatty acids were derived from acetate *in vivo* (Rittenberg and Bloch, 1944). In fact, the pathway of fatty acid biosynthesis resembles, in several important respects, the biosynthesis of polyketides. The subject of fatty acid biosynthesis has recently been reviewed (Lynen, 1967a, b). The essential chemical transformations are shown in Fig. 6.

1. Conversion of acetate into malonate (carboxylation).

$$CH_3COSCoA + CO_2 \longrightarrow \begin{array}{l} COOH \\ | \\ CH_2COSCoA \end{array}$$

(22) (23)

2. Priming reaction.

$$CH_3COSCoA + HS\text{-enzyme} \longrightarrow CH_3COS\text{-enzyme} + CoASH$$

3. Chain lengthening reaction.

$$\begin{array}{c} \begin{array}{l} COOH \\ | \\ CH_2COS\text{-enzyme} \end{array} \\ + \\ CH_3COS\text{-enzyme} \end{array} \longrightarrow CH_3COCH_2COS\text{-enzyme} + CO_2$$

4. 1st reduction.

$$CH_3COCH_2COS\text{-enzyme} \longrightarrow CH_3CH(OH)CH_2COS\text{-enzyme}$$

5. Dehydration

$$CH_3CH(OH)CH_2COS\text{-enzyme} \longrightarrow CH_3CH{=}CHCOS\text{-enzyme}$$

6. 2nd reduction

$$CH_3CH{=}CHCOS\text{-enzyme} \longrightarrow CH_3CH_2CH_2COS\text{-enzyme}$$

7. Chain lengthening reaction.

$$\begin{array}{c} \begin{array}{l} COOH \\ | \\ CH_2COS\text{-enzyme} \end{array} \\ + \\ CH_3CH_2CH_2COS\text{-enzyme} \end{array} \longrightarrow CH_3CH_2CH_2COCH_2COS\text{-enzyme}$$

FIG. 6. Pathway of fatty acid biosynthesis.

Important features of this scheme in relation to polyketide biosynthesis are as follows:

(i) In a preliminary step, acetate is activated by conversion to the coenzyme A thioester (acetyl-CoA, (22)). Before this can be utilised in the condensation reaction it is carboxylated by a biotin-dependent enzyme to give malonyl-CoA (23) (reaction 1). Both the malonyl and acetyl residues are now transferred in an ester exchange reaction to thiol groups on the enzyme, where they condense together to give an

$$CH_3COSCH_2CH_2NHCOCH_2CH_2NHCOCH(OH)C(CH_3)_2CH_2OP(O)(O^-)OP(O)(O^-)—OCH_2$$

Acetyl-CoA (22)

acetoacetyl unit and carbon dioxide (reaction 3). The formation of intermediate thioesters provides additional activation for the condensation step as discussed previously. The formation of malonyl-CoA also can be seen to augment this activation since malonate esters are well known to be more reactive in condensation reactions than the corresponding acetates.

(*ii*) Throughout the entire sequence of reactions, the substrate is bound to the enzyme by the thioester link. The fatty acid molecule is only liberated from the enzyme when the synthesis is complete.

(*iii*) The first reduction, dehydration and second reduction of the primary condensation product (reactions 4, 5 and 6) take place *before* any further chain extension.

The overall reaction in relation to palmitic acid biosynthesis can be summarised by the following equations:

$$CH_3COOH + CO_2 \longrightarrow HOOC{-}CH_2{-}COOH$$

$$CH_3COOH + 7\ CH_2(COOH)_2 \longrightarrow \underset{\text{Palmitic acid}}{CH_3(CH_2)_{14}COOH}$$

Lynen has presented evidence to show that the fatty acid synthetase system in yeast consists of a multi-enzyme complex which cannot be broken down into individual units without drastic loss of activity (Lynen, 1967a, b). The substrate is visualised as being attached to the central unit (Fig. 7) by means of a flexible molecular arm (suggested to be 4′-phosphopantotheine). Whilst remaining attached to the central enzyme unit in this way, the substrate is transferred from one

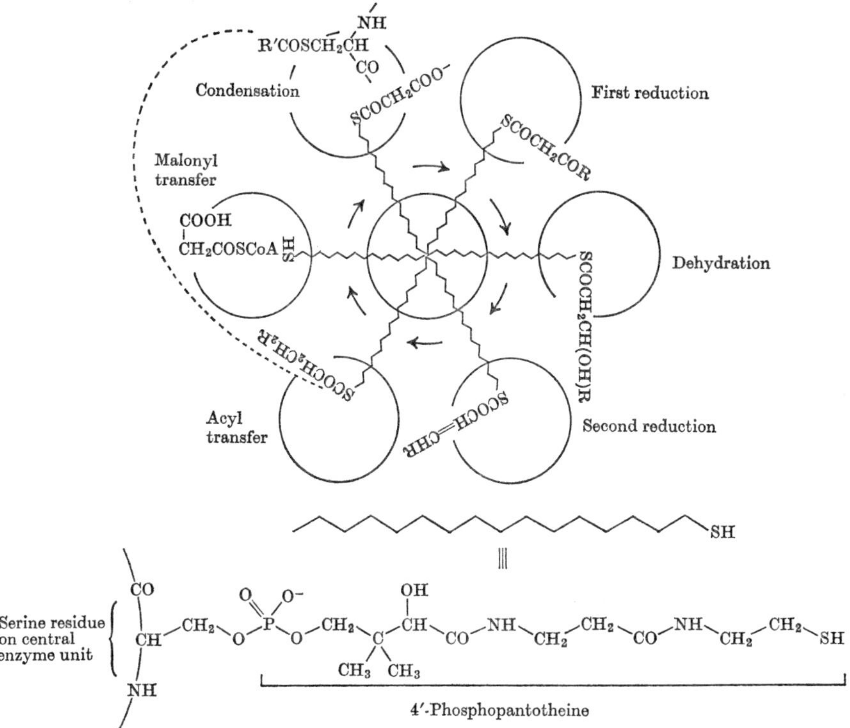

FIG. 7. Diagrammatic representation of yeast fatty acid synthetase (after Lynen).

to another of the surrounding enzyme units on which the reactions are carried out. When the reduction–dehydration–reduction sequence is complete, the molecule is transferred to a peripheral thiol group and a new malonyl unit is attached to the central thiol group. This condenses with the partially completed fatty acid molecule in the chain elongation step and the entire sequence of reactions is repeated. The whole operation is very reminiscent of a modern computer-controlled machine tool where the work piece is passed from tool to tool, each one carrying out its specific operation, until the job is complete.

The major chemical difference between fatty acid and polyketide biosynthesis is that in the latter the intermediate reduction steps by which oxygen is eliminated are often omitted, so that the sites of the original carboxyl carbon atom of acetate are marked by oxygen functions, usually in the form of phenolic hydroxyl or methoxyl groups.

Confirmation of Birch's hypothesis by tracer experiments with ^{14}C-labelled acetate rapidly followed the initial publication. The patterns of incorporation of labelled acetate into some representative compounds are illustrated in Fig. 8.*

Penicillium urticae (Birch, Massy-Westropp and Moye, 1955a,b; Birch, Cassera and Rickards, 1961)

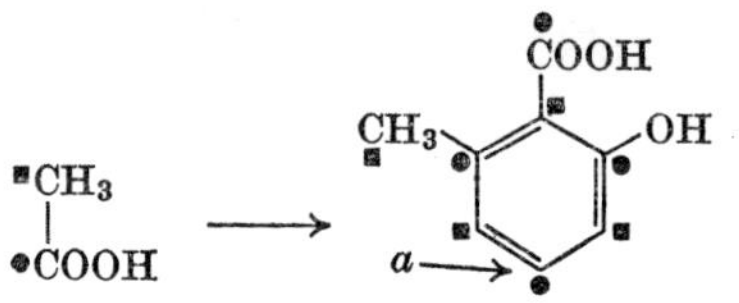

CH_3–COOH → 6-Methylsalicylic acid (24)

Penicillium islandicum (Gatenbeck, 1958, 1960)

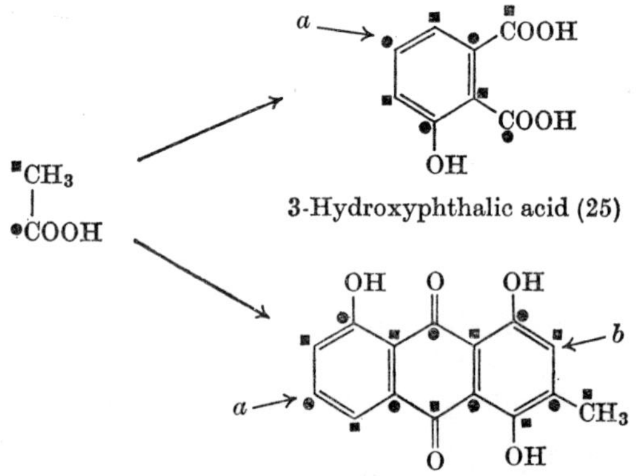

CH_3–COOH → 3-Hydroxyphthalic acid (25)

CH_3–COOH → Islandicin (26)

Fig. 8. Biogenesis of some typical polyketides.

* The following code, due to Bentley (Bentley, 1962) will be used to designate the origin of various labelled atoms. ■ Carbon from acetate methyl, ● carbon from acetate carboxyl, ◧ carbon from acetate methyl or malonate methylene, ◐ carbon from acetate carboxyl or malonate carboxyl, ▲ carbon from the methyl group of methionine or other C_1 compounds such as formate or formaldehyde. Other symbols will be defined for each Figure in which they are used.

Several examples of typical secondary transformations can be seen in these metabolites. In each compound, one position (arrows *a*), derived from the carboxyl group of acetate, lacks an oxygen function. This can probably be attributed to reduction and dehydration of the poly-β-keto intermediate (Birch and Donovan, 1953; Lynen and Tada, 1961).

In 3-hydroxyphthalic acid (25), an acetate methyl group has been oxidised to a carboxyl function and in islandicin (26), the carboxyl group of the terminal acetate unit has been lost (arrow *b*).

The simple polyacetate hypothesis was soon modified by the suggestion that acetate itself acted as the 'starter' molecule but that subsequent condensations to extend the chain involved malonate units as in fatty acid biosynthesis (Bu'Lock and Smalley, 1961; Lynen and Tada, 1961). Experimental support for this proposal was obtained *inter alia* from feeding experiments in *Penicillium urticae*

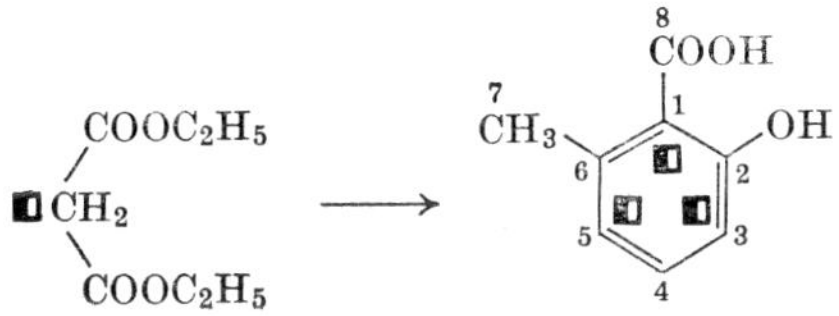

FIG. 9. The incorporation of labelled malonate into 6-methylsalicylic acid.

with [2-^{14}C]-diethyl malonate. This was incorporated readily into 6-methylsalicylic acid (Fig. 9), but whereas carbon atoms 7 and 6 carried only 8 per cent of the total activity, C(3), C(5) and C(1) carried 30, 29 and 30 per cent respectively. This pattern of incorporation is entirely consistent with a biogenetic pathway involving one acetate 'starter' molecule and three malonate condensing units (Bu'Lock, Smalley and Smith, 1962).

When labelled acetate was fed, the situation was reversed and the 'starter' unit was more highly labelled than the rest of the molecule, although the differentiation was much less pronounced (Birch, Cassera and Rickards, 1961). Comparable results were obtained almost simultaneously by Lynen and Tada (1961).

In general, some randomisation of the label is always found, particularly when [^{14}C]-acetate is fed, since carboxylation of acetyl-CoA to malonyl-CoA seems to occur readily in many organisms. The reverse reaction, decarboxylation of malonate to acetate also takes place but is usually less important, as in the feeding experiments with *P. urticae* just described.

An important advance in firmly establishing the acetate pathway came in 1959 when it was shown that acetate, doubly labelled with ^{14}C and ^{18}O was incorporated into orsellinic acid (27) in *Chaetomium cochliodes* with retention of both isotopes (Fig. 10). Since it was a premise of the original acetate hypothesis that oxygen substituents in polyketides marked the positions of former acetate carboxyl groups, this experimental verification was highly desirable (Gatenbeck and Mosbach, 1959). ^{18}O was diluted with respect to ^{14}C by a factor of 3·6 in the carboxyl group of orsellinic acid (27), the corresponding figure for the phenolic hydroxyl groups being 1·9, indicating that some exchange with ^{16}O from the medium had taken place. The relative dilution of ^{18}O in the carboxyl and hydroxyl groups by a factor of

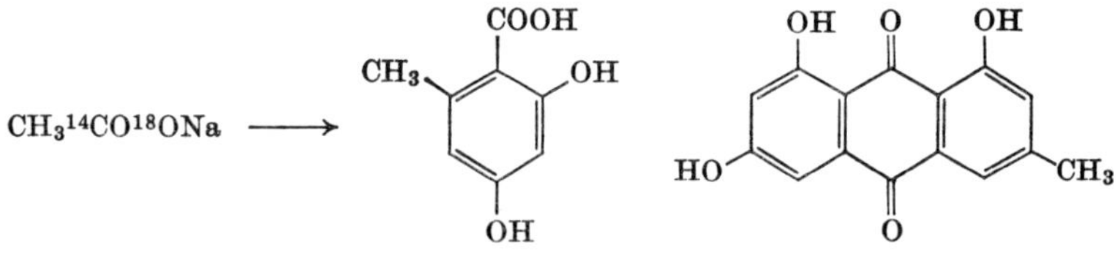

$CH_3{}^{14}CO^{18}ONa$ ⟶

Orsellinic acid (27)

Emodin (28)

FIG. 10. The incorporation of ^{18}O-labelled acetate into polyketides.

almost precisely two to one was regarded as being consistent with the hydrolysis of an intermediate coenzyme A thioester, which would result in the introduction into the carboxyl group of additional ^{16}O from the medium. Essentially similar results were obtained from studies of ^{18}O incorporation into islandicin (26), and emodin (28) in *Penicillium islandicum* (Gatenbeck, 1960b).

Important secondary modifications of polyketides are achieved by *O*- and *C*-methylation. The methyl donor *in vivo* is methionine (29), although in many organisms there appears to be a biosynthetic pool of one-carbon compounds which are to some extent equivalent so that other substrates such as bicarbonate and formate can sometimes serve as precursors of *C*- and *O*-methyl groups, apparently by conversion into methionine.

Before methionine can function in transmethylation reactions it must be activated by conversion to *S*-adenosylmethionine (30). The transmethylation process can then be regarded as a substitution

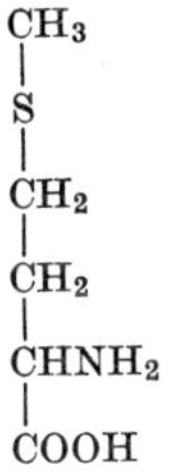

CH_3—S—CH_2—CH_2—$CHNH_2$—COOH

Methionine (29)

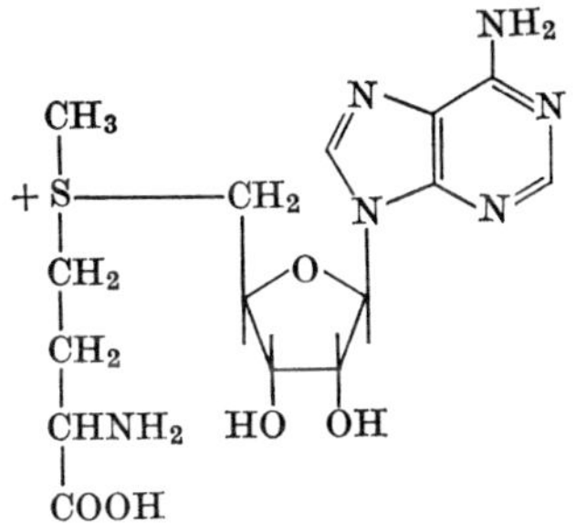

S-adenosylmethionine (30)

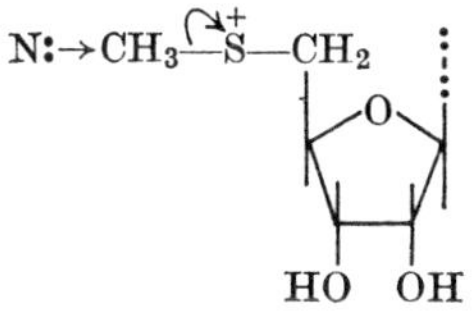

(31)

N = Nucleophilic centre in acceptor molecule

FIG. 11. Mechanism of biological methylation by *S*-adenosylmethionine.

reaction on carbon through attack by a nucleophilic centre in the acceptor molecule, as shown in Fig. 11 (31).

A laboratory analogy for this process has been described by Challenger and his collaborators (Challenger, Bywood, Thomas and

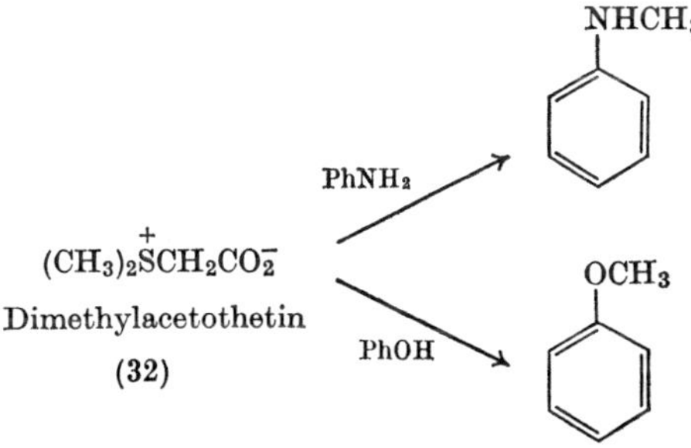

$(CH_3)_2\overset{+}{S}CH_2CO_2^-$

Dimethylacetothetin

(32)

PhNH₂ → $NHCH_3$ (N-methylaniline)

PhOH → OCH_3 (anisole)

FIG. 12. Methylation by dimethylacetothetin.

Hayward, 1957). Dimethylacetothetin (32) methylates both aniline and phenol at 140° to give the corresponding *N*- and *O*-methylated products (Fig. 12). By analogy with these reactions, methylation *in vivo* would be expected to occur at comparable sites, which appears to be generally true since methylation almost invariably occurs at potentially anionic positions derived from the methyl group of acetate, or at phenolic or enolic hydroxyl groups.

The role of methionine as a methyl donor in polyketide biosynthesis was first demonstrated with mycophenolic acid (33). The nuclear

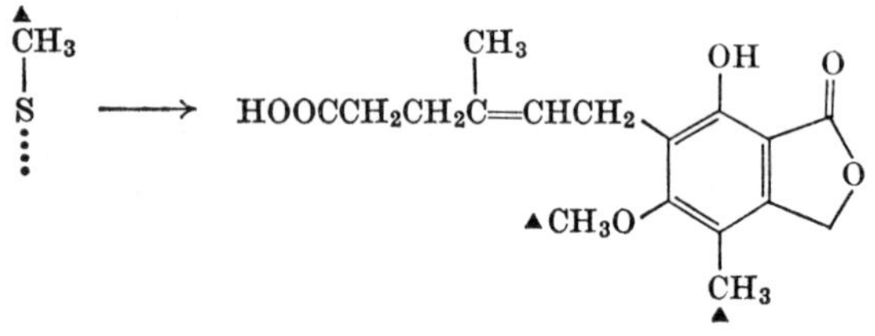

Mycophenolic acid (33)

C- and *O*-methyl groups were equally labelled when [$^{14}CH_3$]-methionine was fed to *Penicillium brevi-compactum* (Birch, English, Massy-Westropp, Slaytor and Smith, 1958).

A novel variation on the normal pattern of methylation was found in barnol (34) from *Penicillium barnense* (Mosbach and Ljungcrantz, 1964). The nuclear methyl group at C(3) was derived from methionine, but surprisingly, so was the methyl group of the ethyl side-chain. Propionic acid, obtained from barnol by Kuhn-Roth oxidation, was degraded in a stepwise manner to show that C(1) and C(2) (C(2) and C(8) of barnol (34)) were derived from acetate carboxyl and methyl respectively and that C(3) (C(10) in barnol) was derived from the methyl group of methionine.

Mechanistically this reaction is best explained in terms of a quinone methide intermediate. Hydroxylation of orsellinic acid (27) or a related compound (step *a*, Fig. 13) would furnish the intermediate (35). The subsequent oxidation step to the paraquinone methide (36) is shown as proceeding by proton removal from the *para* hydroxyl group with concomitant redistribution of electrons and transfer of a hydride ion to nicotinamide adenine dinucleotide (NAD) (partial formula shown). This, however, is only one of several possible mechanisms which have been discussed recently by Turner (Turner, 1964). Ionisation of the quinone methide (36) would produce the anionic species (37) required for the methylation step.

Two other major reactions must be accomplished before the

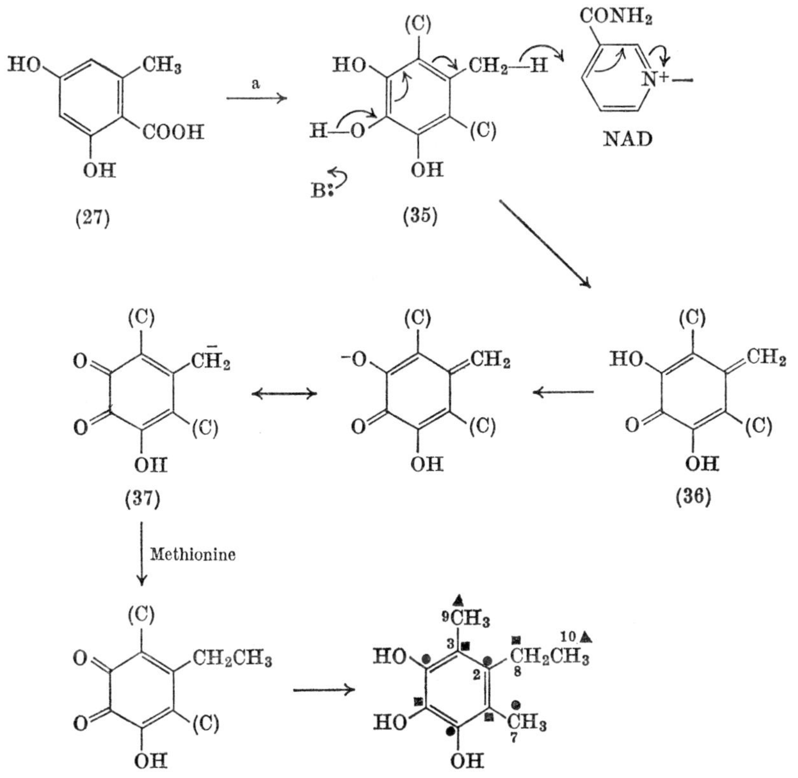

Barnol (34)

FIG. 13. Biosynthesis of barnol.

biosynthesis of barnol is complete; introduction of the nuclear methyl group C(9) and reduction of the terminal acetate carboxyl group to give the C(7) methyl function.

Methionine has also recently been shown to provide the central methylene unit in the methylenebisphloroglucinol derivatives of ferns such as *p*-aspidin (41) from *Dryopteris* species (Penttila, Kapadia and Fales, 1965). It was shown that methylaspidinol (38), labelled in the nuclear methyl group at C(2), was also incorporated, and it was suggested that this took place by oxidation to the quinone methide (39) followed by Michael addition of the anion of butyrylfilicinic acid (40).

The overall reaction strongly resembles the condensation of

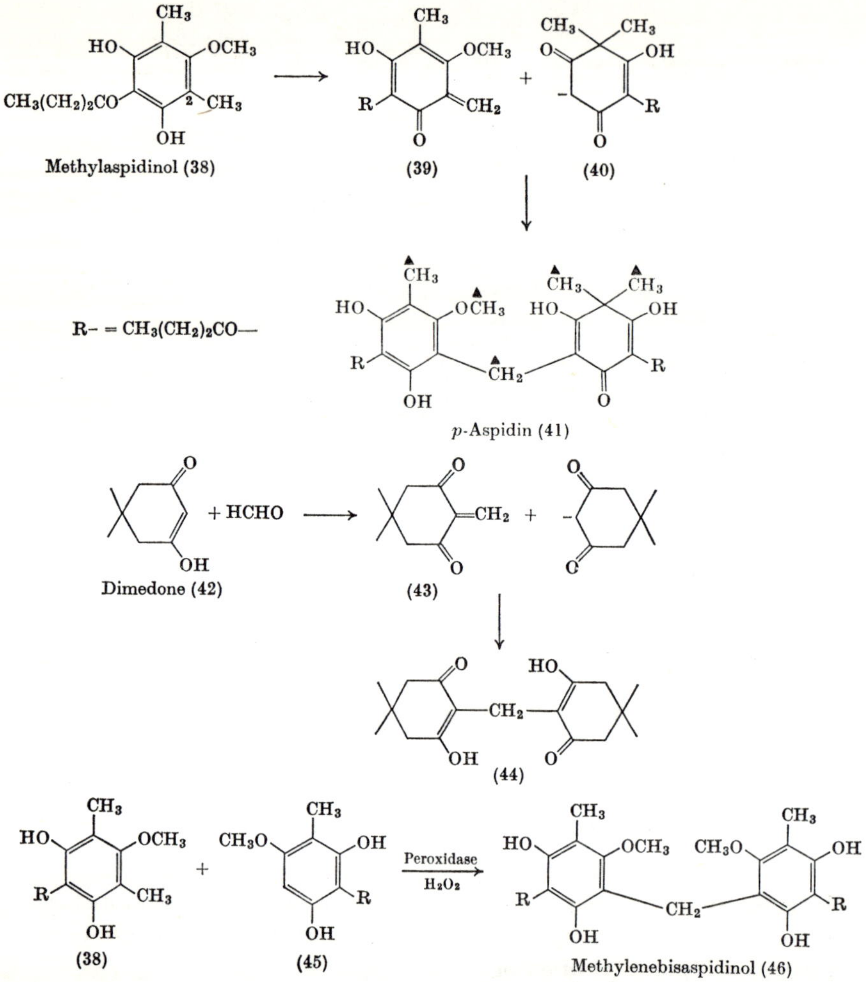

R– = $CH_3(CH_2)_2CO$—

Methylaspidinol (38) (39) (40) *p*-Aspidin (41) Dimedone (42) (43) (44) (38) (45) Methylenebisaspidinol (46)

Peroxidase H_2O_2

FIG. 14(a). The biosynthesis of methylenebisphloroglucinol derivatives.

formaldehyde with dimedone (42 → 44, Fig. 14(a)) and related resorcinol or dihydroresorcinol derivatives. Whereas the keto-methylene intermediate (43) in this condensation is formed directly, the formation of the corresponding compound (39) from methylaspidinol (38) requires an additional oxidation step. The enzyme involved is probably peroxidase since incubation of methylaspidinol

(38) and aspidinol (45) with this enzyme, in the presence of hydrogen peroxide, gave methylenebisaspidinol (46) (Penttila and Fales, 1966a).

In addition to the central methylene group, the nuclear *C*- and *O*-methyl groups in *p*-aspidin are also methionine-derived (Penttila *et al.*, 1965). *gem*-Dimethyl groups of the type found in *p*-aspidin, although not common, are fairly widespread in nature.

Until recently, the basic ring system in these compounds was reasonably assumed to have a normal polyketide origin, although this had not been experimentally demonstrated. The hydroxylation pattern is quite consistent with this assumption, but it is typical of the state of flux in our knowledge of biosynthetic processes that a recent report suggests that a novel biochemical chain-extension mechanism is involved instead of the normal acetate–malonate pathway (Gordon, Penttila and Fales, 1968). Gordon *et al.* have administered [1-^{14}C]-sodium butyrate to *Dryopteris marginalis* in order to examine the possibility that the polyketide chain might have been initiated by butyryl-CoA rather than acetyl-CoA (cf.

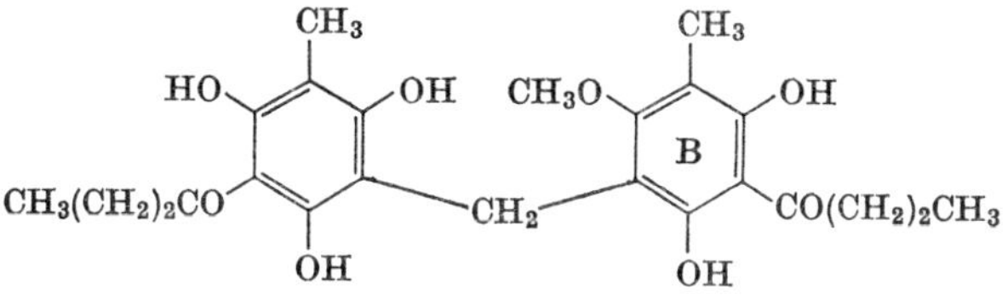

CH_3 CH_3
HO OH CH_3O OH
B
$CH_3(CH_2)_2CO$ CH_2 $CO(CH_2)_2CH_3$
OH OH

Margaspidin

$CH_3(CH_2)_2COSCoA$ (A) ⟶ $CH_3CH{=}CHCOSCoA$ $\xrightarrow{CO_2}$ $CH_2CH{=}CHCOSCoA$ (with COOH on CH_2) (B)

(A) + (B) ⟶ $CH_3(CH_2)_2COCH_2CH{=}CHCOSCoA$ + CO_2 $\xrightarrow[\text{Cyclisation, Hydroxylation.}]{\text{Malonyl CoA}}$

HO OH
$CH_3(CH_2)_2CO$
OH
⟶ Margaspidin

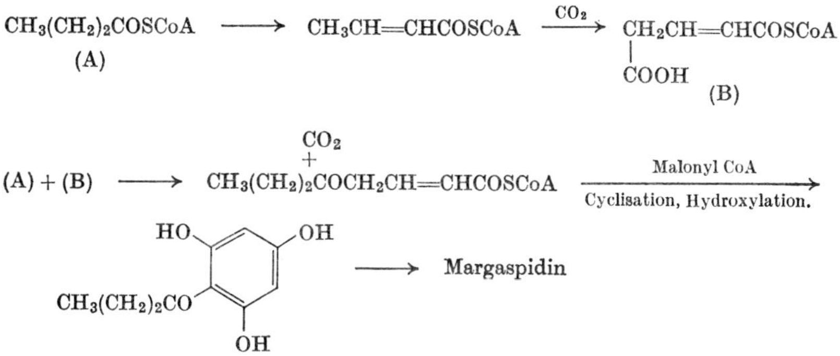

FIG. 14(b). Hypothetical scheme for methylenebisphloroglucinol biosynthesis from butyrate.

section H.5). Degradation of the resulting labelled margaspidin (Fig. 14(b)) showed that the activity was distributed equally between the two halves of the molecule, that the carboxyl groups of the butyryl side chains contained 50 per cent of the total activity, and that the propyl component was inactive. These results indicated that the side chain was indeed derived from an intact butyryl 'starter' molecule. However, 50 per cent of the activity of the margaspidin remained unaccounted for. Controlled degradation of the ring systems by ozonolysis showed that the entire activity of ring B was located at the carbon atom bearing the methoxyl group. The authors were forced to conclude from these results that in the construction of this molecule, two butyryl residues had been joined head to tail, the resulting C_8 unit condensing with a further C_2 unit (probably from malonyl-CoA). A possible pathway for these conversions is illustrated in Fig. 14(b); a close analogy for the dehydrogenation and carboxylation steps is found in a related sequence in leucine metabolism (del Campillo-Campbell, Dekker and Coon, 1959).

Regardless of whether the proposed coupling mechanism is correct, the demonstration that two intact C_4 units are involved in the construction of the butyrylphloroglucinol system is of great interest and certainly warrants careful study and further experimental verification.

Further examples of secondary transformations of a polyketide are revealed in the biosynthesis of the *Aspergillus flavus* metabolite flavipin (48). The biogenesis of flavipin via the acetate–malonate pathway has been firmly established (Pettersson, 1965a); the addi-

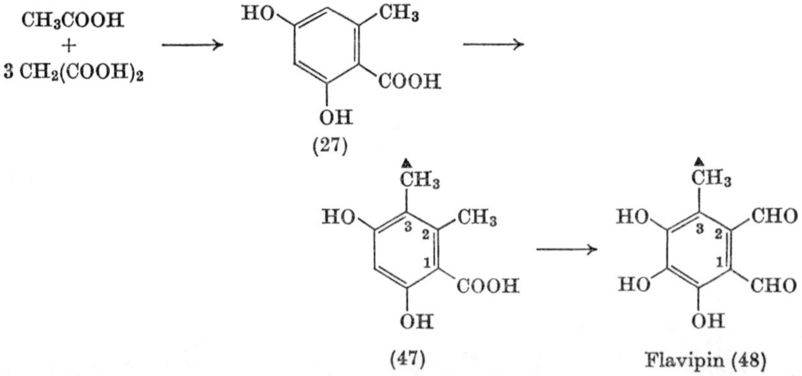

CH_3COOH + 3 $CH_2(COOH)_2$ ⟶ (27) ⟶ (47) ⟶ Flavipin (48)

FIG. 15. Biosynthesis of flavipin.

tional methyl group at C(3) is derived from methionine. Isolation of orsellinic acid (27) and 2,4-dihydroxy-5,6-dimethylbenzoic acid (47) from the same cultures indicated that these might be precursors and this was established by the direct incorporation of orsellinic acid (6·5 per cent) and the benzoic acid (47) (4·5 per cent) (carboxyl-labelled) into flavipin (Pettersson, 1965b) (Fig. 15).

It is interesting to note that the C(1) and C(2) formyl groups in flavipin (48) are produced by reduction of a carboxyl group and oxidation of a methyl group respectively.

2. *More Complex Polyketides*

A number of compounds of greater complexity than those so far considered appear to be derived via the acetate–malonate pathway. With these, additional problems in the elucidation of the biosynthetic pathway can arise. For example, the monomeric unit of the dimeric metabolite duclauxin (49) (from *Penicillium duclauxi*) could arise from a poly-β-keto precursor in at least three ways, as shown in Fig. 16 (Sankawa, Taguchi, Ogihara and Shibata, 1966; Shibata, 1967).

Routes *b* and *c* require, in addition to *O*-methylation, the insertion of two and four extra *C*-methyl groups respectively, whereas route *a* requires only *O*-methylation. Accordingly, the methoxyl methyl group in duclauxin (49), derived from [^{14}Me]-methionine, should contain 100, 33 or 20 per cent of the total activity depending on which of the three routes, *a*, *b* or *c*, respectively, is followed. The degradation necessary to resolve this point was carried out by cleavage of the methoxyl group with hydriodic acid, the methyl iodide formed being trapped as triethylammonium iodide. Ninety-six per cent of the total activity of duclauxin was found in this derivative, showing that pathway *a* was the correct one. Kuhn-Roth oxidation followed by Schmidt degradation of the resulting acetic acid gave methylamine and carbon dioxide which were also labelled in a manner consistent with the operation of route *a* (Fig. 16). It is of interest that this pathway involves the elimination of one carbon atom from the polyketide chain.

In addition to the problem of deciding between different types of folding in a poly-β-keto precursor, it is sometimes necessary to consider whether more than one polyketo chain might be involved. For example, the biphenyl derivative alternariol (50) from *Alternaria*

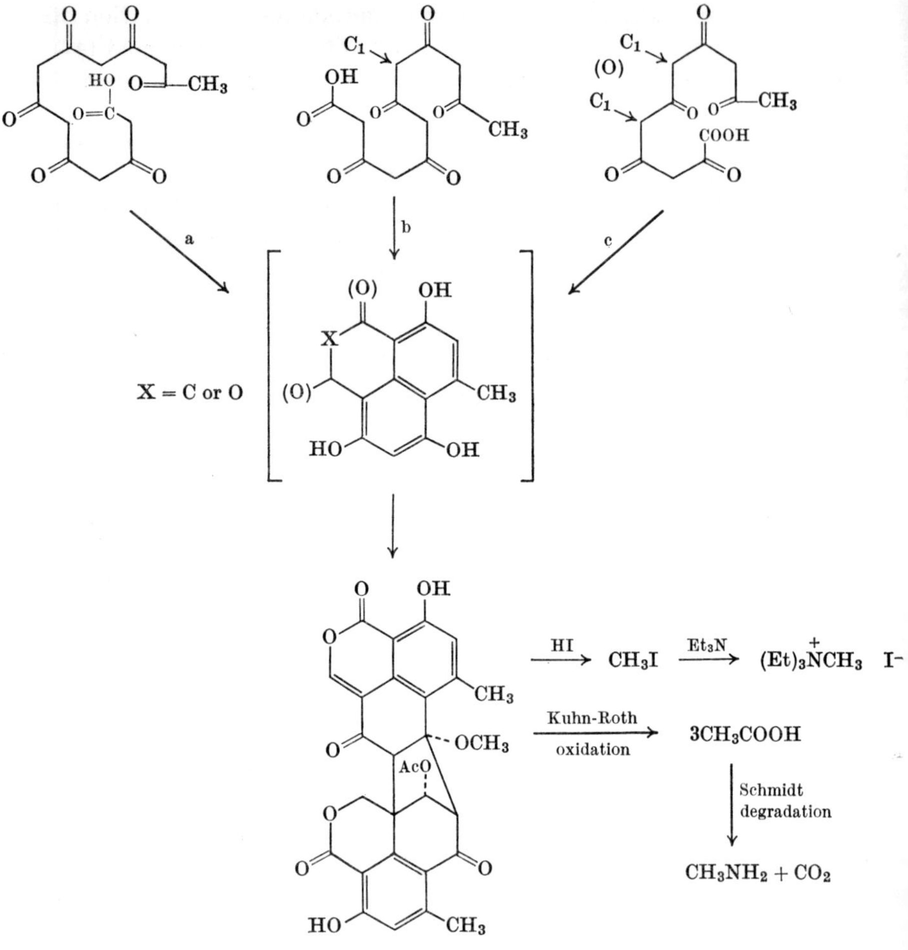

FIG. 16. Possible pathways for the biosynthesis of duclauxin.

tenuis, could be derived either from two separate chains or from a single C_{14} chain, as shown in Fig. 17. (The methyl and carboxyl groups are included to clarify the suggested pathway and do not imply any knowledge of the real identity of these precursors.)

Although the polyketide nature of alternariol was not in doubt (Thomas, 1961), the available evidence did not allow a distinction

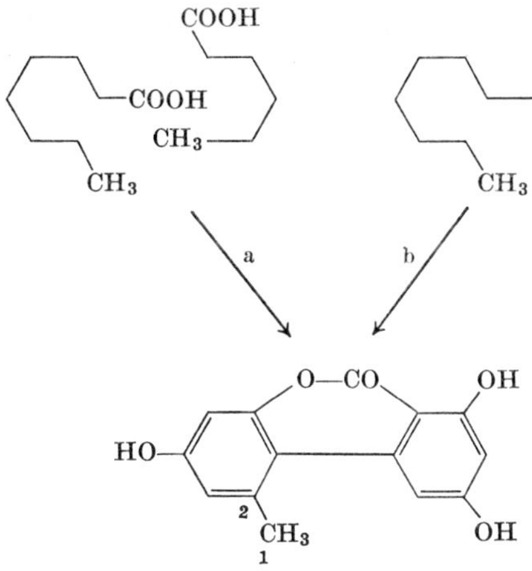

Alternariol (50)

FIG. 17. Possible pathways for the biosynthesis of alternariol.

to be made between these two possibilities. Gatenbeck and Hermodsson, using a partially purified enzyme system from *A. tenuis*, which was able to incorporate [1-^{14}C]-acetate into alternariol, found 83 per cent of the total activity in the acetic acid* (from C(1) and C(2)) obtained by Kuhn-Roth oxidation (Gatenbeck and Hermodsson, 1965). This result provides strong support for the formation of alternariol from a single poly-β-keto precursor (route *b*, Fig. 17). It is worth while commenting on the high relative incorporation into the methyl-terminal end of the polyketide chain. The high figure of 83 per cent compared with 2·3 per cent found for the lactone carbonyl position, which may be considered typical of the acetate carboxyl-derived positions in alternariol, contrasts strongly with the results obtained

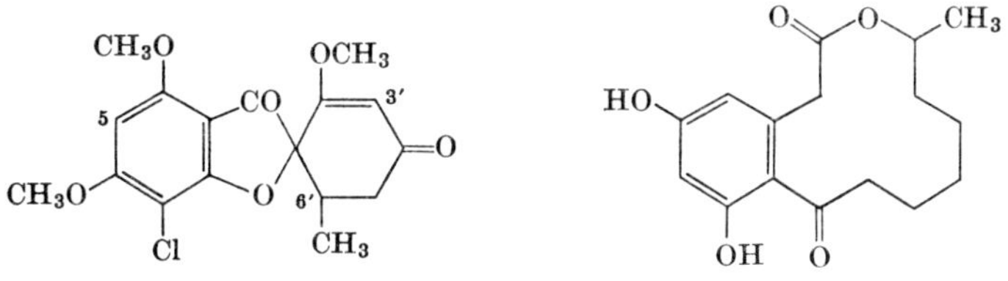

Griseofulvin (51)

Curvularin (52)

* This result was not obtained directly but was calculated from the observed specific activities of alternariol and the carbon dioxide produced in the Kuhn-Roth oxidation.

when intact organisms are used, when a much smaller difference is observed, usually of the order of 5 per cent excess. For example, the 'starter' acetate units in griseofulvin (51) (Birch, Massy-Westropp, Rickards and Smith, 1958) and curvularin (52) (Birch, Musgrave, Rickards and Smith, 1959) after feeding [1-^{14}C]-sodium acetate, contained only 5–8 per cent activity in excess of the random value. The differentiation was increased to 12–15 per cent when the precursor was administered together with unlabelled diethyl malonate (Birch *et al.*, 1959).

A biogenetic pathway for citromycetin (53) from two polyketide chains has been suggested on similar evidence (Fig. 18) (Gatenbeck and Mosbach, 1963).

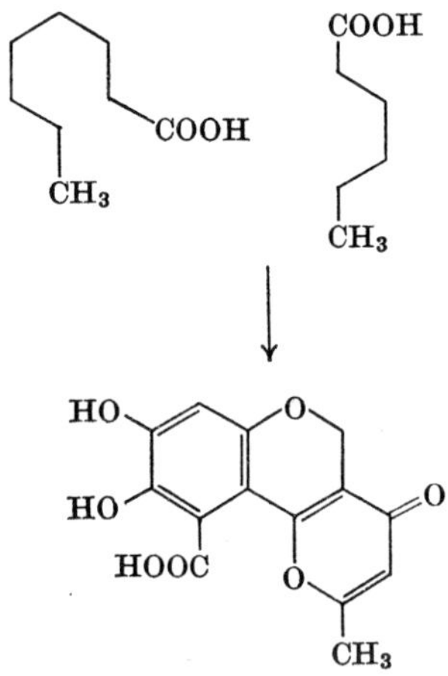

COOH
COOH
CH_3
CH_3
HO
HO
O
O
HOOC
O
CH_3
Citromycetin (53)

FIG. 18. The biosynthesis of citromycetin from two separate polyketide chains.

The figures just quoted serve to emphasise the difficulty inherent in the problem of deciding how many 'starter' molecules are involved in the biosynthesis of a polyketide since with large molecules one is trying to obtain significant differences between small relative activities.

The biosynthesis of sulochrin (54) in *Aspergillus terreus* has recently been the subject of a careful study designed to reveal whether one or two separate polyketide chains are utilised (Curtis, Harries, Hassall, Levi and Phillips, 1966; Curtis, Hassall and Pike, 1968).

Degradation of ^{14}C-labelled sulochrin (54) from feeding experiments with [2-^{14}C]-acetate and [2-^{14}C]-malonate gave a clear indication that two polyketide units are involved. The relative activities of C(17) and C(11) in sulochrin derived from [2-^{14}C]-acetate were 1·33:1, whereas in the corresponding experiment with [2-^{14}C]-malonate the ratio was 0·62:1, showing that C(10) and C(17) represented an acetate 'starter' molecule. A similar difference was observed when the activity of C(14) was compared with the average activity of the acetate methyl derived positions (C(1), C(3) and C(5)) in ring A. With [2-^{14}C]-acetate as precursor the ratio was 1·14:1, which, although lower than the corresponding value for ring B, was still large by comparison with the differences usually observed.

Mutant strains of *Aspergillus terreus* produce a number of closely related compounds (55)–(60) (some of which are also found in other

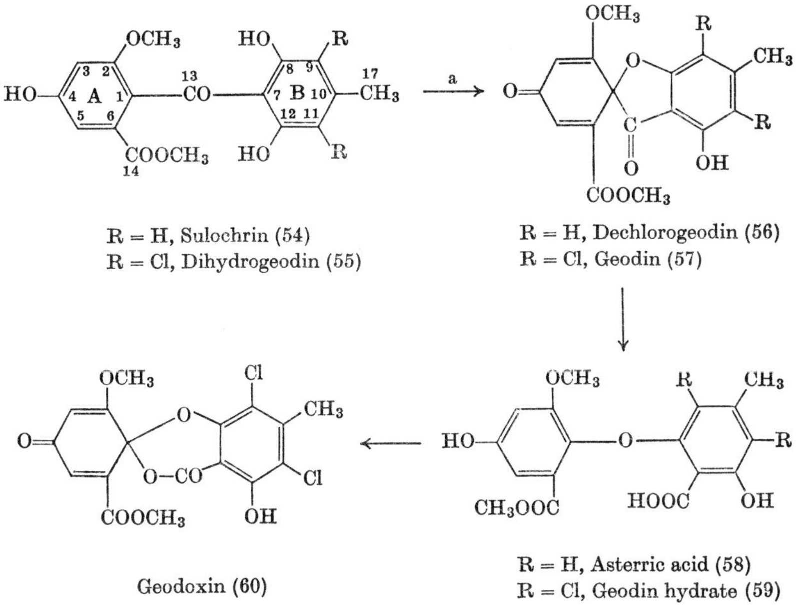

R = H, Sulochrin (54)
R = Cl, Dihydrogeodin (55)

R = H, Dechlorogeodin (56)
R = Cl, Geodin (57)

Geodoxin (60)

R = H, Asterric acid (58)
R = Cl, Geodin hydrate (59)

FIG. 19. The biosynthesis of geodoxin and related metabolites.

microorganisms), comprising a chlorinated and an unchlorinated series. Studies with blocked mutants (Hassall, 1965) have provided evidence for the biogenetic sequence illustrated in Fig. 19 which is supported by conversion studies in *A. terreus* (Rhodes, McGonagle and Somerfield, 1962). The cyclisation step *a* has been demonstrated enzymatically by the conversion of dihydrogeodin (55) into geodin (57) with a cell-free extract from *Penicillium estinogenum* (Komatsu, 1957). This reaction has also been effected *in vitro* using alkaline potassium ferricyanide (Curtis, Hassall, Jones and Williams, 1960). Laboratory analogies for the remaining steps are also known. The conversions of geodin (57) into geodin hydrate (59) (Barton and Scott, 1958) and of dechlorogeodin (56) into asterric acid (58) (Curtis *et al.*, 1960) were readily effected by acid catalysed hydration, and the final step, cyclisation of geodin hydrate (59) to geodoxin (60) was achieved using lead dioxide (Hassall and Lewis, 1961).

The biosynthesis of griseofulvin (51) appears to follow a closely similar pathway from the benzophenone (61) (Fig. 20). Conversion

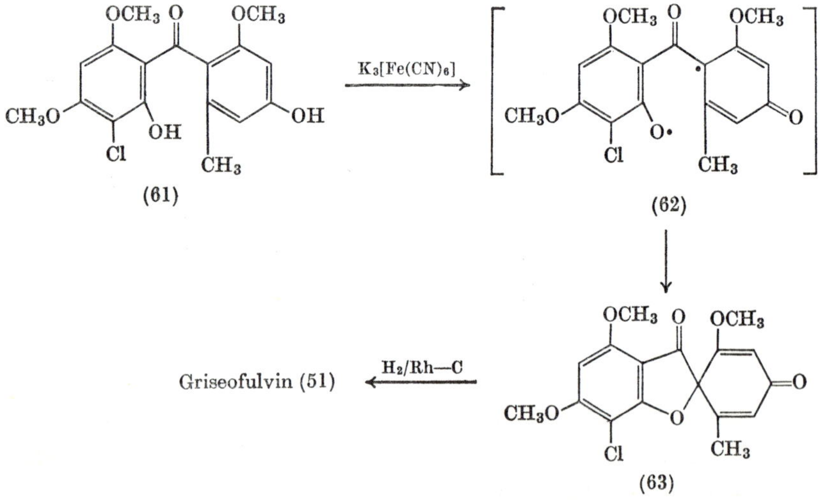

FIG. 20. Griseofulvin synthesis.

studies have indicated that the chlorination step takes place at the benzophenone stage (Rhodes, Boothroyd, McGonagle and Somerfield, 1961). The analogy for the oxidative cyclisation step was provided by the dihydrogeodin (55)–geodin (57) conversion mentioned above. The final hydrogenation step has been demonstrated directly by conversion of ^{14}C-labelled dehydrogriseofulvin (63) into griseofulvin (51) in *Penicillium patulum* (Hassall and Scott, 1961).

The laboratory synthesis of griseofulvin first reported by Scott and his collaborators (Day, Nabney and Scott, 1961) closely parallels the *in vivo* pathway. The benzophenone (61) was converted in 50–60 per cent yield into dehydrogriseofulvin (63) by potassium ferricyanide oxidation, presumably through the diradical intermediate (62). Reduction to griseofulvin was effected in 8 per cent yield by partial hydrogenation over a rhodium-charcoal catalyst.

An interesting study, with the aid of nuclear magnetic resonance spectroscopy, of the incorporation of [2-^{13}C]-sodium acetate into griseofulvin (51) in *Penicillium urticae* has recently been reported. By making use of the large ^{13}C–H coupling constant and without degradation of the molecule, Tanabe and Detre were able to show that C(3′), C(5) and the C(6′) methyl group were all derived from the methyl group of acetic acid (Tanabe and Detre, 1966).

The rye-infecting fungus, ergot (*Claviceps purpurea*), produces a number of related pigments, the ergochromes, which are dimeric compounds derived from reduced xanthone monomer units as in ergochrome AB (64). The ergochromes are accompanied in the fungus by anthraquinones (65) and (66), which prompted Franck to propose the biosynthetic pathway shown in Fig. 21 (Franck, Hüper, Gröger and Erge, 1966). The cleavage step, *a*, is analogous to the well-known Baeyer-Villiger conversion of ketones to esters and lactones (Hassall, 1957).

Feeding experiments with [2-^{14}C]-acetate in *Claviceps purpurea* gave labelled ergochrome AB (64). Partial degradation revealed an incorporation pattern which was in quantitative agreement with the scheme proposed, assuming that the intermediate anthraquinone was derived from acetate in the expected manner (as in 67).

The biosynthetic scheme proposed for the ergochromes has obvious relevance to the problem of sulochrin biosynthesis in that oxidative cleavage of the anthraquinone (67) at the B/C ring junction (route *b*, Fig. 21) would give the sulochrin (54) skeleton with the correct oxygenation pattern (cf. Gatenbeck, 1960c), whereas cleavage at the A/B junction (route *a*, Fig. 21) would lead to the ergochrome monomer structure.

There are, however, significant differences in the incorporations of [2-^{14}C]-acetate which militate against this interpretation; these appear in the activities of the methyl and carbomethoxy carbon atoms in ergochrome A/B (64) which were lower (10·6 per cent and 7·7 per cent of the total activity respectively) than the average value for the labelled positions of ring C (12·2 per cent). This assumes that only

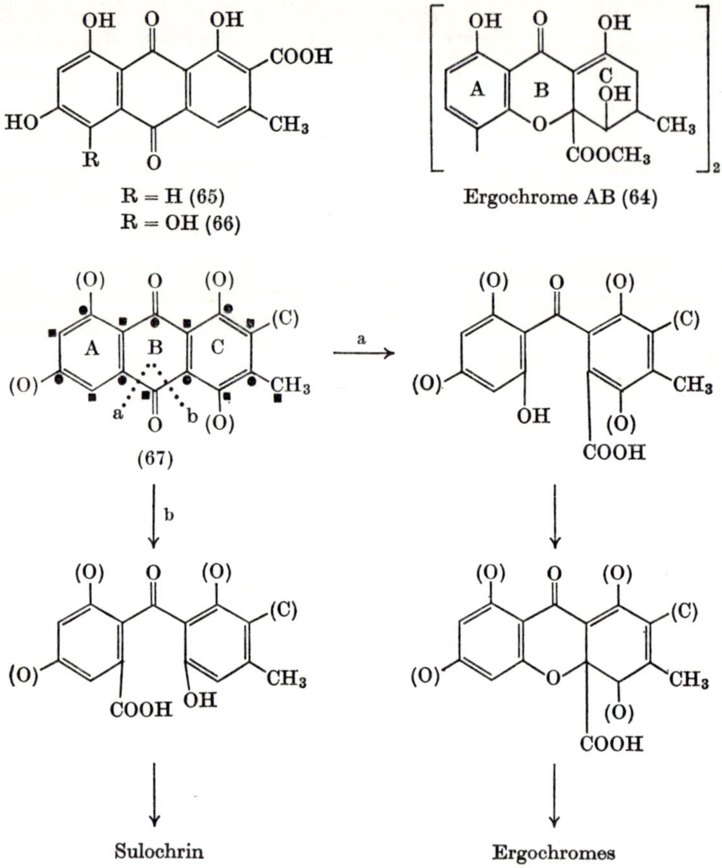

R = H (65)
R = OH (66)

Ergochrome AB (64)

(67)

Sulochrin

Ergochromes

FIG. 21. Possible pathway of ergochrome and sulochrin biosynthesis.

three of the ring C positions were labelled, but granted this assumption, the results can be seen to differ radically from the corresponding values for sulochrin (54) which indicate a much higher activity for the corresponding methyl and carbomethoxy carbon atoms relative to the other labelled positions. The results for ergochrome A/B are also unusual in that the methyl group of ring C, which should represent the 'starter' acetate unit, has a lower activity than the average rather than the slightly higher value which is usually found.

A related anomaly was found by Gatenbeck and Bentley in the incorporations of [1-^{14}C]-acetate and [2-^{14}C]-malonate into javanicin

(68) (Gatenbeck and Bentley, 1965). Kuhn-Roth oxidation of javanicin from the acetate feeding gave C(14, 13) and C(11, 3) as acetic acid and hypoiodite oxidation gave C(14) as iodoform. The activities of these degradation products showed that C(13) and C(11) were derived from the carboxyl group of acetic acid. When [2-^{14}C]-malonate-derived javanicin was degraded, however, it was found that whereas C(3) carried 11·5 per cent of the total activity, C(14), with 14·2 per cent was even more active. It was mentioned previously that carboxylation of acetate to malonate takes place readily in most microbiological systems, but that the reverse reaction, decarboxylation of malonate to acetate, appears to be less important, so that a marked differentiation between malonate and acetate-derived positions is usually

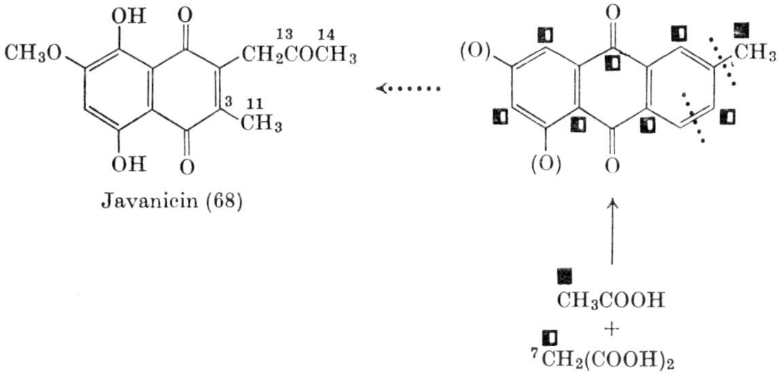

Javanicin (68)

FIG. 22. A hypothetical pathway for javanicin biosynthesis.

observed when the former is used in feeding experiments. A route through a normal acetate-derived anthraquinone, with cleavage of ring C and loss of the 'starter' acetate methyl group (Fig. 22) would provide an explanation for these results; the theoretical activity of C(14) would then be 14·3 per cent of the total. However, it was pointed out by the original authors that javanicin production was extremely slow and that under these conditions an abnormally high proportion of the malonate may have been converted into acetate before incorporation.

The results just described, together with others of a similar kind, suggest that the pattern of incorporation of acetate may, by itself, not be at all reliable as an aid to deciding whether more than one polyketide unit is used in the biosynthesis of a metabolite. In this respect, malonate usually gives a much clearer picture and should be the preferred precursor in such experiments. Because of the uncertainty in conclusions based solely on the results of acetate incorporation patterns, the available evidence does not eliminate the possibility that sulochrin and the ergochromes might have a basically similar biogenesis. The separate question as to whether these arise by ring cleavage of anthraquinone intermediates is still a matter for speculation which can best be resolved by feeding experiments with the suspected intermediates in suitably labelled form.

It would be much easier to decide questions about starter units if the postulated poly-β-keto precursors could be isolated directly, but with one exception (section H.4) this has not so far been possible. It has often been suggested that these intermediates have no existence other than as enzyme-bound entities which are not released until the synthesis is essentially complete. This question can only be settled by experiments using the techniques of enzymology. However, there are enough metabolites which are either known, or suspected of being biosynthesised from biogenetically discrete intermediates, to justify attempts at investigating some of the intervening steps by tracer methods.

The fungal metabolites sclerotiorin (69) and rotiorin (70) from

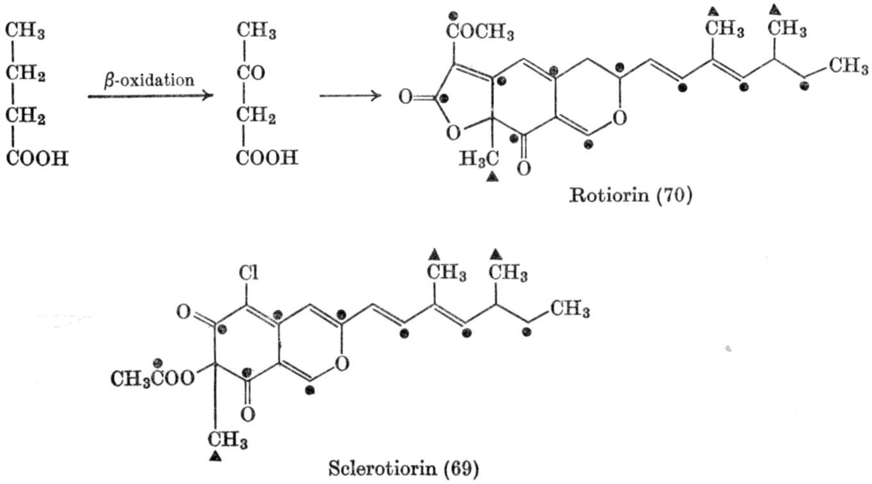

Rotiorin (70)

Sclerotiorin (69)

FIG. 23. The biogenesis of sclerotiorin and rotiorin.

Monascus purpureus also appear to have a clear biogenesis from acetate and malonate (Birch, Fitton, Pride, Ryan, Smith and Whalley,

1958; Birch, Cassera, Fitton, Holker, Smith, Thompson and Whalley, 1962) with subsequent introduction of three methyl groups from methionine. The acetoacetyl unit forming the lactone ring in rotiorin (70) appears to be derived from two molecules of acetate without the intervention of malonate. Since specifically labelled butyrate was incorporated intact into this C_4 unit, it was suggested that this was oxidised to acetoacetate by an enzyme system similar to that involved in the β-oxidation of fatty acids (Fig. 23) (Holker, Staunton and Whalley, 1964).

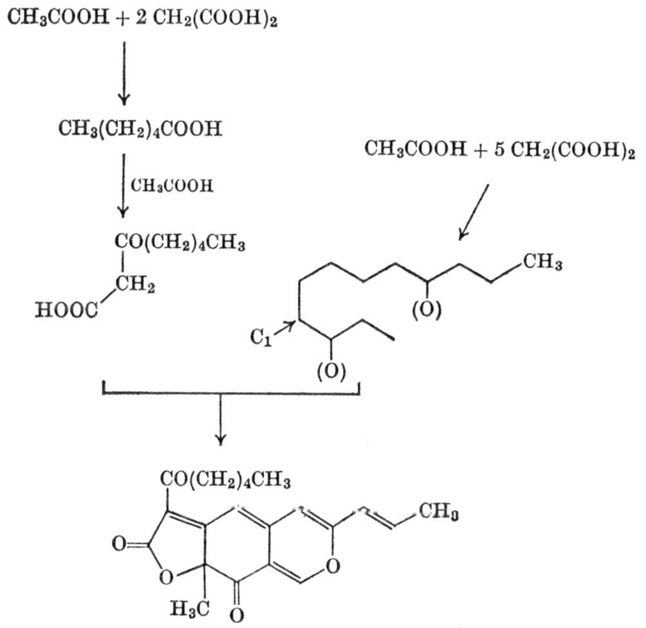

$CH_3COOH + 2\ CH_2(COOH)_2$

$CH_3(CH_2)_4COOH$

CH_3COOH

$CO(CH_2)_4CH_3$

CH_2

HOOC

$CH_3COOH + 5\ CH_2(COOH)_2$

CH_3

(O)

C_1

(O)

$CO(CH_2)_4CH_3$

CH_3

O

O

O

H_3C

O

Rubropunctatin (71)

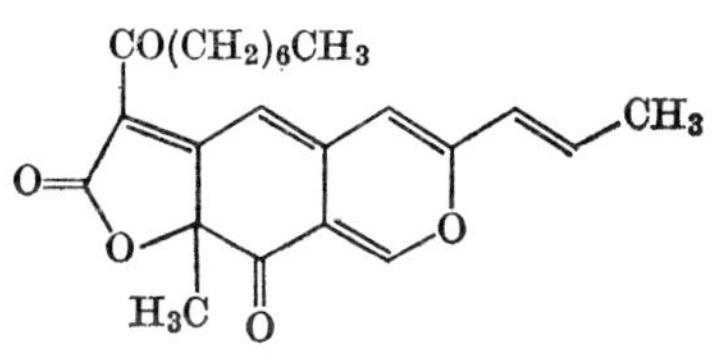

$CO(CH_2)_6CH_3$

CH_3

O

O

O

H_3C

O

Monascorubin (72)

Fig. 24. The biogenesis of rubropunctatin and monascorubin.

The co-metabolites rubropunctatin (71) and monascorubin (72) are synthesised in a slightly different manner since it has been shown that the β-oxo-lactone systems are derived by condensation of hexanoate and octanoate respectively, with acetate. Although these C_6 and C_8 units are elaborated by the normal acetate–malonate pathway of fatty acid biosynthesis, the final chain extensions to β-oxo-octanoate and β-oxo-decanoate respectively, do not involve malonate. The suggested biosynthetic pathway to rubropunctatin is illustrated in Fig. 24 (Hadfield, Holker and Stanway, 1967).

3. *Benzenoid Polyacetylenes*

A vast number of polyacetylenic compounds have been isolated from fungi and higher plants and the biogenesis from acetic acid of both acyclic and cyclic representatives has been firmly established. The four typical benzenoid polyacetylenes shown in Fig. 25, for example, have been shown to be derived from acetate by appropriate feeding experiments with *Chrysanthemum frutescens* and *Coreopsis lanceolata* (Bohlmann and Jente, 1966; Bohlmann, 1967).

Jones and his collaborators have recently brought the list up to date by confirming the polyacetate origin of compound (73) which occurs at the extraordinary level of 0·5 per cent in the tubers of certain Dahlia hybrids (Fairbrother, Jones and Thaller, 1967). This compound, like the triacetylenic analogue (74) consists of seven acetate residues, the carboxyl group of the terminal unit having been lost.

Bu'Lock has developed a unified scheme for the biosynthesis of acyclic polyacetylenes in which oleic acid occupies a fundamental position and is suggested to be converted into polyacetylenes by successive dehydrogenation steps (Bu'Lock, 1966). Experimental support for this pathway has been obtained in the acyclic series (Fairbrother *et al.*, 1967; Bu'Lock and Smith, 1967).

Very recently, Bohlmann and his collaborators have shown that the benzenoid polyacetylene (74) has a similar biogenesis from oleic acid, probably via the intermediates shown in Fig. 25 (Bohlmann, Jente, Lukas, Laser and Schulz, 1967; Bohlmann, Bonnet and Jente, 1968). All three precursors, suitably labelled, were converted into the polyacetylene (74) in *Coreopsis lanceolata*. (The acids were administered as the methyl esters and the alcohol as the acetate.) The suggested cyclisation mechanism is also illustrated in Fig. 25 (dotted arrows). This mechanism involves an aldol condensation in the normal way

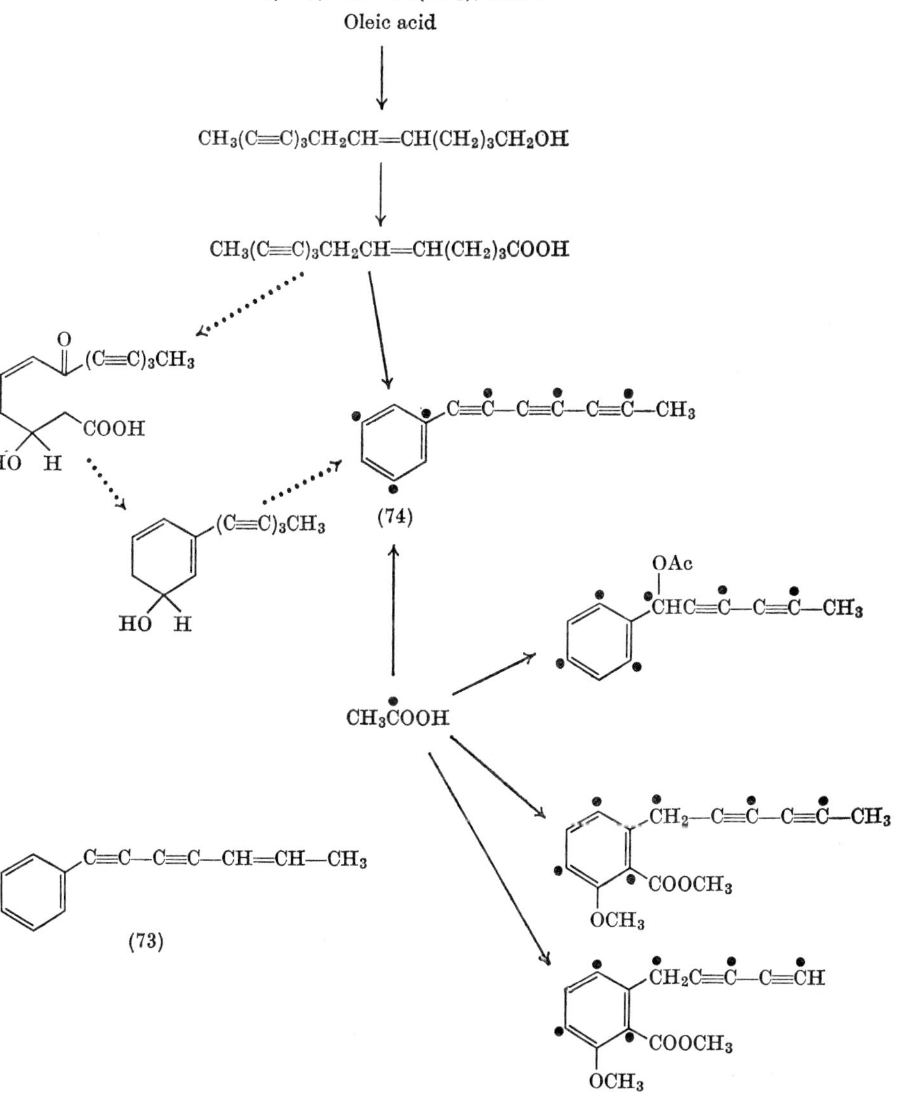

FIG. 25. Biosynthesis of benzenoid polyacetylenes.

associated with cyclic polyketide biosynthesis, with the significant difference that the ketonic intermediate must be produced by oxidation of a non-oxygenated precursor. This pathway also results in the overall loss of a five-carbon section from the carboxyl-terminal end of oleic acid.

4. *Tropolones*

The fungal tropolones stipitatic (75) and stipitatonic (76) acids (from *Penicillium stipitatum*), puberulic (77) and puberulonic (78)

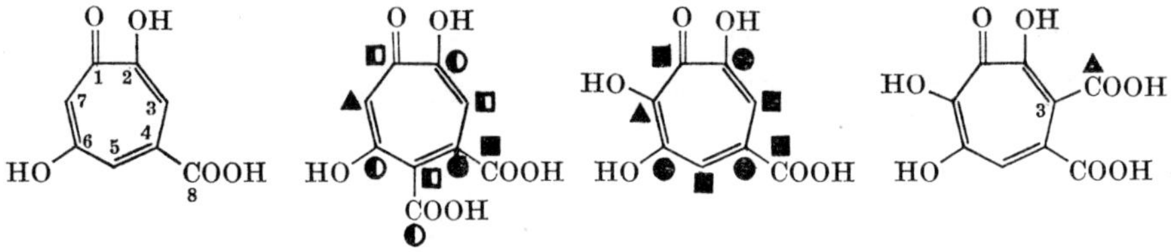

Stipitatic acid (75) | Stipitatonic acid (76) | Puberulic acid (77) | Puberulonic acid (78)

acids (from *Penicillium cyclopodium* and *P. aurantio-virens*) have a common biogenesis from acetate. The ring system is constructed from a polyketide chain which has been expanded by the insertion of an additional one-carbon unit (Bentley, 1962, 1963b; Andrew and Segal, 1964). Stipitatic acid (75) is derived from stipitatonic acid (76) by decarboxylation (Bentley, 1963b).

Stipitatonic (76) and puberulonic (78) acids differ biogenetically in that the carboxyl groups of the former are derived from acetate and represent the terminal groups of the postulated poly-β-keto precursor, whereas the carboxyl group at C(3) in the latter is not derived from acetate but appears to be provided by the one-carbon pool (for a summary of the evidence, see Bentley, 1963b).

Feeding experiments with *P. stipitatum* have shown that stipitatic acid (75) is derived from one acetate 'starter' molecule and three malonate units (Bentley, 1963b), but perhaps the most significant biogenetic feature is the derivation of C(7) from formaldehyde or formic acid (Bentley, 1958, 1963b). This was revealed by Baeyer-Villiger oxidation of [^{14}C]-HCHO-derived stipitatic acid (18 per cent incorporation), to give malonic and *cis*-aconitic (79) acids (Fig. 26). The *cis*-aconitic acid was essentially inactive, whereas decarboxylation of the malonic acid revealed that all of the activity resided in the methylene carbon atom derived from C(7) in stipitatic acid.

Degradation in this series is complicated by the equivalence of various positions under many of the reaction conditions employed, giving rise to products in which some or all of the carbon atoms have multiple origins (cf. Fig. 26). By comparison, the outcome of the

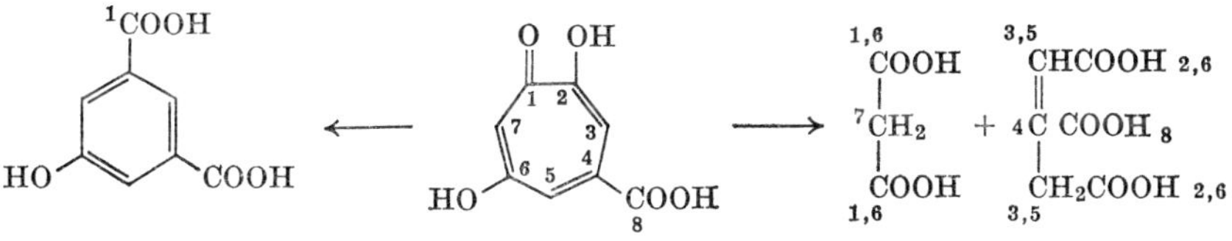

(80) (75) (79)

FIG. 26. The degradation of stipitatic acid.

alkaline (benzilic acid) rearrangement of stipitatic acid (75) to 5-hydroxyisophthalic acid (80), proved to be surprisingly straightforward. In this useful rearrangement, only C(1) was extruded in spite of the fact that keto-enol tautomerism might have been expected to make C(1) and C(2) equivalent (Bentley, 1963a; Andrew and Segal, 1964).

Many compounds have been put forward as possible precursors of the tropolone acids, including orsellinic acid (27), shikimic acid (section I), triacetic lactone (83A), cyclopolic (81) and cyclopaldic (82) acids. The last two compounds are also produced by strains of

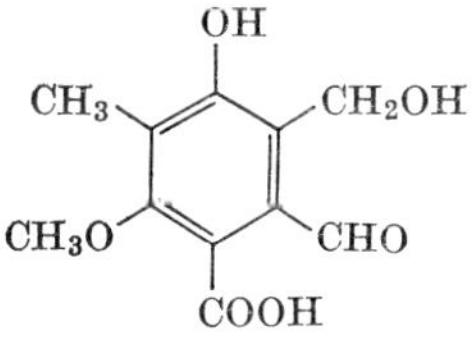

Cyclopolic acid (81)

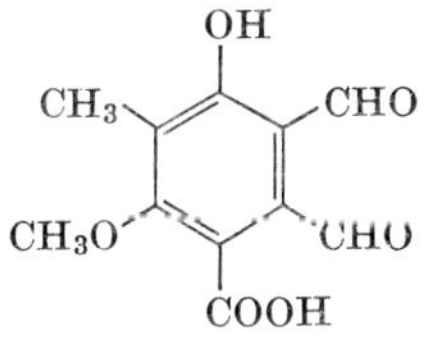

Cyclopaldic acid (82)

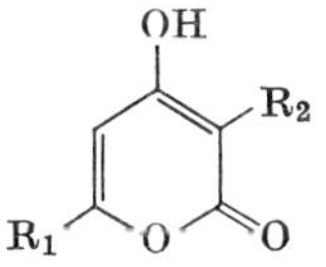

R_1	R_2	
CH_3	H	(83A)
CH_3	CH_3	(83B)
CH_3COCH_2	H	(83C)

Penicillium cyclopodium (Birkinshaw, Raistrick, Ross and Stickings, 1952). The isolation of methyl triacetic lactone (83B) from a strain of *Penicillium stipitatum* has been reported (Acker, Brenneisen and Tanenbaum, 1966), and the same metabolite, together with triacetic lactone (83A) and tetraacetic lactone (83C), was produced by a *P. stipitatum* culture in which tropolone biosynthesis had been suppressed by the addition of ethionine (Bentley and Zwitkowits, 1967a).

The discovery of compound (83C) is extremely interesting as it is a form of the hypothetical poly-β-keto precursor of orsellinic acid (27) (page 84). This is the first reported isolation of such a non-cyclic polyketide precursor.

The formation of compounds (83A–C) when tropolone biosynthesis was suppressed suggested that the former might be intermediates, and indeed, incorporation of activity from the [^{14}C]-labelled lactones into the tropolones in normal cultures was observed (Bentley and Zwitkowits, 1967b). However, the pattern of incorporation showed quite conclusively that the labelled lactones had not been incorporated intact but had undergone degradation to acetate, and that it was the labelled acetate which had been utilised in tropolone biosynthesis.

The incorporation of a one-carbon fragment into stipitatic acid is suggestive of a ring expansion step in the biosynthetic pathway and this has been a feature of a number of hypothetical schemes (Bentley, 1963b). Moreover, a one-carbon insertion step is made more attractive by the existence of laboratory syntheses based on analogous reactions. In their synthesis of stipitatic acid, for example, Johnson and his collaborators (Bartels-Keith, Johnson and Taylor, 1951) converted

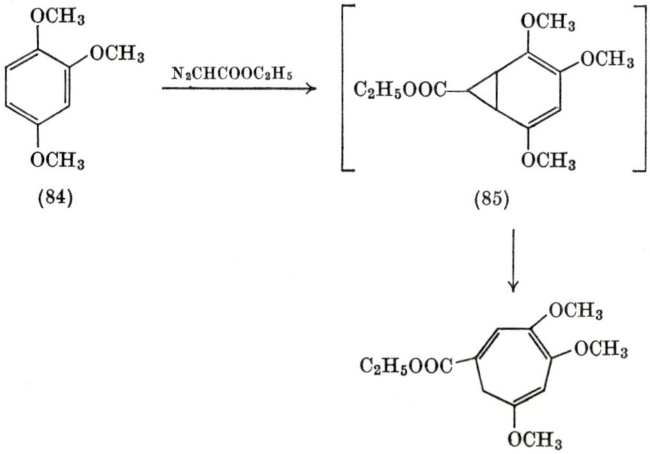

FIG. 27. The ring expansion step in the synthesis of stipitatic acid.

1,3,4-trimethoxybenzene (84) into the ester (86) by ring expansion with ethyl diazoacetate at 150° (Fig. 27). This reaction is presumed, by analogy with closely related reactions (see Bartels-Keith *et al.*,

1951), to proceed via the norcarane intermediate (85). The analogy with the possible biosynthetic pathway would be more pertinent if a biological equivalent of ethyl diazoacetate were known. However, formation of the cyclopropane ring system by insertion of a methionine-derived methylene group into an olefinic double bond is a well-established reaction *in vivo* (section L) and although there is as yet no evidence that this reaction occurs with aromatic substrates, it does give some support to the concept of a biosynthetic ring expansion reaction.

The well-known *in vivo* rearrangement of methylmalonic acid to

$$\begin{array}{c} COR \\ | \\ CH{-}CH_3 \\ | \\ COR \end{array} \longrightarrow \begin{array}{c} COR \\ | \\ CH_2 \\ | \\ CH_2 \\ | \\ COR \end{array} \qquad R = SCoA$$

FIG. 28. The methylmalonate-succinate conversion.

succinic acid (Fig. 28) is formally a chain expansion of the type under consideration (Beck and Ochoa, 1958) and may possibly proceed via an intermediate cyclopropyl species (Ingraham, 1962).

Bentley has critically reviewed the various theories relating to tropolone biosynthesis (Bentley, 1963b) and has shown that suggested pathways which start with the insertion of a hydroxymethyl group into an orsellinic acid type of precursor are not consistent with the observed labelling pattern in stipitatic acid. A number of other plausible routes have been suggested, but the available evidence does not allow a distinction to be made between them (Bentley, 1963b).

5. *'Starters' other than Acetate*

It is perhaps not surprising that continued exploration of fatty acid and polyketide biosynthesis should have disclosed metabolites with 'starter' units derived from compounds other than acetic acid. Among the best authenticated examples are the optically active D(+)-12-methyltetradecanoic (87) and D(+)-14-methylhexadecanoic (88) acids from *Bacillus subtilis*. L-Isoleucine was specifically incorporated into the terminal five-carbon unit of both these acids without change in configuration at the β-carbon atom, probably through the formation of α-methylbutyryl-CoA (Fig. 29), followed by successive condensations with malonyl-CoA (Lennarz, 1961; Kaneda, 1966).

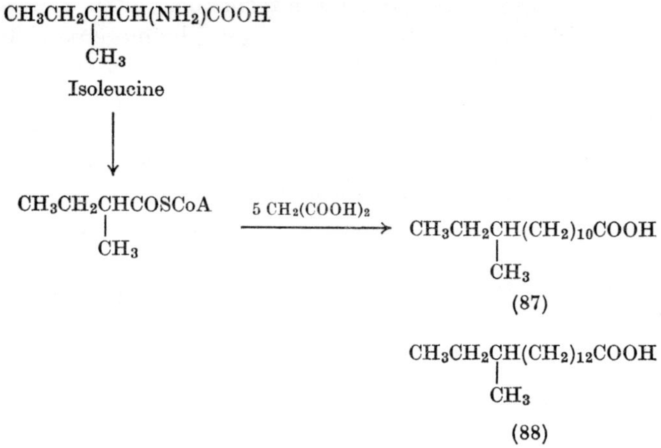

$CH_3CH_2CH(CH_3)CH(NH_2)COOH$

Isoleucine

↓

$CH_3CH_2CH(CH_3)COSCoA$ $\xrightarrow{5\ CH_2(COOH)_2}$ $CH_3CH_2CH(CH_3)(CH_2)_{10}COOH$ (87)

$CH_3CH_2CH(CH_3)(CH_2)_{12}COOH$ (88)

FIG. 29. The biosynthesis of branched chain fatty acids.

Tasmanone (89) from various *Eucalyptus* species probably belongs to this class of polyketide. The nuclear and methoxyl methyl groups have been shown to be derived from methionine in appropriate tracer experiments (Birch, Willis, Hellyer and Salahud-Din, 1966), and although direct evidence is lacking, the basic ring system is probably derived from isobutyryl-CoA (from valine (Coon, Robinson and Bachawat, 1956)) and three malonate units as shown in Fig. 30. The

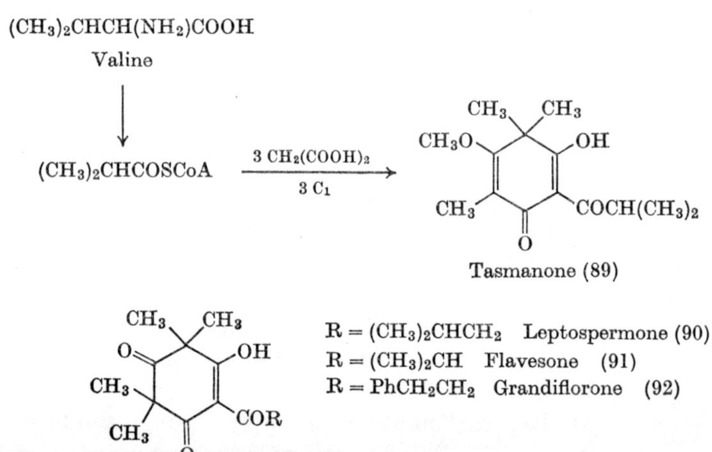

FIG. 30. The probable biogenesis of tasmanone and related metabolites.

related compounds leptospermone (90), flavesone (91) and grandiflorone (92) (Hellyer and Pinhey, 1966) probably have a similar biogenesis from isovaleryl-CoA, isobutyryl-CoA and the coenzyme A ester of a phenylpropane acid (possibly cinnamic acid), respectively.

In the biosynthesis of the macrolide antibiotics, a parallel with fatty acid biosynthesis has been found in which propionyl-CoA serves as the 'starter' molecule and chain elongation proceeds by condensation with methylmalonyl-CoA (Grisebach and Hofheinz, 1964). Metabolites formed in this way have branching methyl groups at alternate

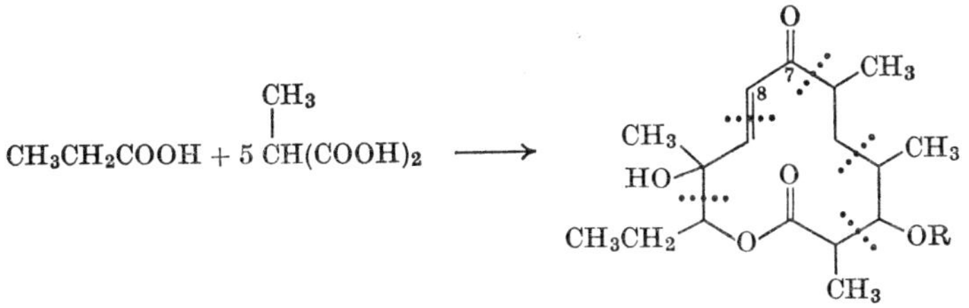

$$CH_3CH_2COOH + 5\ CH(CH_3)(COOH)_2 \longrightarrow$$

Methymycin (93)
(R = Desosamine)

positions in the chain. This pathway has been conclusively demonstrated in, for example, methymycin (93) biosynthesis, by feeding suitably labelled propionic acid to *Streptomyces venezuelae* (Birch, Pride, Rickards, Thompson, Dutcher, Perlman and Djerassi, 1960). The odd two-carbon unit, C(7, 8) may be derived from acetate or may represent a methylmalonyl residue in which the side chain methyl group has been lost (Bentley, 1962). The same type of chain extension has been demonstrated in the branched chain portions of the phthienoic acids from *Mycobacterium tuberculosis* (Gastambide-Odier, Delaumeny and Kuntzel, 1966).

It is curious that no fully carbocyclic analogues of the macrolides have been discovered. The nearest related compound, biogenetically,

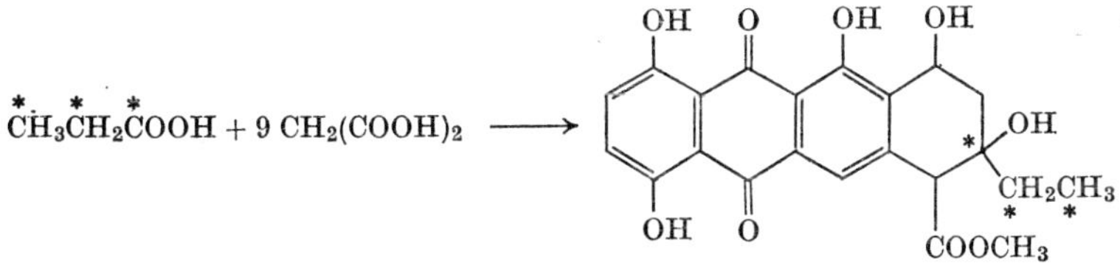

$$\overset{*}{C}H_3\overset{*}{C}H_2\overset{*}{C}OOH + 9\ CH_2(COOH)_2 \longrightarrow$$

ε-Pyrromycinone (94)

is ε-pyrromycinone (94), the aglycone of various antibiotics (pyrromycin, the cinerubins and rutilantins). Incorporation experiments with labelled acetic and propionic acids demonstrated the derivation of this metabolite from one 'starter' molecule of propionic acid and nine of malonic acid (inferred from the results of acetate incorporation) as shown (Ollis, Sutherland, Codner, Gordon and Miller, 1960).

6. *Benzoquinones*

A number of simple benzoquinones have been shown to be biosynthesised via the acetate–malonate pathway. Since this route leads initially to products in which the phenolic hydroxyl groups are *meta* disposed, the formation of a *para*-benzoquinone nucleus must inevitably involve an additional hydroxylation step.

Incorporation studies with ^{14}C-labelled acetate and malonate revealed the basic polyketide nature of aurantiogliocladin (95), the major secondary metabolite produced by *Gliocladium roseum* (Bentley and Lavate, 1965). In an attempt to define the intermediate stages, Pettersson investigated the incorporation of a number of ^{14}C-labelled aromatic precursors and found that orsellinic acid (27) and 2,4-dihydroxy-5,6-dimethylbenzoic acid (96) were incorporated with efficiencies of 0·07 and 0·9 per cent respectively (Pettersson, 1965c). On the basis of these results, the pathway for aurantiogliocladin

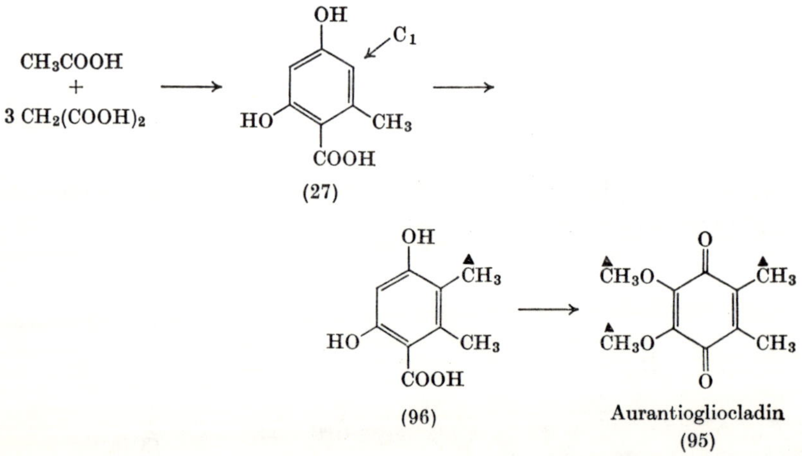

CH_3COOH + 3 $CH_2(COOH)_2$ → (27) → (96) → Aurantiogliocladin (95)

FIG. 31. The biosynthesis of aurantiogliocladin.

biosynthesis shown in Fig. 31 was proposed. The investigations of Birch and his collaborators showed that the three additional methyl groups were provided by the C_1 pool as expected (Birch, Fryer and Smith, 1958).

The number of possible permutations and combinations of nuclear hydroxyl and methoxyl groups, in even the simple fungal quinones, is rather large, as can be seen from the selection of six toluquinones (97)–(102) isolated from *Aspergillus fumigatus* (Pettersson, 1963a;

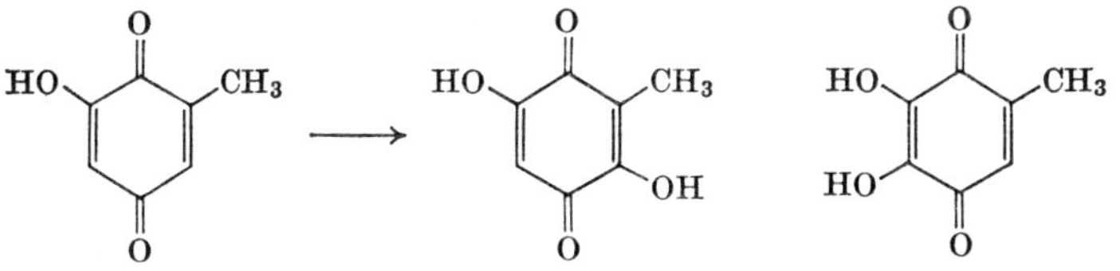

(97) → (98) (99)

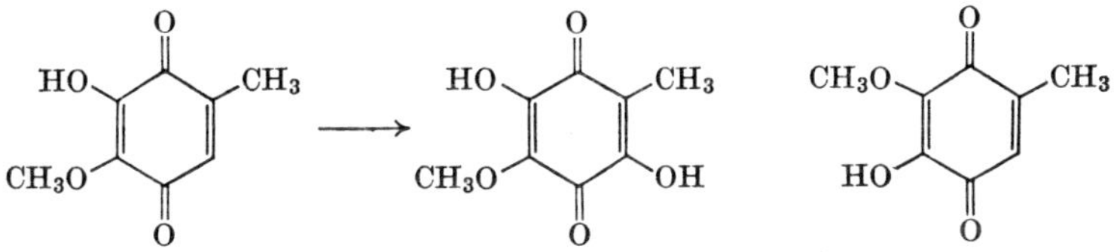

Fumigatin (100) → Spinulosin (101) (102)

1964a). There is evidence that orsellinic acid (27) is a common precursor of these metabolites (Pettersson, 1963b; 1964a) but information on the chronology of the various hydroxylation and methylation steps is still only fragmentary. It does appear, however, that hydroxylation *ortho* to the methyl group takes place at a late stage, since the conversions of 3-hydroxy-2,5-toluquinone (97) into 3,6-dihydroxy-2,5-toluquinone (98), and of fumigatin (100) into spinulosin (101) have been established. The available evidence suggests that the various hydroxylation and methylation steps occur at the hydroquinone or fully aromatic level and that oxidation to the quinone is the ultimate step in the pathway. Evidence has been presented to show that the final oxidation is non-enzymatic, a rare feature which is undoubtedly due to the special redox nature of the hydroquinone–quinone system (Pettersson, 1964b, c).

If one had a choice of organism to use in a biosynthetic investigation, it is doubtful if members of the beetle family would spring to mind as the most agreeable or convenient to work with. In practice, however, they appear to be quite amenable to experimentation. Meinwald and his collaborators have been able to carry out tracer experiments with the tenebrionid beetle *Eleodes longicollis* which contains in its defensive

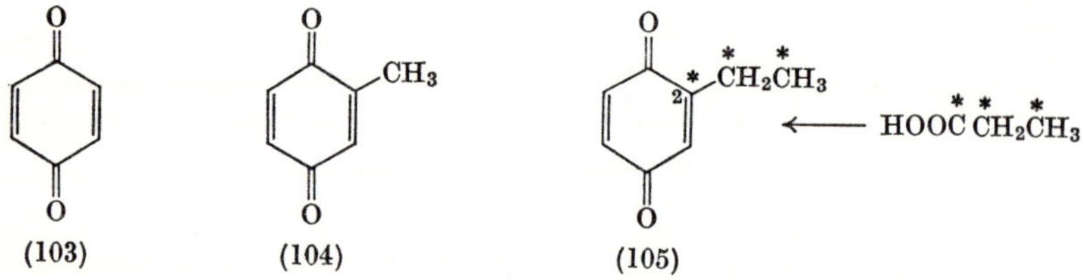

FIG. 32. Benzoquinones from arthropod defensive secretions.

secretion benzoquinone (103), toluquinone (104) and 2-ethylbenzoquinone (105). Formation of benzoquinone (103) required the preformed ring of an aromatic precursor (phenylalanine), but toluquinone (104) and 2-ethylbenzoquinone (105) were formed by the normal polyketide pathway, with the difference that in biosynthesis of the latter, propionate served as the starter molecule. [1-^{14}C]-Propionate labelled the ring carbon atom C(2) almost exclusively, whereas the labels from [2-^{14}C]- and [3-^{14}C]-propionate were incorporated specifically into the ethyl side-chain (Fig. 32) (Meinwald, Koch, Rogers and Eisner, 1966).

The polyketide nature of the interesting quinone epoxide terreic acid (106) from *Aspergillus terreus* has recently been established (Read and Vining, 1968). Incorporation of atmospheric oxygen into

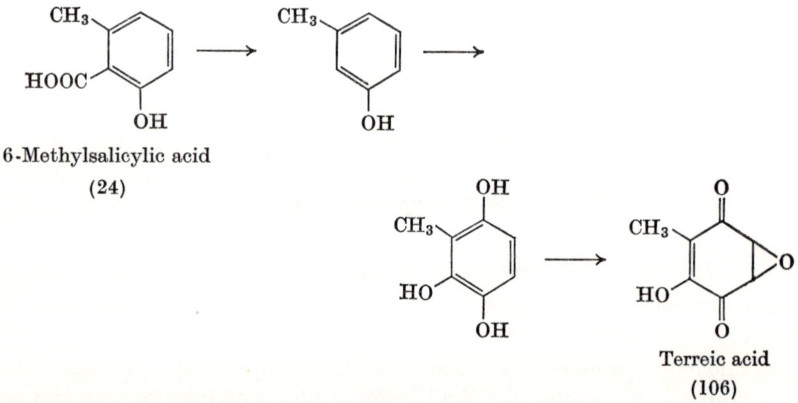

FIG. 33. The probable pathway of terreic acid biosynthesis.

the epoxide group was directly demonstrated in studies with $^{18}O_2$. This latter result rules out orsellinic acid as a precursor since this would require removal of a phenolic hydroxyl group from the aromatic ring—a process which has been observed in a few cases but which does not appear to be very common (El-Basyouni, Chen, Ibrahim, Neish and Towers, 1964; Scheline, Williams and Witt, 1960; Dacre and Williams, 1962; Booth, Murray, DeEds and Jones, 1955). The failure of ^{14}C-labelled orsellinic acid to be incorporated supported this conclusion; on the other hand, 6-methylsalicylic acid was incorporated without prior degradation. These results are consistent with a route from 6-methylsalicylic acid as shown in Fig. 33, with *m*-cresol as a probable intermediate (cf. patulin biosynthesis, section H.10).

7. *Tetracyclines*

Later stages in the elaboration of the tetracycline antibiotics such as 7-chlorotetracycline (107) are understood in considerable detail. At present the basic skeleton is considered to have a normal polyketide origin with reservations concerning the 'starter' component (see below). The early stages in the pathway were investigated using radiotracer techniques (Miller, McCormick and Doerschuk, 1956; Snell, Birch and Thomson, 1960), but the later stages, illustrated in the scheme for 7-chlorotetracycline (107) biosynthesis (Fig. 34) have all been elucidated by extended investigations involving large numbers of *Streptomyces* mutants (McCormick, 1965). A key feature of the pathway fell into place with the discovery of fully aromatic intermediates such as 4-hydroxy-6-methylpretetramid (108) (McCormick and Jensen, 1965), which was efficiently converted into 7-chlorotetracycline (107) *in vivo*. The later stages are characterised by the various oxidative transformations illustrated in Fig. 34, by methylation at C(6) (as in (107)), and at the amino function of ring A, and by the introduction of chlorine. The diversion into the 5-hydroxy series takes place at the penultimate stage as shown (Mitscher, Martin, Miller, Shu and Bohonos, 1966).

It is a significant feature of this scheme that the carboxamide function is already present in the earliest defined intermediates, the pretetramids (as in (108)). Feeding experiments with [^{14}C]-bicarbonate (Gatenbeck, 1962) demonstrated that almost the entire activity in the oxytetracycline (109A) formed was located in this group. This led Gatenbeck to suggest that the polyketide chain was

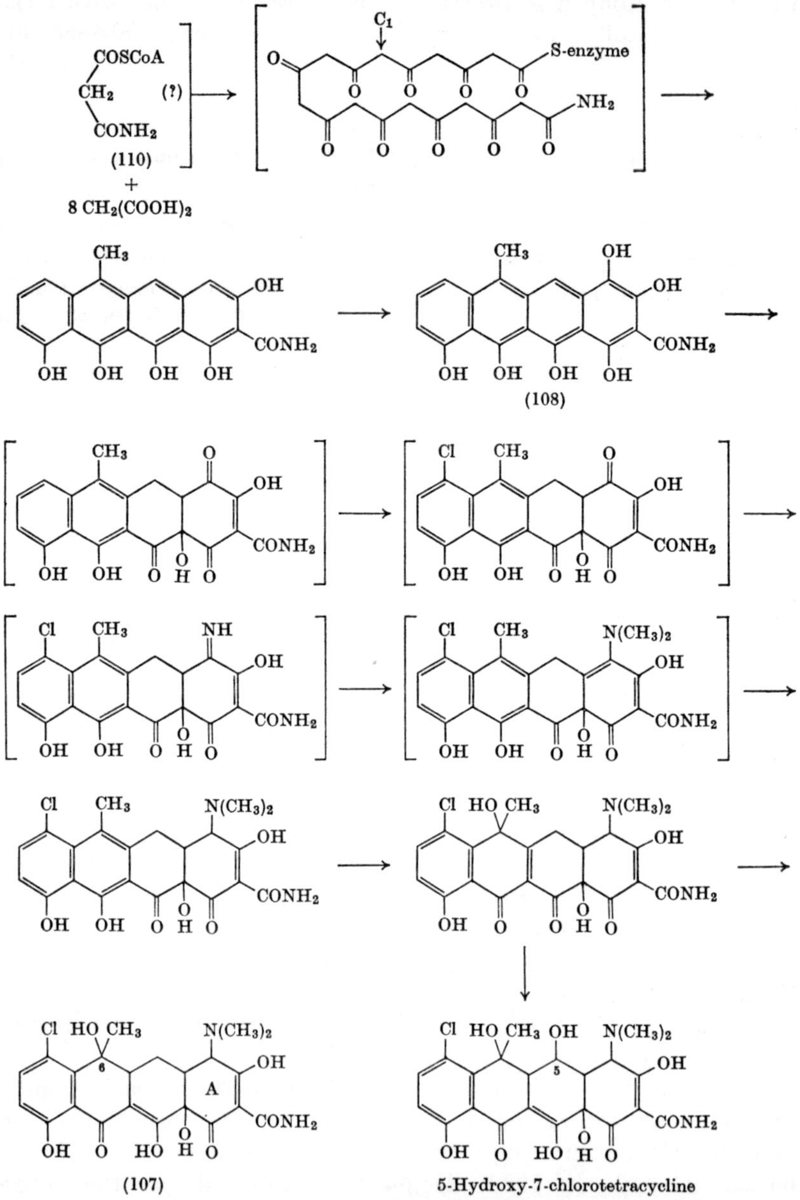

$CH_2(COSCoA)CONH_2$ (110) + 8 $CH_2(COOH)_2$ (?) →

(108)

(107)

5-Hydroxy-7-chlorotetracycline

FIG. 34. Tetracycline biosynthesis.

initiated by malonamyl-CoA (110), subsequent condensations with malonyl-CoA leading to the fully aromatic tetracyclic intermediates. Although there is some negative evidence which tends to refute this idea (McCormick, 1965), an investigation of the incorporation of suitably labelled malonamyl-CoA, the experiment which might be definitive, does not appear to have been carried out as yet.

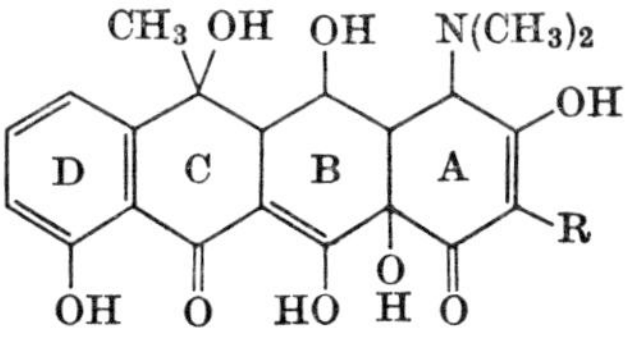

(109A) R = —$CONH_2$
(109B) R = —$COCH_3$

The supposition that the carboxamide group marks the ‘starter’ unit is supported by the discovery of co-metabolites such as 2-acetyl-decarboxamidotetracycline (109B) in which the ‘starter’ position is clearly indicated by the acetyl group (McCormick, 1965).

Two other explanations were offered by Gatenbeck to account for the specific incorporation of [^{14}C]-bicarbonate. The first was that succinate, derived from pyruvate by carboxylation, was a precursor. This proposal could be tested by a detailed degradation of ring A in a suitably labelled tetracycline, but this unfortunately appears to be technically very difficult. The second suggestion, that carboxylation of a complex intermediate was involved, again precludes experimental verification. The suggestion that malonamyl-CoA might function as the ‘starter’ in the polyketide chain is possibly relevant to the problem of javanicin biosynthesis discussed above.

It has been pointed out by McCormick in a recent detailed review of tetracycline biosynthesis (McCormick, 1965) that because of the lack of specificity of some of the enzymes involved, pathways such as that shown in Fig. 34, which are largely based on precursor conversion studies with blocked mutants, must be tentative to some degree. There is little doubt, however, of the validity of the basic pathway.

8. *Phenol Coupling Reactions*

The probable intervention of oxidative coupling of phenolic substrates in the biosynthesis of griseofulvin and related compounds has been mentioned already. The conclusion that similar processes, which almost certainly involve free radical intermediates, are very widespread in nature, rests largely on the structural analysis of natural products. Many structures can be seen as *in vivo* expressions of oxidative coupling reactions which have been extensively investigated in the laboratory. Recent reviews are available which describe these studies (Hassall and Scott, 1961; Scott, 1965; 1967).

A reinvestigation of the formation of Pummerer's ketone (110) from *para*-cresol (111) by ferricyanide oxidation led Barton *et al.* to suggest its formation through the coupling of the two radical species illustrated in Fig. 35 (Barton, DeFlorin and Edwards, 1956). This concept was

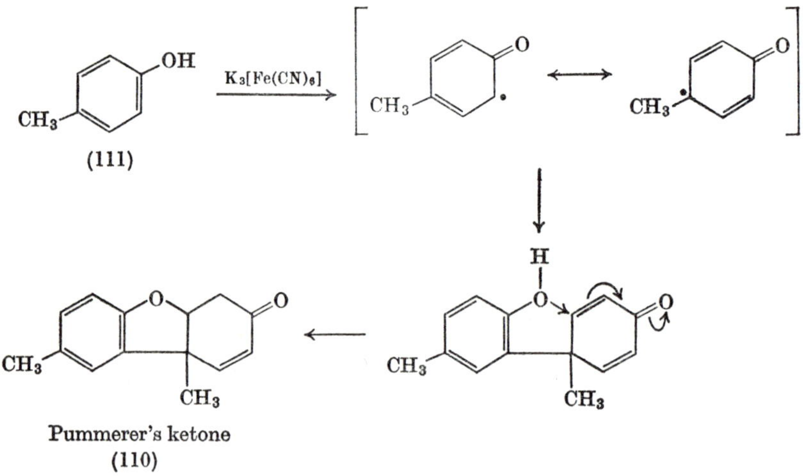

FIG. 35. The synthesis of Pummerer's ketone by phenol coupling.

developed to provide an ingenious synthesis of the ubiquitous lichen metabolite usnic acid (112), which was obtained in two steps from methylphloracetophenone (113) as shown (Fig. 36). Ferricyanide oxidation of (113) gave the intermediate (115) in 15 per cent yield; this was converted in good yield to usnic acid diacetate (114) by treatment with acetic anhydride containing a little sulphuric acid.

In spite of the elegance of this synthesis and its obvious implications for usnic acid biosynthesis, incorporation experiments were lacking until recently. Lichens are symbiotic combinations of algae and fungi; as a consequence their laboratory culture creates special difficulties and this undoubtedly has deterred investigators. Ten years after the

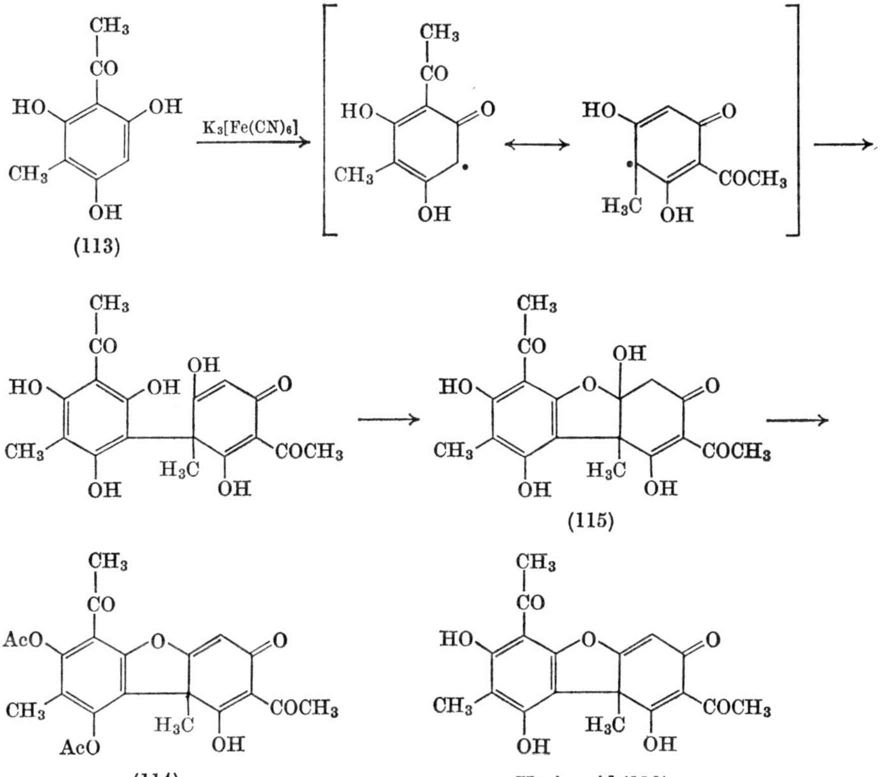

Fig. 36. The synthesis of usnic acid.

publication of the original laboratory synthesis, however, Shibata and his collaborators have reported incorporation experiments with labelled precursors in a number of lichen species (Taguchi, Sankawa and Shibata, 1966). Both ^{3}H- and ^{14}C-labelled methylphloracetophenone (113) were incorporated with efficiencies ranging from 0·022 to 0·018 per cent (^{3}H) and 0·07 to 0·28 per cent (^{14}C).

Usnic acid derived from methylphloracetophenone (113), labelled with ^{14}C in the methyl carbon atom of the acetyl side chain (Fig. 37), was degraded with sodium hypoiodite to give iodoform which contained all the activity of the metabolite. Incorporation experiments with [2-^{14}C]-acetate, [2-^{14}C]-malonate and [^{14}C]-formate clearly demonstrated that the acetate–malonate pathway was involved with

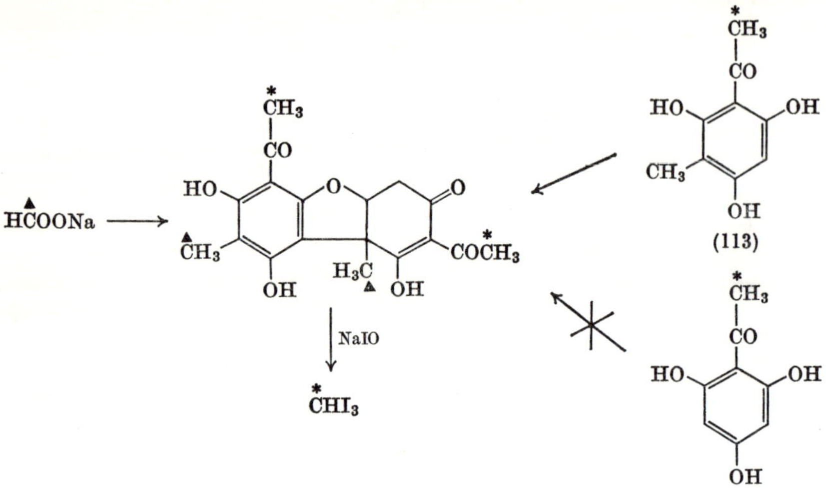

Fig. 37. Experiments on usnic acid biosynthesis.

the two nuclear methyl groups coming from the C_1 pool. [^{14}C]-Phloracetophenone itself was not incorporated showing that introduction of these methyl groups must take place at an early stage (Fig. 37).

Oxidative phenol coupling reactions have also been studied in isolated enzyme systems. The most frequently used system is horseradish peroxidase–hydrogen peroxide, which catalyses a variety of different coupling reactions including that of *meta*-cresol to Pummerer's ketone (Fig. 35) (Westerfield and Lowe, 1942) and of methylphloracetophenone (113) to usnic acid hydrate (Penttila and Fales, 1966b). The large number of such transformations which have been carried out with isolated enzyme systems leave no doubt that such processes are important in nature, but very few have been demonstrated by direct tracer experiments (Scott, 1965). This is undoubtedly due in part to the fact that the substrates involved are usually quite complex; thus their synthesis in labelled form often represents a major undertaking. The most striking tracer investigations into such reactions have been in the benzylisoquinoline alkaloid series. This work has been reviewed recently (Barton, 1964; Battersby, 1967).

As phenol coupling reactions usually do not generate new carbocyclic rings, but generally serve to couple together preformed cyclic systems, they will not be discussed further here, except to point out their probable role in the biosynthesis of a few well-known metabolites such as 4,5,4′,5′-tetrahydroxy-1,1′-binaphthyl (116) and 3,10-dihydroperylene-4,9-quinone (117) from *Daldinia concentrica*, which co-occur with 1,8-dihydroxynaphthalene (118) (Allport and Bu'Lock, 1960), the sesquiterpene dimer gossypol (119) from cotton seed (Edwards and Cashaw, 1957) and picrolichenic acid (120) from the lichen *Pertusaria amara* (Davidson and Scott, 1960).

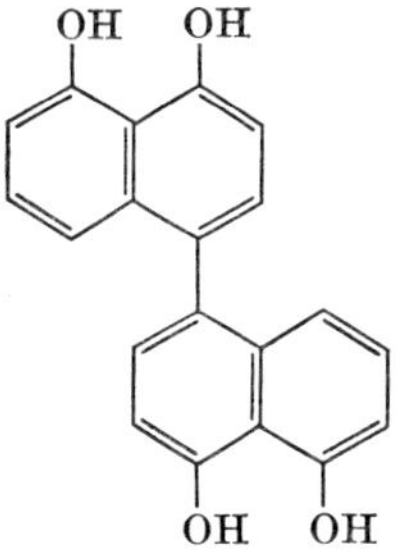

(116) 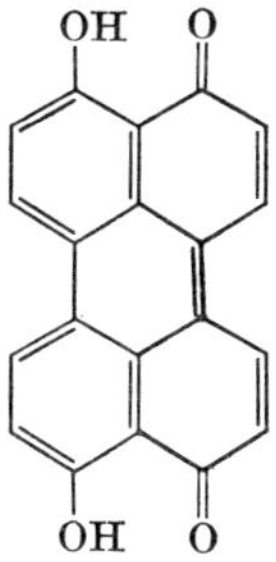 (117) 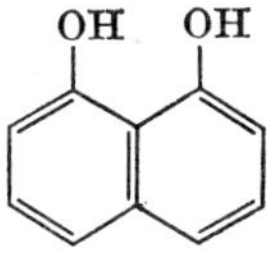 (118)

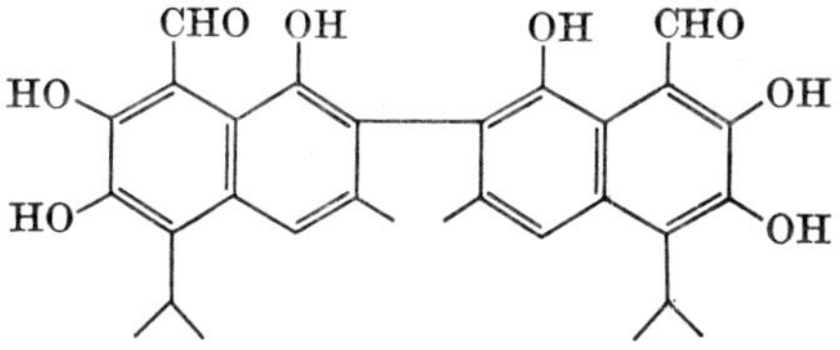

Gossypol (119) 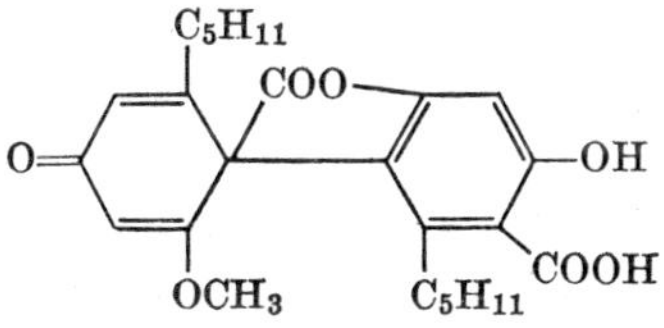 Picrolichenic acid (120)

It is noteworthy that although the sequence (118) → (117) (presumably via (116)) has been demonstrated in *Daldinia concentrica*

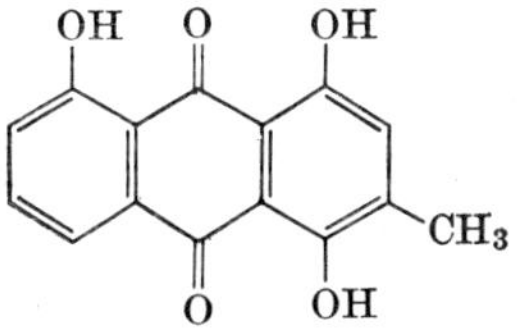

Islandicin (26)

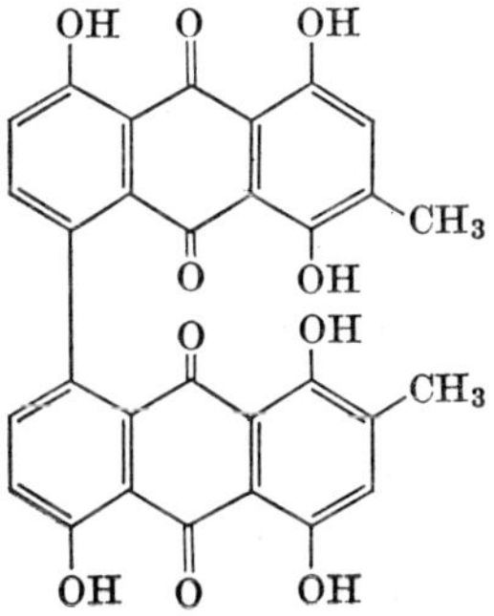

Iridoskyrin (121)

(Allport and Bu'Lock, 1960), experiments with *Penicillium islandicum*, which produces islandicin (26) and the corresponding dimer iridoskyrin (121), seem to indicate that these metabolites are formed from glucose by independent routes (Gatenbeck, 1960a). This finding serves to underline the need for experimental verification of even the most plausible hypothetical pathways.

9. *Terrein*

Terrein (122), a metabolite of *Aspergillus terreus*, appears to be derived from a modified polyketide precursor. The asymmetry in the

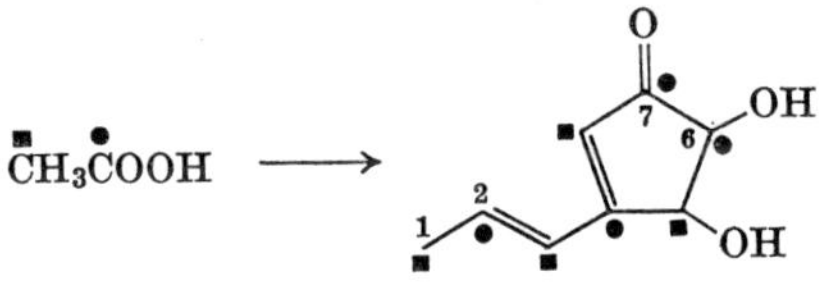

Terrein (122)

incorporation of [^{14}C]-acetate was found at C(6) and C(7) which were both derived, with nearly equal efficiency, from the carboxyl carbon of acetate. The incorporation of [2-^{14}C]-malonate paralleled that of [2-^{14}C]-acetate, but with both precursors the activity in all the labelled positions was nearly equal so that it is not clear that C(1) and C(2) represent a 'starter' acetate unit (Birch, Cassera and Jones, 1965). It was suggested that the biosynthesis of terrein involved ring contraction of a six-membered ring precursor.

10. *Oxidative Transformations of Carbocyclic Polyketides*

Oxidative ring cleavage is an important process in the metabolic degradation of aromatic compounds. A great deal is known about these reactions in micro-organisms both from the chemical and enzymatic points of view (Evans, 1963). The two modes of ring cleavage of catechol (Fig. 38) serve to illustrate the type of transformation involved.

A number of tracer investigations have provided examples of carbocyclic polyketides serving as precursors of oxygen-heterocyclic compounds by pathways involving ring cleavage. One such metabolite which has been investigated is penicillic acid (123) from various

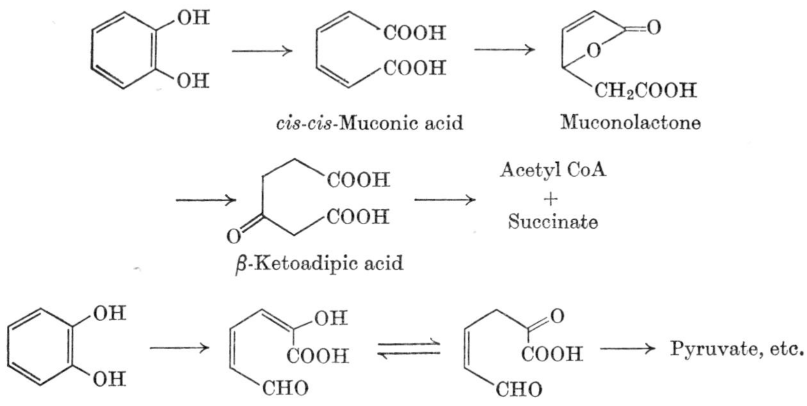

OH OH → COOH COOH → O O CH_2COOH

cis-cis-Muconic acid Muconolactone

→ COOH COOH O → Acetyl CoA + Succinate

β-Ketoadipic acid

OH OH → OH COOH CHO ⇌ O COOH CHO → Pyruvate, etc.

FIG. 38. Oxidative degradation of catechol by micro-organisms.

Penicillium species. ^{14}C-labelled orsellinic acid (27), obtained from a culture of *Chaetomium cochliodes* which had been grown in the presence of [1-^{14}C]-acetate, was converted into penicillic acid (123) in *Penicillium barnense* (Mosbach, 1960). This conversion involves ring cleavage, decarboxylation, oxidation and methylation, as shown in Fig. 39. The pattern of incorporation of [^{14}C]-acetate (both labels)

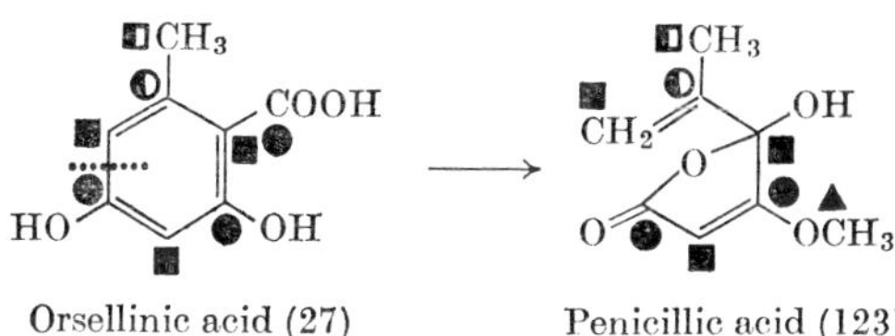

FIG. 39. The biosynthesis of penicillic acid from orsellinic acid.

and [2-^{14}C]-malonate into penicillic acid is entirely consistent with this pathway (Bentley and Keil, 1962).

An analogous ring cleavage is involved in the conversion of 6-methylsalicylic acid (26) into patulin (124) in *Penicillium patulum* (Tanenbaum and Bassett, 1959, 1960). The previously generally accepted pathway via gentisaldehyde (129, R = H) is illustrated in Fig. 40, pathway *a* (Bentley, 1962). The discovery of new metabolites of *Penicillium patulum*, notably *m*-cresol (125, R = H), *m*-hydroxybenzyl alcohol (126) and toluquinol (127) prompted Scott to suggest the possible alternative pathways *b* or *b'* (Fig. 40) (Scott and Yalpani,

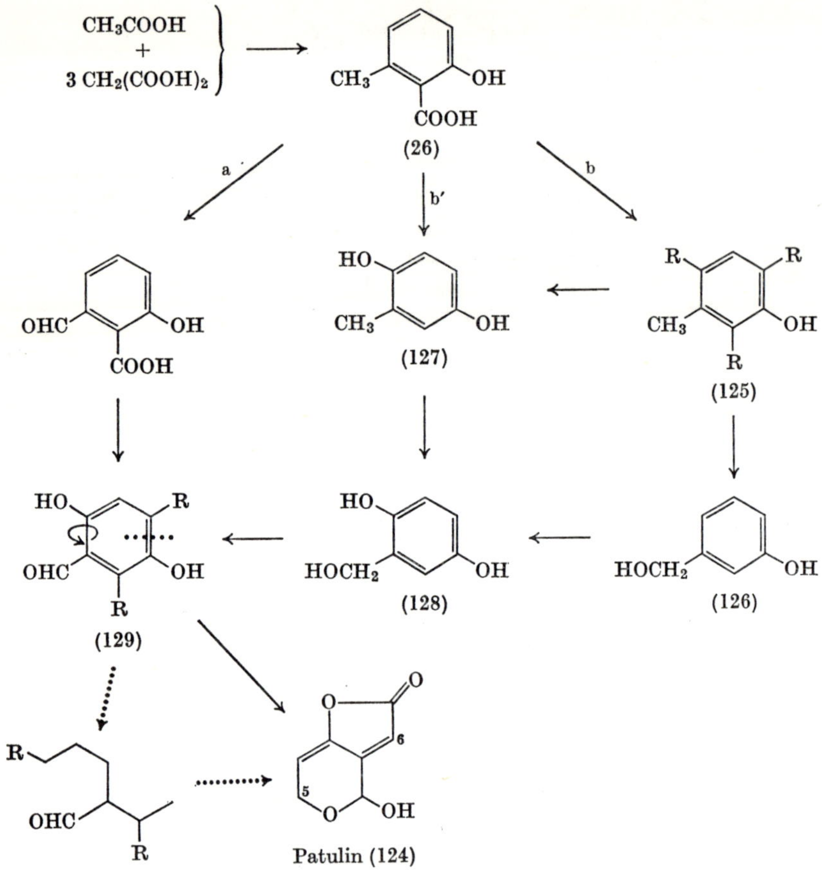

FIG. 40. Biosynthesis of patulin.

1967). By making elegant use of mass spectrometry, Scott and Yalpani were able to demonstrate the conversion of [2,4,6-^{2}H]-3-methylphenol (125, R = ^{2}H) into patulin (124) with dilution of the deuterium content to 30 per cent of that of the precursor. It was also shown that the deuterium was located in the predicted positions (C(5) and C(6) in 124)). In addition, toluquinol (127), gentisyl alcohol (128) and gentisaldehyde (129, R = H), isolated from the same culture, showed a deuterium content corresponding to 71·0, 65·6 and 66·7 per cent of that of the precursor, respectively, providing strong support for the proposed pathway *b′*. This experiment provides convincing proof of the utility and convenience of mass spectrometry in biosynthetic investigations, provided reasonably high incorporations can be attained.

11. *Model Reactions*

The initial postulation of poly-β-keto precursors in polyketide biosynthesis naturally stimulated research into the synthesis of model compounds. Investigations on these lines are hindered by the lack of stability of such compounds and by the formation of complex mixtures from them on treatment with basic reagents. In spite of these difficulties, a number of interesting investigations have been carried out

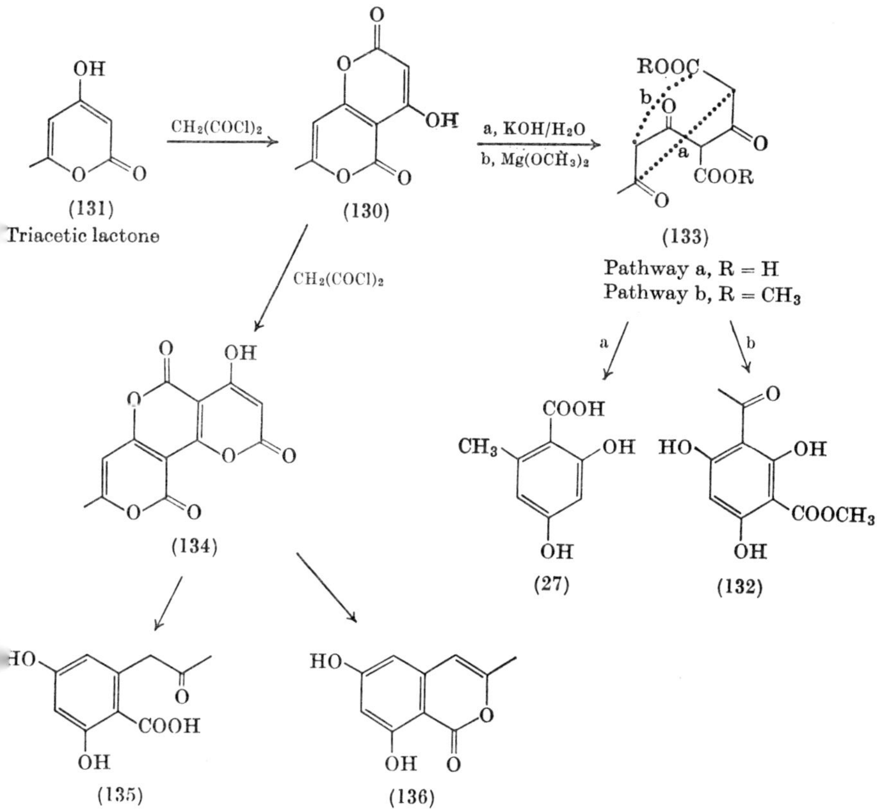

FIG. 41. The base-catalysed cyclisation reactions of polypyrones.

which have thrown light on the extent to which purely 'chemical' factors can affect the various possible modes of cyclisation.

Scott and his collaborators have developed methods for the synthesis of polypyrones based on successive acylations with malonyl chloride. In this way, the pyrone (130) was obtained from triacetic acid (131) (Money, Qureshi, Webster and Scott, 1965). The base-catalysed ring opening and recyclisation of this compound, which can be regarded as a 'protected' polyketide, shows an interesting dependence on the nature of the base employed. Treatment with aqueous potassium hydroxide gave orsellinic acid (27) whereas methanolic magnesium methoxide gave the acylphloroglucinol (132). The formation of these products by aldol condensation (pathway *a*) and Claisen condensation (pathway *b*) of the acyclic intermediate (133) is illustrated in Fig. 41. Further acylation of the pyrone (130) gave the tricyclic product (134). Treatment of this with methanolic alkali gave the known microbial metabolites *C*-acetyl orsellinic acid (135) and 3-methyl-6,8-dihydroxyisocoumarin (136) amongst other products. Discrimination between the various pathways of cyclisation of the tripyrone (134) in relation to the type of base employed, however, was much less marked than was the case with the dipyrone (130) (Comer, Money and Scott, 1967).

Recently, a synthetic approach to the 'free' poly-β-keto acids as in (137) have been developed (Harris and Carney, 1966). 1,3,5-Triketones

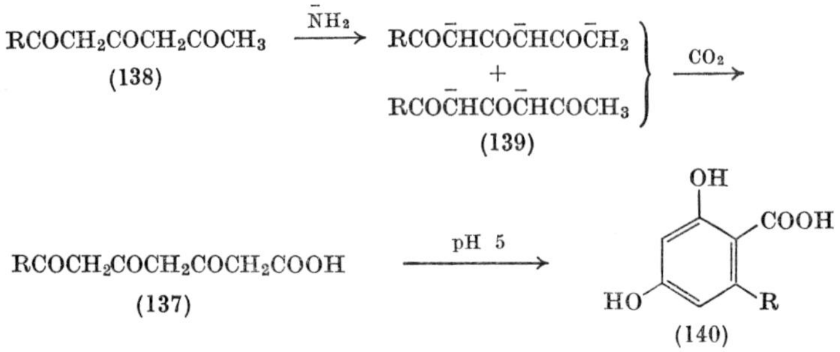

$$\underset{(138)}{RCOCH_2COCH_2COCH_3} \xrightarrow{\bar{N}H_2} \left.\begin{array}{c} RCO\bar{C}HCO\bar{C}HCO\bar{C}H_2 \\ + \\ RCO\bar{C}HCO\bar{C}HCOCH_3 \\ (139) \end{array}\right\} \xrightarrow{CO_2}$$

$$\underset{(137)}{RCOCH_2COCH_2COCH_2COOH} \xrightarrow{pH\ 5} (140)$$

FIG. 42. The synthesis and base-catalysed cyclisation of poly-β-keto acids.

(138) (Fig. 42), when treated with three or more equivalents of alkali amide in liquid ammonia, gave mixtures of the di- and tri-anions (139). These were carboxylated with carbon dioxide in ether or tetrahydrofuran solution to give the required acids (137). Cyclisation of the

triketoacids was effected at room temperature in aqueous solution at pH 5. Excellent yields of β-resorcylic acids (140) were obtained as in the conversion (90 per cent) of the acid ((137), $R = nC_5H_{11}$) into olivetol carboxylic acid ((140), $R = nC_5H_{11}$), a common lichen metabolite. Product dependence on the type of base used was also noted in these investigations.

The dependence of the nature of the product on the amount of base

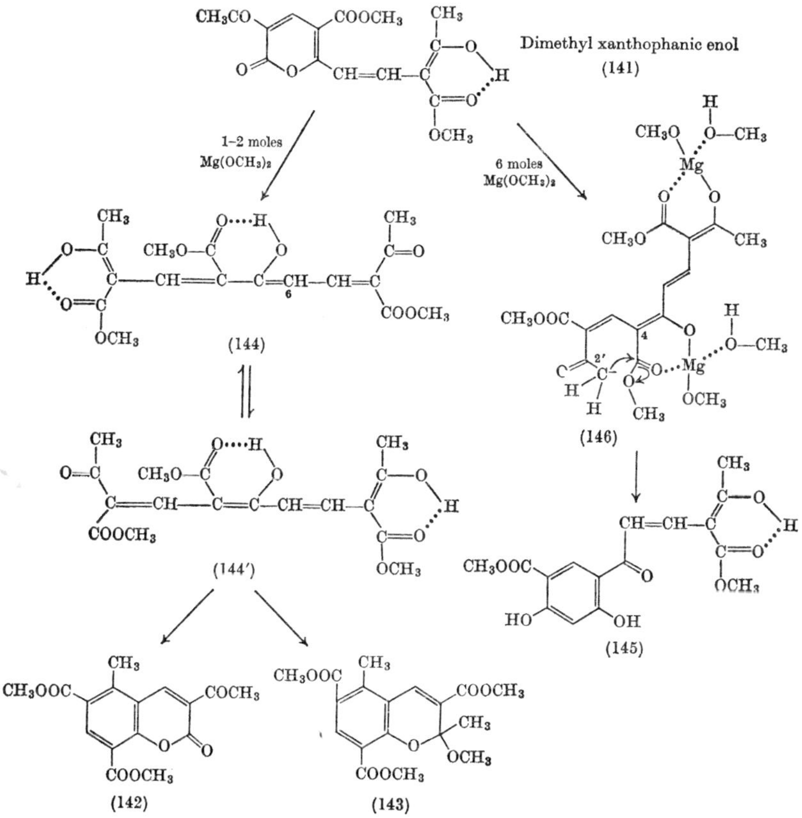

FIG. 13. The effect of base concentration on the cyclisation of polyketonic compounds.

used in similar cyclisation reactions has been interpreted by Crombie in terms of the conformational constraint and inhibition of enolisation introduced by chelation with the metal ion (Crombie, Games and

Knight, 1966). For example, dimethyl xanthophanic enol (141) (Fig. 43) with a limited amount of magnesium methoxide (1–2 moles) gave compounds (142) and (143). These conversions were interpreted as proceeding by opening of the pyrone ring in (141) and cyclisation by attack of the C(6) anion formed adjacent to the acidic and least complexable enolic hydroxyl group in the two possible tautomeric intermediates (144) and (144′).

With an excess of magnesium methoxide (6 moles), a different result was obtained, the resacetophenone derivative (145) being formed in 78 per cent yield. This was explained by assuming the formation of the magnesium-containing chelated intermediate (146), in which reaction by enol ionisation was blocked. It was considered that in this intermediate (146) the geometry of the molecule was controlled so that cyclisation by attack of the C(2′) anion on the C(4) ester group was favoured, giving the required product.

This work is of considerable interest in that it goes some way towards revealing the possible chemical basis of enzymatic control in polyketide cyclisation processes.

12. *Summary*

It will be evident that most of the information discussed above is of a biogenetic rather than a biosynthetic nature. The mechanisms of the various steps in polyketide biosynthesis are still largely conjectural and very little information is available concerning the various reactions at the enzyme level. It is also evident that biogenetically, even apparently straightforward polyketides (such as the methylene-bisphloroglucinol constituents of ferns), may show a surprising divergence from the normal pathway.

There has been a great deal of speculation about the possible mechanism of polyketide biosynthesis. Lynen has discussed the biosynthesis of 6-methylsalicylic acid in terms of an enzyme model related to the yeast fatty acid synthetase system, but in which certain of the intermediate reduction-dehydration steps are omitted (Lynen, 1967a). This scheme has the considerable merit of being based on an established mechanism and is made all the more plausible by the basic biogenetic similarity of the polyketides and fatty acids. Richards and Hendrickson favour a significantly different mechanism on the grounds that polyketides in general are formed without loss of the oxygen functions of the linear poly-β-keto precursor, whereas in fatty acid biosynthesis, elimination of the oxygen substituents is the rule rather than the exception (Richards and Hendrickson, 1964). These authors regard this difference as indicating a discontinuity of mechanism in the transition from fatty acid to polyketide biosynthesis. This assumption is in itself arguable, since a number of polyketides feature quite extensive loss of oxygen (cf. curvularin (52), sclerotiorin (69) and rotiorin (70)). Richards and Hendrickson have proposed a model in which malonyl units are individually bound as thioesters, to a linear succession of thiol groups on the enzyme surface. Acetyl-CoA (or some other 'starter' molecule) then attacks the end of the chain, initiating a sequence of condensation reactions leading to the poly-β-keto intermediate, which remains bound to the enzyme surface, stabilised by hydrogen bonding to the released thiol groups. The entire enzyme-substrate complex is then coiled in such a way as to permit internal condensations leading to the final product. There are, however, rather serious objections to this model. As applied to tetracycline biosynthesis, for example, the model required the close juxtaposition of no less than nine thiol groups (presumably from cysteine residues in the polypeptide chain) at the active centre of the enzyme. In view of the ready susceptibility of thiol groups to oxidation with the formation of disulphide bridges, such an arrangement would be highly unstable and therefore rather unlikely.

Birch has recently discussed a number of ways in which chemical control of the various transformations in polyketide biosynthesis might be exerted at the enzymatic level (Birch, 1966).

The recent discovery of the hypothetical orsellinic acid precursor, tetraacetic lactone (page 105) has reopened the question as to the necessity for the cyclisation to be completed before dissociation of the enzyme-substrate complex. It would now appear possible that there are discrete stages in the construction of polyketides, in which prefabricated units composed of a few acetate residues are coupled together. The evidence from the pathways of methylenebisphloroglucinol biosynthesis in ferns, and of citromycetin, rotiorin and sclerotiorin biosynthesis, would support this hypothesis. The structure of curvularin (52), for example, is suggestive of such a pathway, in which a preformed C_8 fatty acid molecule acts as 'starter' for the extension of the chain by four more acetate–malonate units. This pathway would still be consistent with the available biogenetic evidence.

All of these uncertainties can be resolved by a study of the enzymology of polyketide biosynthesis, and it is probable that further clarification will have to wait on the results of such investigations.

I. The Shikimic Acid Pathway

The essential amino acids, phenylalanine (147), tyrosine (148) and tryptophan (149) serve as precursors of a wide range of aromatic compounds. The biosynthetic pathway to these from D-glucose is shown in Fig. 44. The various steps in this sequence were elucidated over a number of years largely through the efforts of Davis and Sprinson and their collaborators. Several detailed reviews covering these investigations are available (Davis, 1955; Sprinson, 1960; Bohm, 1965).

The key role of shikimic acid (150) in this sequence was reported by Davis in 1951 (Davis, 1951). Experimentally, the investigations depended on the isolation of many auxotrophic (nutritionally dependent) mutants of the bacterial species *Escherichia coli*. *E. coli* mutants were obtained which required the full complement of aromatic amino acids, phenylalanine, tryptophan, tyrosine and *p*-aminobenzoic acid as well as *p*-hydroxybenzoic acid, for growth. In a systematic investigation of the effectiveness of added substrates in replacing this requirement, sixty-six compounds were tested, and of these, shikimic acid (150) was found to be an effective substitute for all four amino acids (Davis, 1951). Quinic acid (151), which is widely distributed in nature, was not an effective substitute. These experiments indicated that the multiply-deficient mutant lacked an enzyme required for the synthesis of shikimic acid and that the latter was a precursor of all four amino acids.

Shikimic acid was first isolated by Eykman from the Japanese plant *Illicium anisatum* (*I. religiosum*) (Eykman, 1885). This species is particularly rich in shikimic acid (Lane, Koch, Leeds and Gorin, 1952). The revelation of its central role in aromatic amino acid biosynthesis has stimulated an intensive search among higher plants for shikimic acid, with the result that it is now known to be very widespread. Bohm lists 159 species in which it has been detected (Bohm, 1965). The name is derived from the Japanese for *Illicium anisatum*, Shikimi-no-ki or Ashikimi.

The various steps in the biosynthetic pathway (Fig. 44) were not elucidated in chronological order but have fallen into place as they have been discovered, rather like an elaborate jig-saw. Prephenic

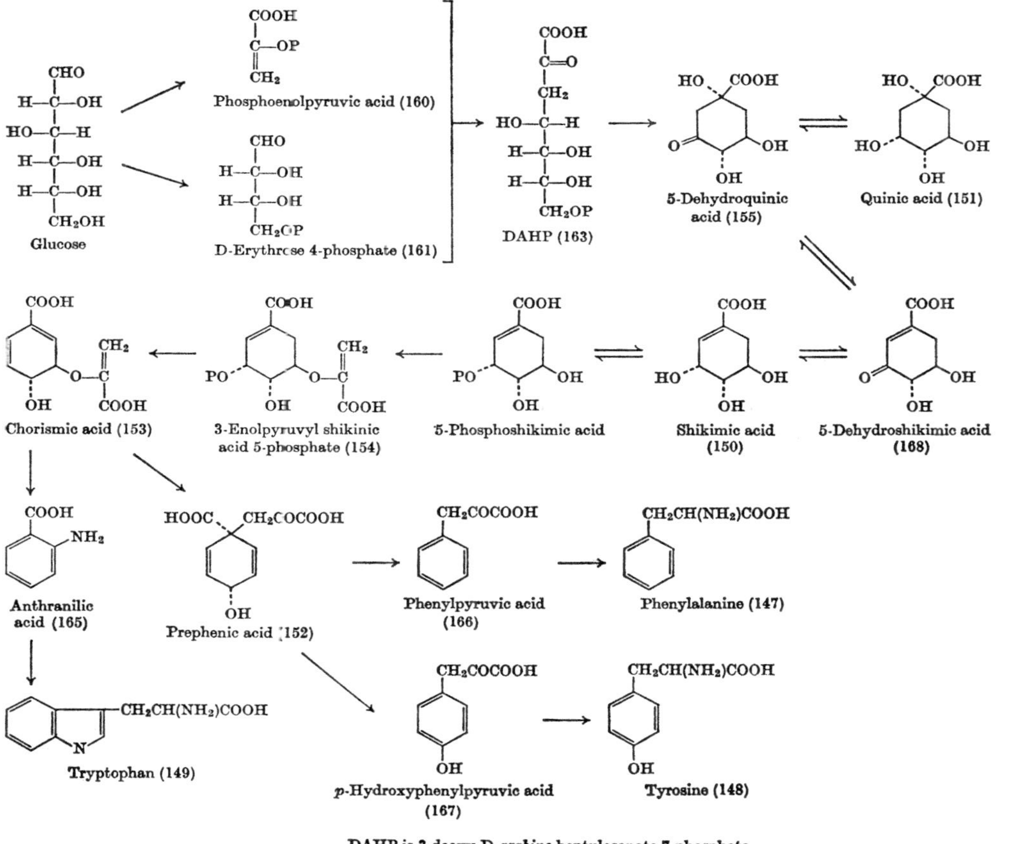

FIG. 44. The glucose–shikimic acid–phenylalanine–tyrosine pathway.

acid (152) for example, was isolated and characterised in 1954 (Weiss, Gilvarg, Mingioli and Davis, 1954), whereas chorismic acid (153) and 3-enolpyruvylshikimate 5-phosphate (154) were only revealed as intermediates in 1964 (Gibson, 1964; Gibson and Gibson, 1964; Levin and Sprinson, 1964). The cyclic intermediates from 5-dehydroquinate (155) to chorismate (153) have been defined almost entirely by studies with blocked mutants, whereas the early steps from glucose to 5-dehydroquinate (155) and shikimate (150) were elucidated by a combination of radiotracer and enzymatic techniques.

The conversion of specifically labelled D-glucose into shikimate revealed a complex pattern of incorporation in which the various positions in shikimic acid were found to be derived from two or more positions in D-glucose as shown in Fig. 45 (Srinivasan, Shigeura, Sprecher, Sprinson and Davis, 1956). However, these incorporations

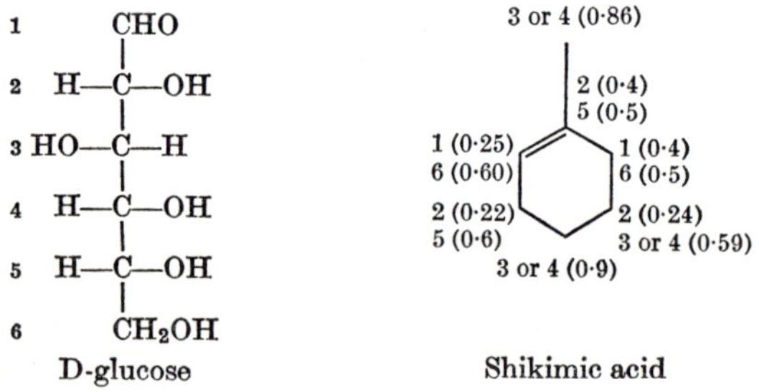

1 CHO
2 H—C—OH
3 HO—C—H
4 H—C—OH
5 H—C—OH
6 CH_2OH
D-glucose

3 or 4 (0·86)
2 (0·4)
5 (0·5)
1 (0·25)
6 (0·60)
1 (0·4)
6 (0·5)
2 (0·22)
5 (0·6)
2 (0·24)
3 or 4 (0·59)
3 or 4 (0·9)
Shikimic acid

(The figures in parentheses give the fraction of the activity from the indicated labelled positions of D-glucose incorporated into each position of shikimic acid.)

FIG. 45. The pattern of incorporation of labelled glucose into shikimic acid.

were explicable in terms of the known transformations involved in the glycolytic and pentose phosphate pathways of glucose metabolism. Sprinson has given a detailed analysis of these results (Sprinson, 1960).

In retrospect, the investigations into the glucose–shikimate conversion constitute an unusual story in that the correct route was discovered as a result of studies involving a postulated intermediate which was subsequently shown not to be directly involved. This intermediate was D-*altro*-heptulose 1,7-diphosphate (156) (Fig. 46).

For structural reasons it was considered that the immediate precursor of the cyclic intermediates might be a seven-carbon sugar.

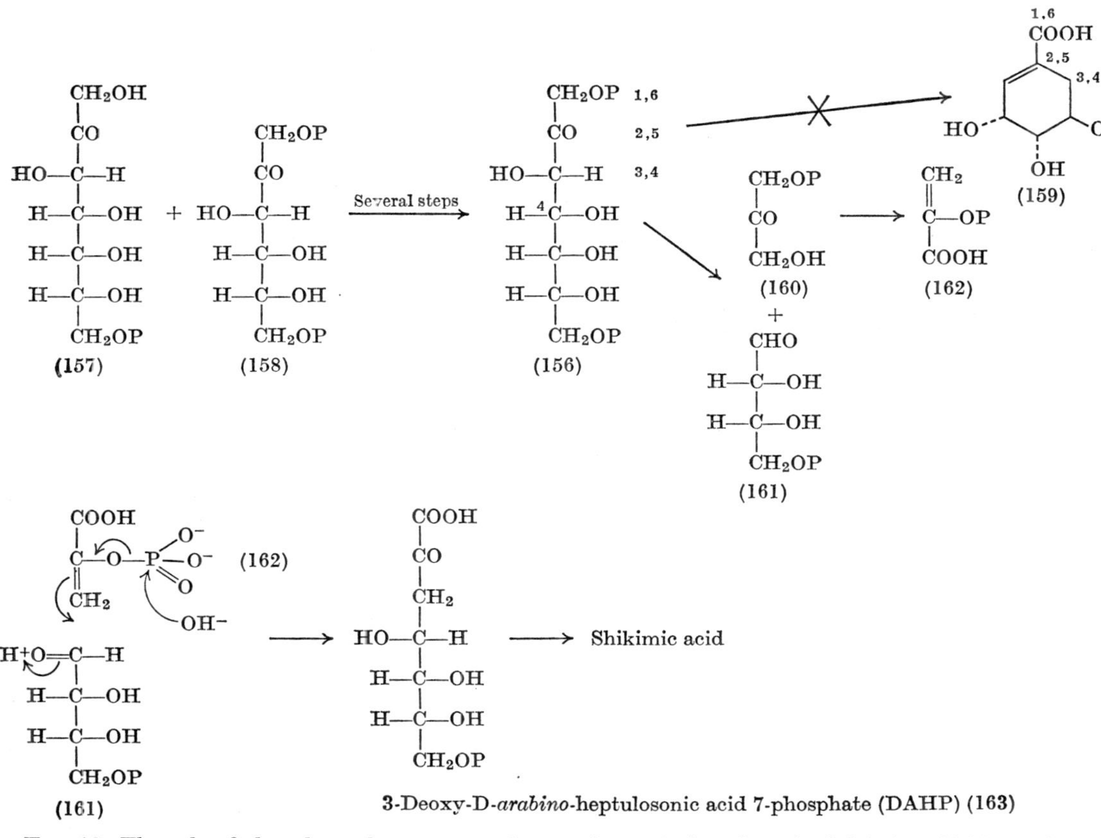

FIG. 46. The role of phosphoenolpyruvate and D-erythrose 4-phosphate in shikimic acid biosynthesis.

Experiments carried out with this in mind showed that D-*altro*-heptulose 7-phosphate (157), an intermediate of the pentose phosphate pathway, was converted into shikimate by bacterial extracts. It was also found that the efficiency of conversion was increased if the substrate was incubated with fructose 1,6-diphosphate (158) (Kalan, Davis, Srinivasan and Sprinson, 1956). Since it was known that these two substrates could undergo a series of fragmentation-recombination reactions to give D-*altro*-heptulose 1,7-diphosphate (156), the latter was incubated with bacterial extracts and found to be nearly quantitatively converted into shikimate (Srinivasan, Sprinson, Kala and Davis, 1956).

In spite of this dramatic result, the observed labelling of shikimic acid derived from specifically labelled glucose clearly demonstrated that D-*altro*-heptulose 1,7-diphosphate (156) was *not* directly cyclised to shikimate. The contribution of the various positions in D-glucose to C(1), C(2) and C(3) of D-*altro*-heptulose 1,7-diphosphate are shown in Fig. 46 (156). These are known from the established metabolic relationship between the two compounds. Direct cyclisation of (156) with this labelling pattern, however, would lead to shikimic acid labelled at C(1), C(2) and C(3) as shown in (159) (Fig. 46), whereas the observed labelling pattern was the reverse of this (Fig. 45). Direct cyclisation would also require epimerisation at C(4) in D-*altro*-heptulose 1,7-diphosphate (156).

Further investigations with *E. coli* extracts into the blocking effects of various enzyme inhibitors led to the formulation for this conversion shown in Fig. 46 (Srinivasan, Katagiri and Sprinson, 1955; 1959). Fission of the heptose diphosphate (156) gives rise to 1,3-dihydroxy-2-propanone phosphate (160) and D-erythrose 4-phosphate (161). The former is oxidised to phosphoenolpyruvate (162) which then condenses with the latter to give, via several intermediate steps, shikimic acid. It was at this stage that it could be seen that D-*altro*-heptulose 1,7-diphosphate (156) was probably not an obligatory intermediate in shikimic acid biosynthesis, since both phosphoenolpyruvate (162) and D-erythrose 4-phosphate (161) can be formed from glucose by pathways not involving (156).

The true pre-cyclic intermediate was shown to be 3-deoxy-D-*arabino*-heptulosonic acid 7-phosphate (DAHP) (163) which arises by condensation of phosphoenolpyruvyl phosphate with D-erythrose 4-phosphate as shown in Fig. 46 (Srinivasan and Sprinson, 1959). An enzyme from an *E. coli* mutant was obtained in partially purified form which was able to catalyse the quantitative conversion of DAHP into 5-dehydroquinate (155) (Srinivasan, Rothschild and Sprinson, 1963). The sequence leading to the first cyclic intermediate was thus complete and D-*altro*-heptulose 1,7-diphosphate (156) could be seen to have been a most informative red herring, providing the clues which led to the identification of DAHP (163) as the true obligatory intermediate. The significance of D-*altro*-heptulose 1,7-diphosphate in carbohydrate metabolism is still not clearly understood. A mechanism for the cyclisation step has been proposed by Srinivasan *et al.* which takes into account the known requirements of the enzyme system for Co^{2+} and nicotinamide adenine dinucleotide (NAD) as cofactors (Fig. 47) (Srinivasan *et al.*, 1963).

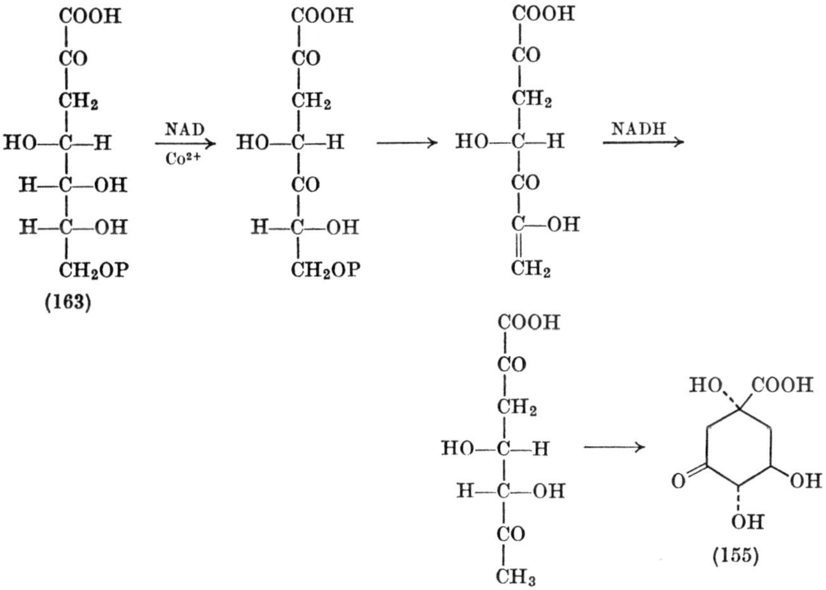

COOH
CO
CH_2
HO—C—H
H—C—OH
H—C—OH
CH_2OP
(163)

NAD, Co^{2+} →

COOH
CO
CH_2
HO—C—H
CO
H—C—OH
CH_2OP

→

COOH
CO
CH_2
HO—C—H
CO
C—OH
‖
CH_2

NADH →

COOH
CO
CH_2
HO—C—H
H—C—OH
CO
CH_3

→

HO COOH
O OH
OH
(155)

FIG. 47. Possible mechanism for the cyclisation of DAHP to 5-dehydroquinic acid.

Several steps in the interconversions of the cyclic intermediates warrant special mention. The sequence from 5-dehydroquinate (155) to chorismate (153) (Fig. 44) involves mechanistically unexceptional transformations, but the rearrangement of chorismate (153) to prephenate (152) is novel from the biochemical point of view. This

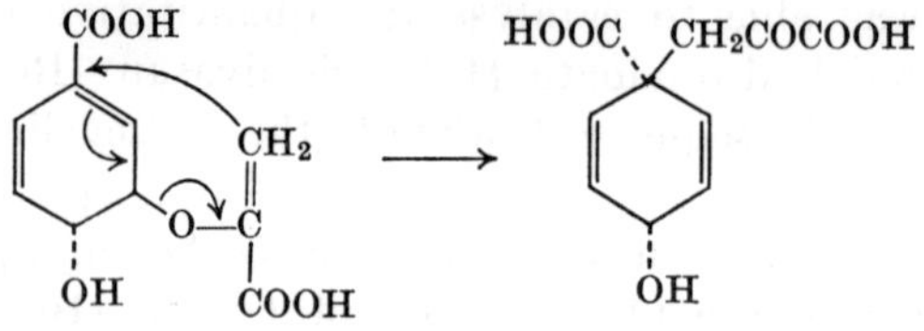

$$\text{(153)} \longrightarrow \text{(152)}$$

Fig. 48. The mechanism of the conversion of chorismic acid into prephenic acid.

transformation (Fig. 48) is analogous to the Claisen allyl vinyl ether rearrangement (Rhoads, 1963) and is readily brought about *in vitro* by warming chorismic acid to 70° at pH 8 for one hour. Below pH 6, prephenate (152) is converted into phenylpyruvate (165) (Gibson, 1964).

Chorismic acid (153) represents a branching point in the pathway, since in the presence of glutamine (164), anthranilic acid (165) the precursor of tryptophan (149) is formed, whereas in the absence of glutamine, prephenate (152) is obtained. A mechanism for the former transformation has been proposed by Levin and Sprinson and is shown in Fig. 49 (pathway *a*) (Levin and Sprinson, 1964).

Prephenate (152) is converted either into phenylpyruvate (166) or into *p*-hydroxyphenylpyruvate (167). The former transformation is best explained in terms of a simple decarboxylation-elimination reaction (Fig. 49, pathway *b*) (Bohm, 1965). The conversion of prephenate into *p*-hydroxyphenylpyruvate has been shown to require NADH as a cofactor and therefore probably involves an intermediate oxidation step followed by reductive decarboxylation as shown in Fig. 49 (pathway *c*) (Sprinson, 1960). The conversion of the phenylpyruvates into the corresponding amino acids phenylalanine (147) and tyrosine (148) by transamination completes the pathway to the aromatic amino acids (Rudman and Meister, 1953).

It should be pointed out that phenylpyruvate (166) and *p*-hydroxyphenylpyruvate (167) are not obligatory intermediates in the biosynthesis of aromatic compounds from shikimate, since aromatisation of earlier intermediates appears to occur in some organisms. For example, a partially purified enzyme system from *Neurospora crassa* has been obtained which catalyses the direct conversion of 5-dehydroshikimate (168) into protocatechuate (169). The mechanism shown in Fig. 50 has been proposed for this transformation on the basis of labelling experiments (Gross, 1958). Further examples of intermediates

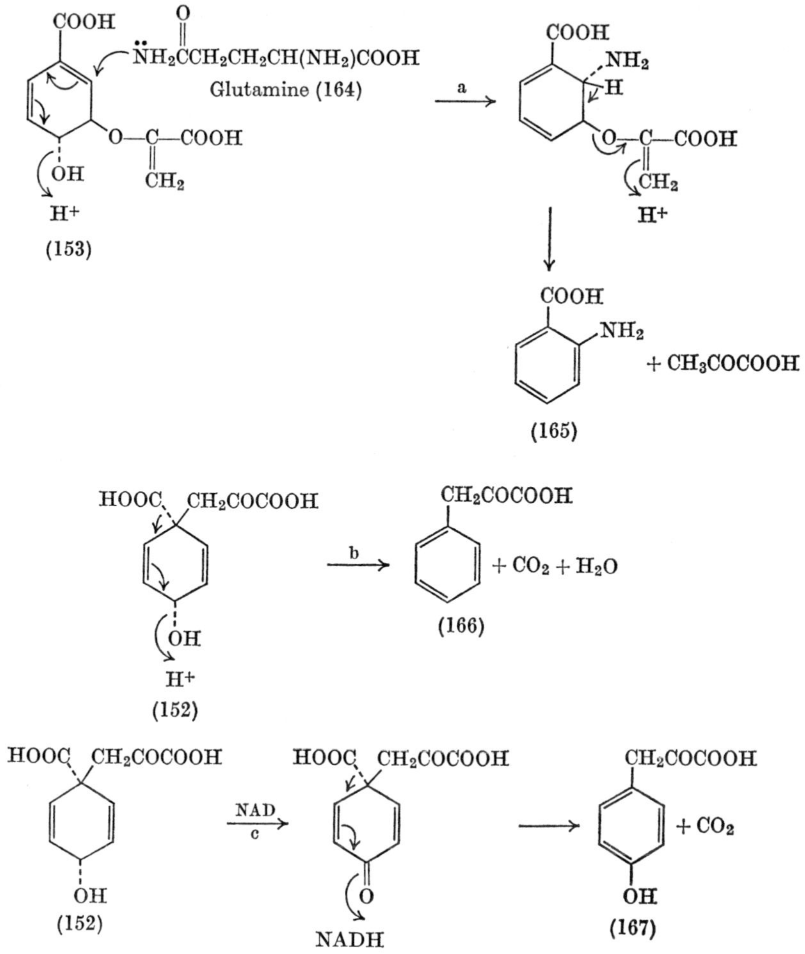

COOH

$\ddot{N}H_2\overset{O}{\overset{\|}{C}}CH_2CH_2CH(NH_2)COOH$

Glutamine (164)

O—C—COOH

CH_2

OH

H^+

(153)

a →

COOH

NH_2

H

O—C—COOH

CH_2

H^+

COOH

NH_2

$+ CH_3COCOOH$

(165)

HOOC $CH_2COCOOH$

OH

H^+

(152)

b →

$CH_2COCOOH$

$+ CO_2 + H_2O$

(166)

HOOC $CH_2COCOOH$

OH

(152)

NAD / c →

HOOC $CH_2COCOOH$

O

NADH

→

$CH_2COCOOH$

$+ CO_2$

OH

(167)

FIG. 49. Mechanisms for the conversions of chorismic and prephenic acids into aromatic compounds.

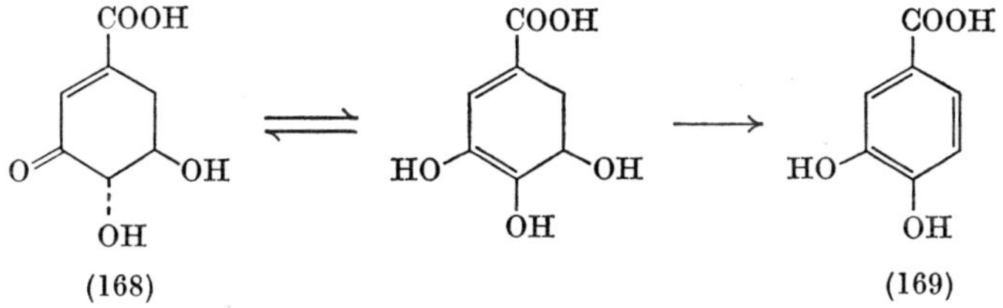

COOH

O OH

OH

(168)

⇌

COOH

HO OH

OH

→

COOH

HO

OH

(169)

FIG. 50. The formation of protocatechuic acid from 5-dehydroshikimic acid.

earlier than prephenate (152) serving as direct precursors of aromatic acids have been discussed by Bohm (Bohm, 1965).

It will be seen from the foregoing account that the shikimic acid pathway was elucidated entirely by experiments with bacterial mutants. The wide distribution of shikimic acid and the results of numerous investigations with labelled compounds, however, leaves little doubt that this pathway is also important in higher plants.

J. Metabolites Derived Directly from Intermediates of the Glucose–Shikimic Acid–Phenylalanine–Tyrosine Pathway

1. *Cinnamic Acids and Coumarins*

Both phenylalanine and tyrosine can furnish the corresponding *trans*-cinnamic (170) and *p*-hydroxy-*trans*-cinnamic (171) acids

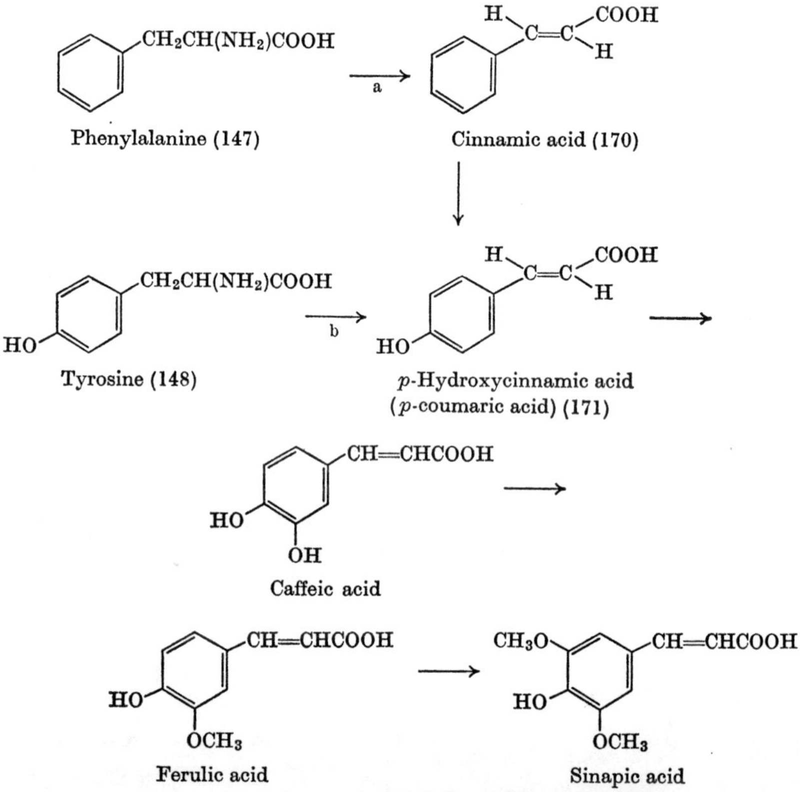

$CH_2CH(NH_2)COOH$

Phenylalanine (147) →(a) Cinnamic acid (170)

$CH_2CH(NH_2)COOH$

HO

Tyrosine (148) →(b) *p*-Hydroxycinnamic acid (*p*-coumaric acid) (171) →

CH=CHCOOH

HO OH

Caffeic acid →

CH=CHCOOH

HO OCH_3

Ferulic acid → Sinapic acid

CH_3O CH=CHCOOH HO OCH_3

Fig. 51. Biosynthesis of cinnamic acids in higher plants.

directly by reactions of the type shown in Fig. 51, pathways *a* and *b*. These reactions are catalysed by enzymes known as deaminases; phenylalanine deaminase appears to be widely distributed in nature but tyrosine deaminase is of more restricted occurrence. A survey of the distribution of these enzymes has been given by Conn (Conn, 1964).

The primary cinnamic acids are elaborated by hydroxylation and methylation reactions. The sequence illustrated in Fig. 51 has been deduced from incorporation experiments in higher plants (McCalla and Neish, 1959; Zenk and Muller, 1964; El-Basyouni, Chen, Ibrahim, Neish and Towers, 1964; Neish, 1964). Although the hydroxylation of phenylalanine is the obligatory route to tyrosine in animals, it does not appear to be as important either in higher plants or in micro-organisms, where both amino acids can be derived directly from shikimate by independent routes, as described previously (Neish, 1964).

The cinnamic acids in turn serve as precursors of the coumarins. The most recent evidence relating to the biosynthesis of coumarin

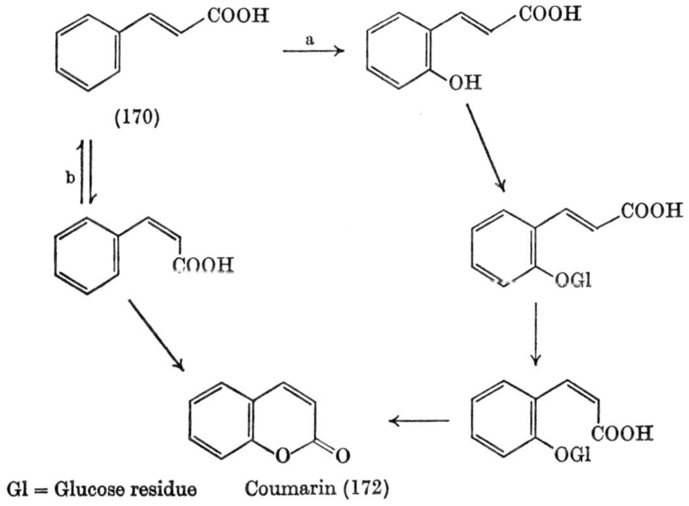

Gl = Glucose residue   Coumarin (172)

FIG. 52. Coumarin biosynthesis.

itself (172) is summarised in Fig. 52 (Brown, 1966). Of the two pathways illustrated, route *a* is the most fully authenticated. Both pathways involve a double bond isomerisation of *trans*-cinnamic acid or a derivative and there has been some debate as to whether this isomerisation is necessarily enzymatic in nature (Edwards and Stoker, 1967).

The production of 7-oxygenated coumarins such as herniarin (173) (Fig. 53) appears to result from a branch in the biosynthetic pathway at the *trans*-cinnamic acid stage (Brown, 1963). *Ortho*-hydroxylation leads to coumarin, whereas *para*-hydroxylation followed by *ortho*-hydroxylation leads to the 7-oxygenated compounds. It was found that in lavender (*Lavandula officinalis*), *o*-hydroxy-*trans*-cinnamic acid was readily converted into coumarin (172) but not readily into

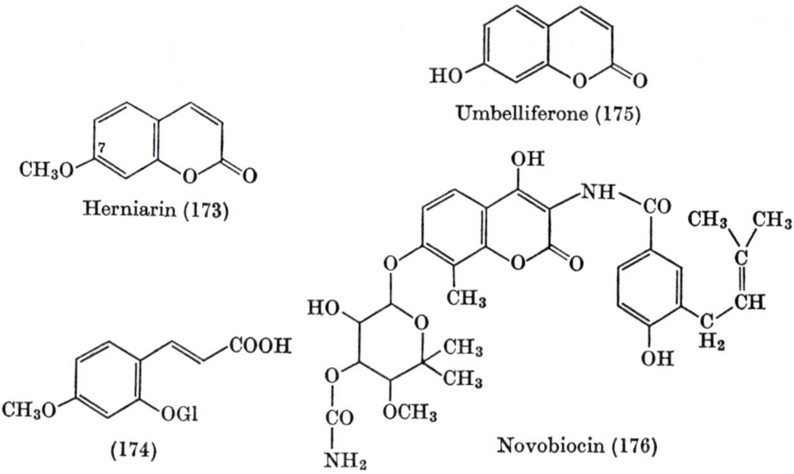

Umbelliferone (175)

Herniarin (173)

(174)

Novobiocin (176)

FIG. 53. Coumarins.

herniarin (173), whereas *p*-hydroxy-*trans*-cinnamic acid was converted with greater efficiency into herniarin (173). *trans*-2-Glucosyloxy-4-methoxycinnamic acid (174) was converted into herniarin (173) with high efficiency, but not into coumarin (172) (Brown, 1963). Similar results were obtained in studies of umbelliferone (175) biosynthesis (Brown, Towers and Chen, 1964).

A different cyclisation mechanism appears to operate in the biosynthesis of novobiocin (176) in *Streptomyces niveus*. It was shown that tyrosine was incorporated into the coumarin nucleus without degradation (Chambers, Kenner, Robinson and Webster, 1960) and that the ring oxygen originated in the carboxyl group of tyrosine (Bunton, Kenner, Robinson and Webster, 1963). The most plausible mechanism for the cyclisation step as suggested by Bunton *et al.* is illustrated in Fig. 54 (pathway *a*) (Bunton *et al.*, 1963). Formation of the spirodienone (177) from *p*-hydroxy-*cis*-cinnamic acid is closely analogous to the geodin hydrate–geodoxin conversion (section H.2) and was suggested to proceed via the formation of a diradical intermediate (178) similar to that invoked to explain the geodin hydrate cyclisation (Hassall and Scott, 1961). The subsequent acid-catalysed step leading to the coumarin structure (179) is an example of the well-known dienone-phenol rearrangement. Several analogies for

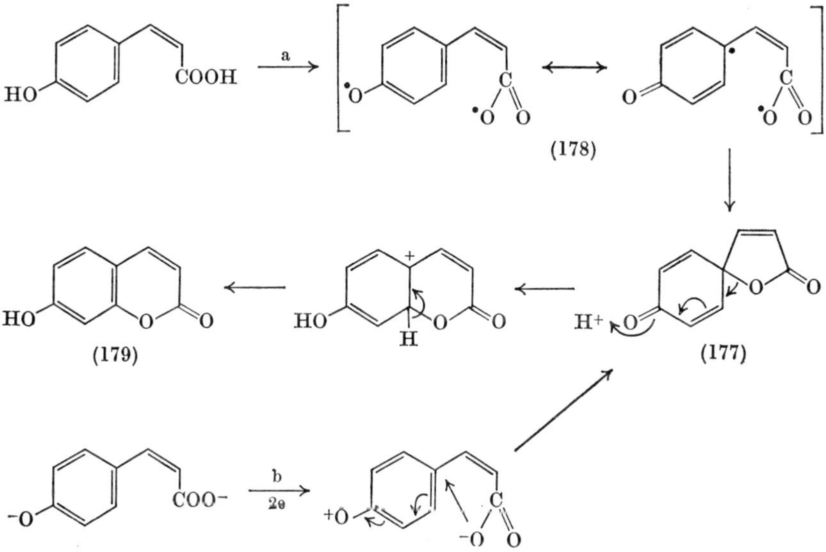

FIG. 54. Possible mechanisms for the cyclisation step in novobiocin biosynthesis.

these reactions are known (see Bunton *et al.*, 1963). It should be pointed out, however, that Barton has criticised the diradical mechanism for the geodin hydrate cyclisation *in vitro* on the grounds that aromatic carboxylic acids are not readily oxidised to free radicals under the reaction conditions employed (Barton, 1964). He has suggested instead that a two electron oxidation of the phenolate anion gives rise to a phenoxonium ion, cyclisation then occurring by attack of a carboxylate *anion*. Extension of this concept to the proposed coumarin cyclisation-isomerisation *in vivo* would lead to the alternative mechanism shown in Fig. 54 (pathway *b*).

A furan ring is present in a number of coumarins such as bergapten (180) and pimpinellin (181). Recent experiments on the incorporation of [2-^{14}C]-mevalonate into the furanocoumarins of *Pimpinella magna* have shown that the furan system arises by isoprenylation of the aromatic ring (probably by electrophilic attack of dimethylallyl pyrophosphate (see below)) followed by oxidative fission of the isopropyl group as shown in Fig. 55 (Floss and Mothes, 1966). This route is supported by the occurrence of isopropylfuranocoumarins and

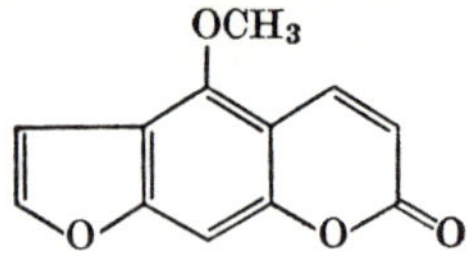

OCH_3

Bergapten (180)

CH_3 OH

$HOCH_2$ COOH

Mevalonic acid

CH_3

CH_3 CH_2OPP

Dimethylallyl pyrophosphate

(O)

OCH_3

CH_3O

Pimpinellin (181)

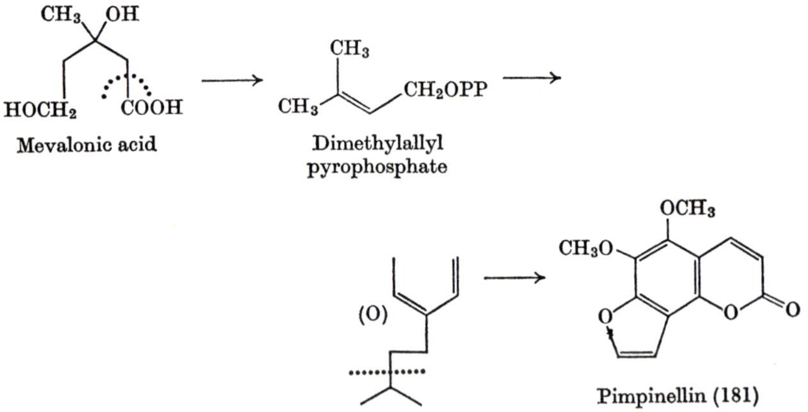

Fig. 55. Biosynthesis of furanocoumarins.

isoprenylcoumarins in plants (Karrer, 1958; Mentzer and Fatianoff, 1964; Mentzer, 1966).

The biosynthesis of the coumarins has been discussed in a recent publication by Brown (Brown, 1966).

2. *Lignin*

The formation of lignin, a complex, polyphenolic, polymeric material composed of phenylpropane monomer units, is quantitatively the most important process, involving phenolic compounds

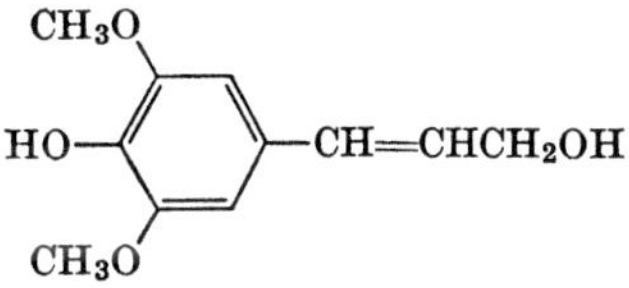

Sinapyl alcohol (182)

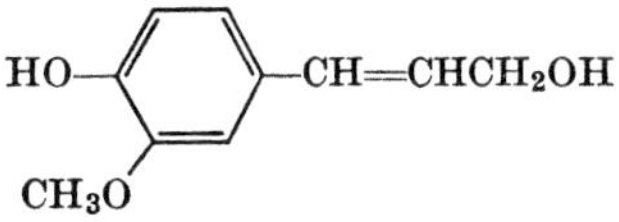

Coniferyl alcohol (183)

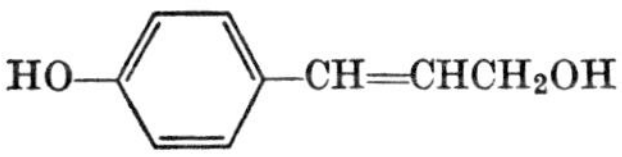

p-Hydroxycinnamyl alcohol (184)

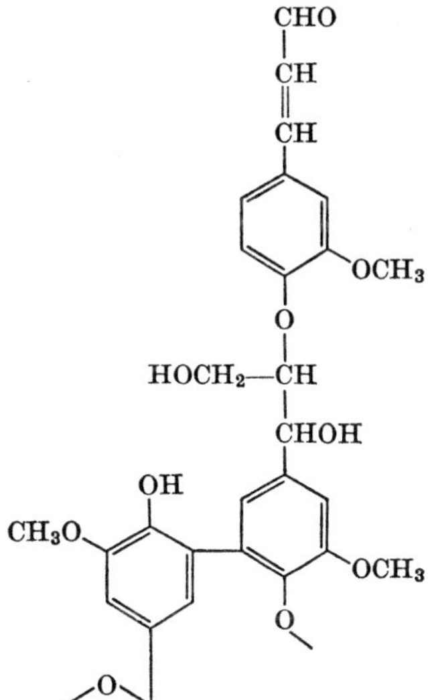

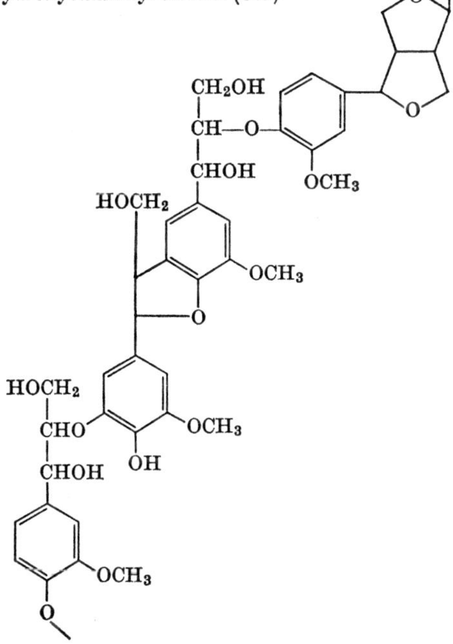

FIG. 56. Hypothetical structure for gymnosperm lignin.

derived from shikimic acid, in plants. The complex, variable, and amorphous character of lignin has precluded any assignments of unequivocal structures and this in turn has hampered attempts to understand its biosynthesis. The weight of the experimental evidence available tends to favour the view that lignin is formed by polymerisation of cinnamyl alcohols via free radical coupling processes. The three most important monomer units are sinapyl alcohol (182), coniferyl alcohol (183) and *p*-hydroxycinnamyl alcohol (184), which could be derived from the corresponding cinnamic acids *in vivo* by reduction of the carboxyl function. *In vitro* experiments on the polymerisation of phenylpropane monomers using the enzymes peroxidase or laccase have given polymeric lignin-like materials and these enzymes are therefore considered to be involved in natural lignification. A hypothetical structure for gymnosperm lignin which illustrates some of the structural features of the polymer is presented in Fig. 56 (Adler, 1957; Isherwood, 1965). Lignin biosynthesis has been discussed in detail in several recent reviews (Isherwood, 1965; Brown, 1964; Freudenberg, 1965; Harkin, 1967).

3. C_6–C_1 *Compounds*

Higher plants possess the ability in varying degrees to hydroxylate a variety of substrates. It is apparent that variously substituted benzoic acids can be obtained either by oxidative degradation of the corresponding cinnamic acids (Zenk and Müller, 1964; El-Basyouni, Chen, Ibrahim, Neish and Towers, 1964; Zenk, 1966) or via hydroxylation and methylation of benzoic acids. Zenk has proposed the scheme shown in Fig. 57 for the biosynthesis of various benzoic acids in plants (Zenk, 1966) (cf. Fig. 51). It has also been found, however, that in *Gaultheria procumbens* and *Primula acaulis*, benzoic acid can be elaborated directly to salicylic (185), *o*-pyrocatechuic (186), gentisic (187) and *p*-hydroxybenzoic (188) acids (Fig. 58) (El-Basyouni *et al.*, 1964). Grisebach and Vollmer similarly found that [3-^{14}C]-cinnamic acid was converted into gentisic (187), *p*-hydroxybenzoic (188), *o*-pyrocatechuic (186), syringic (189), protocatechuic (190) and vanillic (191) acids in *G. procumbens*, although it was not determined whether the hydroxylation and methylation steps took place at the cinnamic acid or the benzoic acid stage (Grisebach and Vollmer, 1964).

From the results so far obtained it would appear that different species are able to carry out a variety of transformations such as

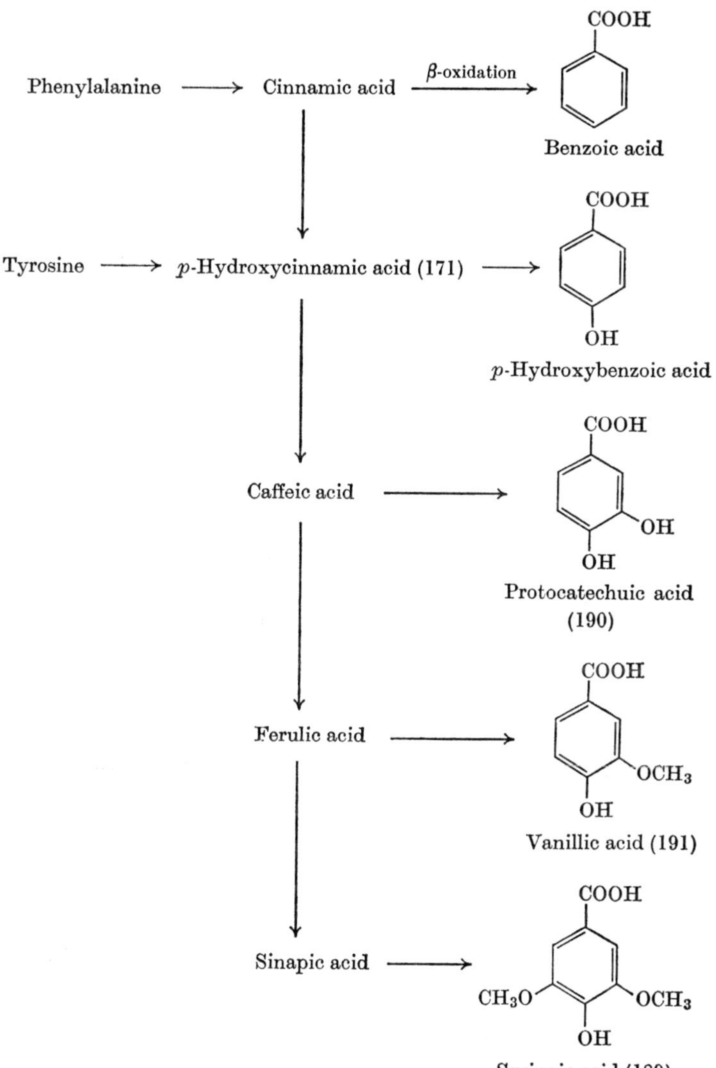

FIG. 57. Biosynthesis of benzoic acids in higher plants.

those described but that there is a considerable difference between individual species in the range of transformations effected. Direct hydroxylation of aromatic systems is, of course, an important

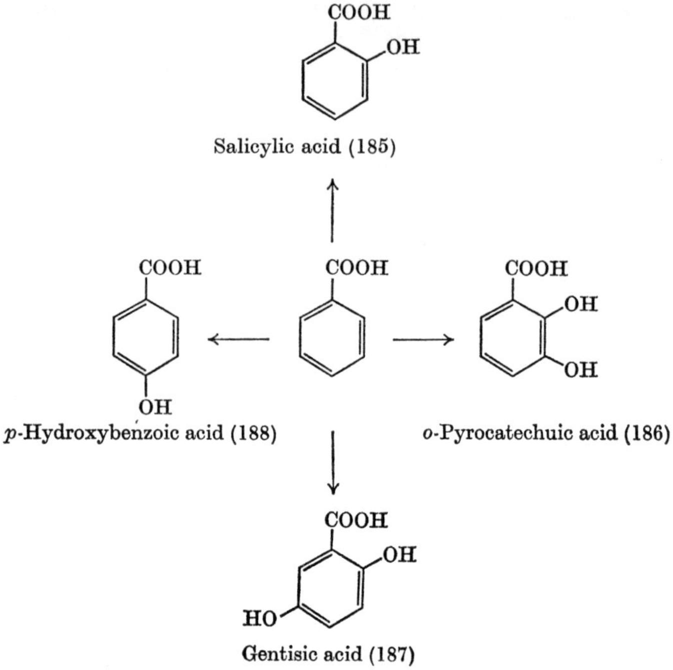

Salicylic acid (185)

p-Hydroxybenzoic acid (188)

o-Pyrocatechuic acid (186)

Gentisic acid (187)

FIG. 58. Metabolism of benzoic acid in higher plants.

preliminary to degradation of aromatic compounds in micro-organisms (Evans, 1963) and animals (Daly and Witkop, 1963).

The possibility that aromatic acids might be derived directly from pre-aromatic intermediates of the shikimic acid pathway has already been mentioned. In this connection it is of interest that [^{14}C]-glucose was found to be a better precursor of gallic acid (192) than [^{14}C]-phenylalanine in *Geranium pyrenaicum* (Conn and Swain, 1961). However, more recently, Zenk has shown that in *Rhus typhina*, [3-^{14}C]-phenylalanine is converted into gallic acid (192) with a significantly higher specific activity than the simultaneously produced benzoic acid (by a factor of about thirty to one). It is also incorporated into gallic acid with about twenty times the efficiency of [2-^{14}C]-glucose. On the basis of these results, Zenk proposed the biosynthetic pathway shown in Fig. 59 for gallic acid biosynthesis in *Rhus typhina* (Zenk, 1964a). On the other hand, it has been shown that gallic acid is obtained directly from 5-dehydroshikimic acid (167) (Fig. 44) in *Phycomyces blakesleeanus* (Haslam, Haworth and

Phenylalanine ⟶ Cinnamic acid ⟶ *p*-Hydroxycinnamic acid ⟶

Caffeic acid ⟶ [CH=CHCOOH; HO, OH, OH] ⟶ [COOH; HO, OH, OH]

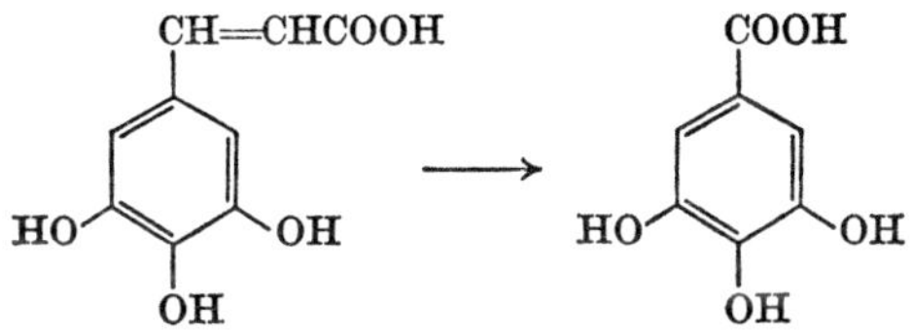

Gallic acid (192)

FIG. 59. Biosynthesis of gallic acid in *Rhus typhina*.

Knowles, 1961). El-Basyouni *et al.* also found that gallic acid (192) was produced by hydroxylation of protocatechuic acid (190) in *Pelargonium hortorum* (El-Basyouni *et al.*, 1964). These results emphasise the fact that the same metabolite can be produced by different routes in different organisms and underline the need for caution in drawing general conclusions about biosynthetic pathways from results obtained with a single species.

Anthranilic acid (165) (Fig. 44) was found to be a precursor of *o*-pyrocatechuic acid (186) in *Claviceps paspali*, but neither salicylic acid (185) nor 3-hydroxyanthranilic acid were incorporated (Gröger, Erge and Floss, 1965).

It has been found that the successive hydroxylation and methylation steps just described are to some extent reversible, since in *Hordeum*, *Oryza* and *Triticum* species, sinapic acid (Fig. 51) was demethoxylated to ferulic (Fig. 51) and/or vanillic (191) acids, and in *Triticum vulgare*, caffeic acid (Fig. 51) was dehydroxylated to *p*-hydroxybenzoic acid (188). The position of the label was not determined in these experiments, however, so that the possibility of degradation followed by resynthesis (as opposed to direct dehydroxylation) cannot be ruled out (El-Basyouni *et al.*, 1964).

As a continuation of the oxidative degradation of the cinnamic

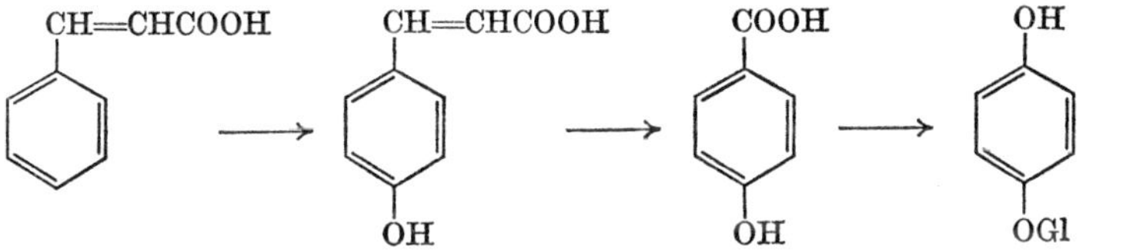

[CH=CHCOOH] ⟶ [CH=CHCOOH; OH] ⟶ [COOH; OH] ⟶ [OH; OGl]

Arbutin (193)

FIG. 60. Biosynthesis of arbutin.

acids, the side chain can finally be eliminated altogether as in the biosynthesis of arbutin (hydroquinone glucoside) (193). *p*-Hydroxybenzoic acid was incorporated into arbutin to the remarkably high extent of 9·8 per cent in *Bergenia crassifolia*, which suggests the scheme illustrated in Fig. 60 for biosynthesis of this compound (Zenk, 1964b).

Zenk has discussed the most recent evidence relating to the biosynthesis of C_6–C_1 compounds (Zenk, 1966).

4. *Terphenyl Derivatives and Related Metabolites*

Various phenylpropane precursors such as [2-^{14}C]-DL-phenyl-lactic acid, [2-^{14}C]-DL-phenylserine, [1-^{14}C]-phenylalanine and uniformly labelled [^{14}C]-shikimic acid were all incorporated into volucrisporin (194) in *Volucrispora aurantiaca*. Since [1-^{14}C]-DL-*m*-tyrosine (195) was also incorporated, it was suggested that volucrisporin was formed by dimerisation of a C_6–C_3 compound, probably a derivative of phenylpyruvic acid (Fig. 61), and that hydroxylation of the aromatic ring took place *before* dimerisation (Read, Vining and Haskins, 1962; Chandra, Read and Vining, 1965).

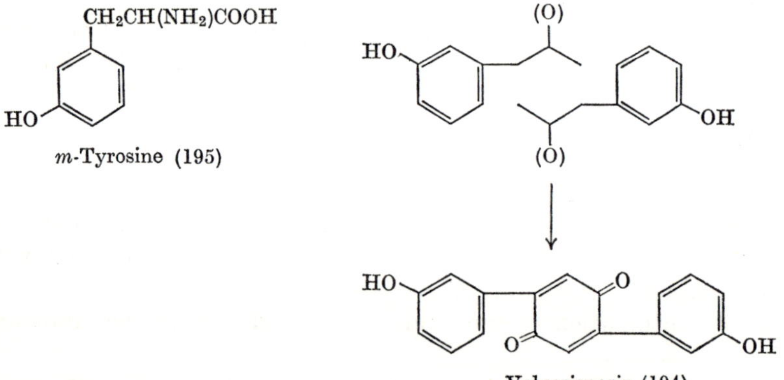

$CH_2CH(NH_2)COOH$ HO

m-Tyrosine (195)

(O) HO OH (O)

HO O O OH

Volucrisporin (194)

FIG. 61. The biosynthesis of volucrisporin.

The lichen metabolites pulvic acid (196) and pulvic dilactone (197), vulpinic acid (198) and calycin (199) almost certainly arise by ring cleavage of the related terphenyl compound polyporic acid (200) as shown in Fig. 62. All four compounds were labelled when [1-^{14}C]-DL-phenylalanine was administered to the fungal symbiont of the lichen *Candellariella vitellina* (Mosbach, 1967). Mosbach has obtained good supporting evidence for this pathway by demonstrating the equal activity of all four labelled positions in vulpinic acid (198) derived from [1-^{14}C]-phenylalanine in *Evernia vulpina* (Mosbach, 1964).

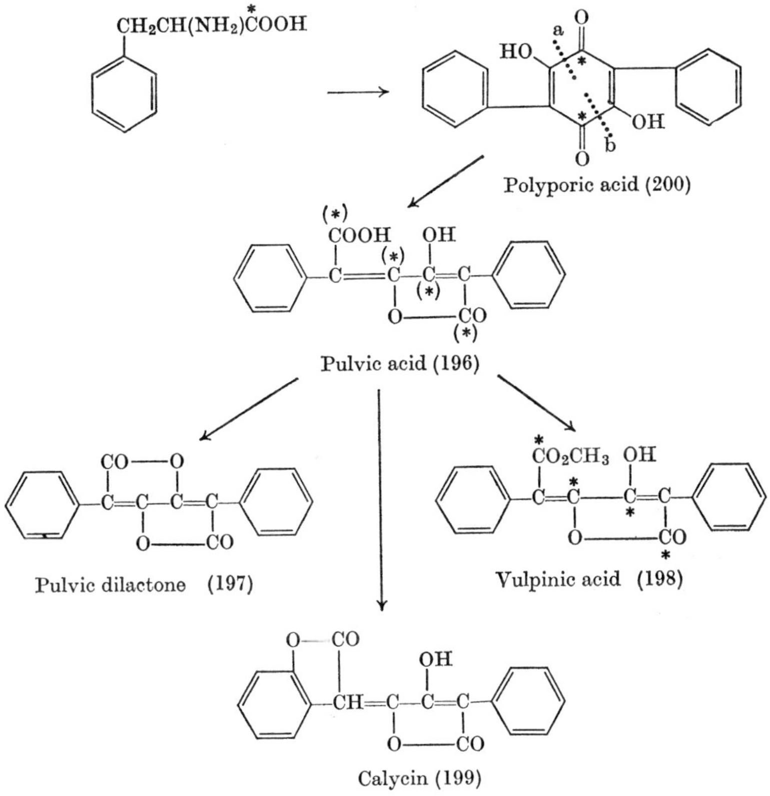

$C_6H_5CH_2CH(NH_2)\overset{*}{C}OOH$

Polyporic acid (200)

Pulvic acid (196)

Pulvic dilactone (197)

Vulpinic acid (198)

Calycin (199)

FIG. 62. The biosynthesis of vulpinic acid and related metabolites.

This is the pattern expected from ring cleavage of a symmetrical intermediate such as polyporic acid (200) at the two equivalent positions *a* and *b* as shown in Fig. 62.

Biosynthesis of xanthocillin (201) requires a different type of dimerisation of a phenylpropane precursor. Grisebach and Achenbach found that tyrosine was a precursor of this metabolite in *Penicillium notatum*. The origin of the carbon atoms of the isocyanide groups was

HO— —CH=C—C=CH— —OH

N N

C C

Xanthocillin (201)

not elucidated; the usual one carbon donors failed to give any incorporations into these positions (Achenbach and Grisebach, 1965).

K. Metabolites with a Mixed Biogenesis

1. *Flavonoids and Neoflavanoids*

The flavonoids constitute one of the most important classes of secondary metabolite for which a mixed biogenesis involving both the shikimate and polyketide pathways has been established. The biosynthesis of the flavonoids has been very fully discussed in a recent review (Grisebach, 1965a).

The current evidence relating to the biogenetic relationships between the various classes of flavonoid is summarised in Fig. 63, in which the substitution pattern of the archetypal series is used. This is a modified version of a recent scheme proposed by Grisebach (Grisebach, 1965b) and takes into account the results on isoflavone biosynthesis described previously (section F) and also accommodates recent work on aurone (204) biosynthesis (Wong, 1966a, b).

The basic carbon skeleton, exemplified in the chalcone structure (205) is considered to arise by a variation of the normal polyketide pathway in which *p*-hydroxycinnamic acid acts as a chain 'starter', condensing with three malonyl-CoA units to give the basic C_6–C_3–C_6 system (Fig. 63). Ring A then arises by cyclisation of the six-carbon polyketide section. The most commonly oxygenated positions in the flavonoids are C(5), C(7) and C(4′) (as in (206)) (Geissman, 1962), a substitution pattern which arises automatically from the proposed biosynthetic pathway.

The central role of chalcones in flavonoid biosynthesis has been attested in a number of investigations (Grisebach, 1965b; Wong, 1965). A temporary uncertainty which arose out of the observation that chalcones can be degraded to cinnamic acids in plants (Patschke, Hess and Grisebach, 1964) has been resolved by the demonstration of the incorporation of an intact, multiply labelled chalcone into biochanin A (10) (page 74) in *Cicer arietinum* and into cyanidin (7)

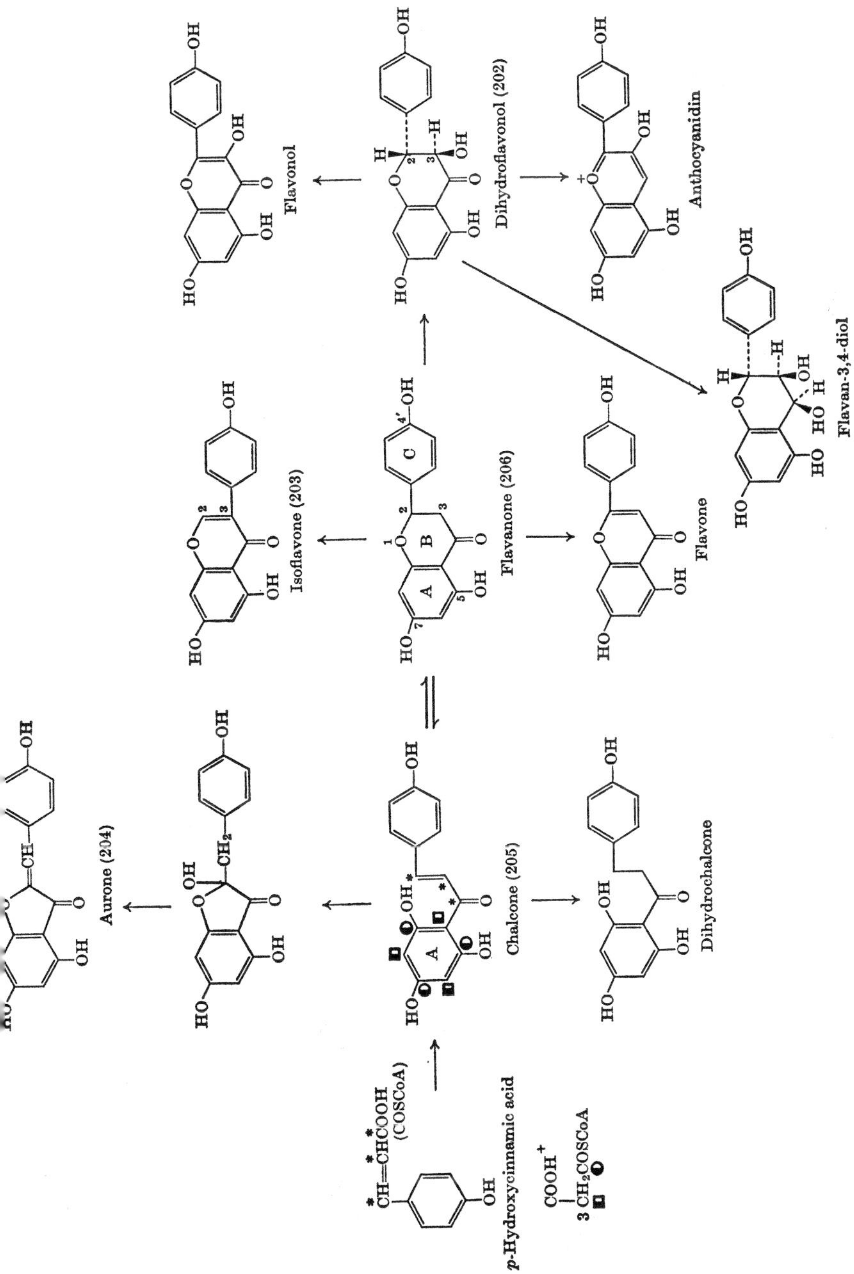

Flavonol

Dihydroflavonol (202)

Anthocyanidin

Isoflavone (203)

Flavanone (206)

Flavone

Flavan-3,4-diol

Aurone (204)

Chalcone (205)

Dihydrochalcone

p-Hydroxycinnamic acid

(COSCoA)

3 $CH_2COSCoA$ / COOH

FIG. 63. Biosynthesis of the flavonoids.

(page 74) in red cabbage seedlings (Patschke, Barz and Grisebach, 1964). The proposed pathway (Fig. 63) also requires the ability of plants to activate *p*-hydroxycinnamic acid by formation of the coenzyme A thioester, and recent experiments in *Cicer arietinum* and parsley (*Apium petroselinum*) indicate that this process does occur (Grisebach, Barz, Hahlbrock, Kellner and Patschke, 1966).

The participation of malonate in the construction of ring A (as in (206)) is inferred by analogy with polyketide biosynthesis, but it has not yet been shown that malonate is a better precursor than acetate. In incorporation experiments with various labelled substrates, no better precursor of ring A than acetate has yet been found (Neish, 1960; Grisebach, 1965a).

Numerous secondary transformations such as hydroxylation, methylation, conjugation with sugars to give glycosides, etc. are involved in the production of the naturally occurring flavonoids. Structurally, the most significant secondary transformation in the scheme shown in Fig. 63 is the migration of ring C from C(2) to C(3) in the formation of the isoflavones (203). For a long time, dihydroflavonols (202) were considered to be likely intermediates in this step, particularly as the usual *trans* stereochemistry at C(2) and C(3) was considered to favour migration by a mechanism involving ionisation

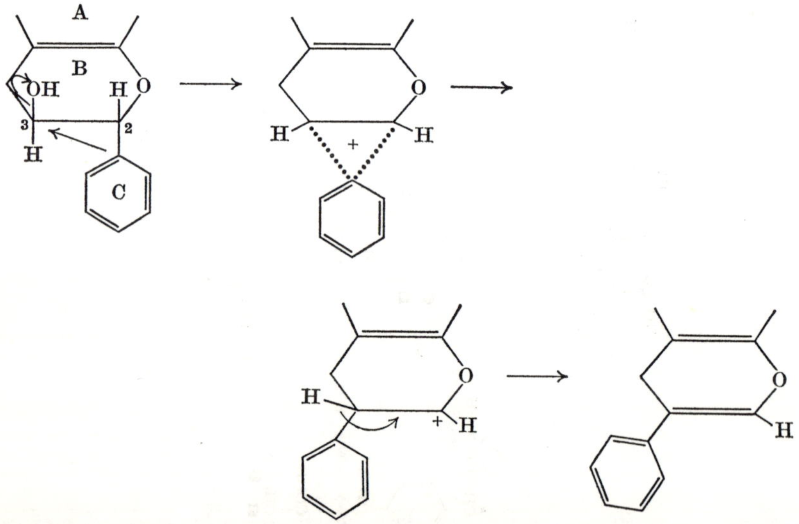

FIG. 64. Possible mechanism for the phenyl migration in isoflavone biosynthesis.

at C(3) with concomitant formation of a bridging phenonium ion, followed by rearrangement and proton loss as shown in Fig. 64 (Grisebach, 1965a). However, the recent results of Grisebach and his collaborators (Barz, Patschke and Grisebach, 1965) and of Wong (Wong, 1965) now require alternative mechanisms to be brought under consideration.

Grisebach and Kellner have recently shown that the chalcone (207) labelled with tritium at C(2) is converted into taxifolin (208) in

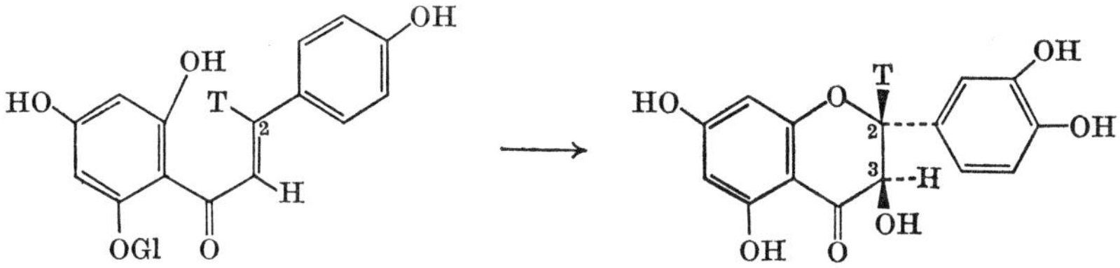

(207)

Taxifolin (208)

Chamaecyparis obtusa, with retention of the label, thus ruling out any intermediate with a double bond between C(2) and C(3). It would appear therefore that dihydroflavonols are derived from chalcones by a pathway involving direct oxidation at C(3) (Grisebach and Kellner, 1965).

A variety of C_6–C_3–C_6 compounds are structurally related to the

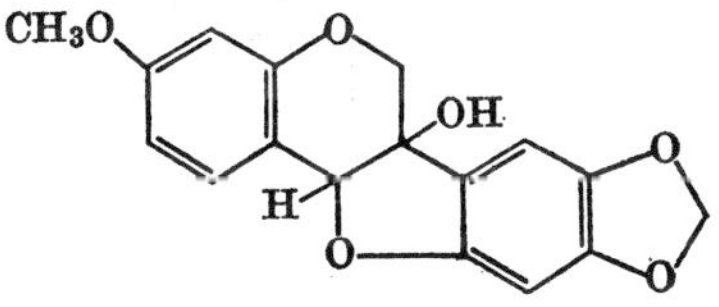

Pisatin (209)

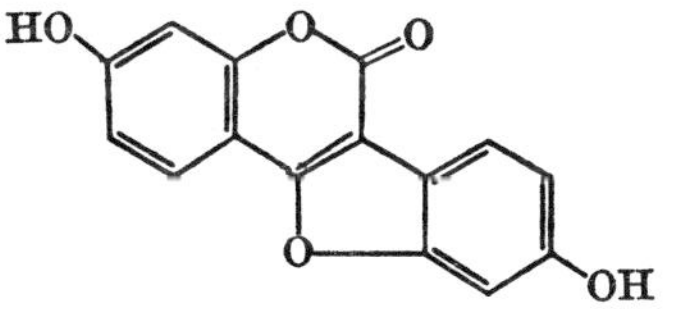

Coumestrol (210)

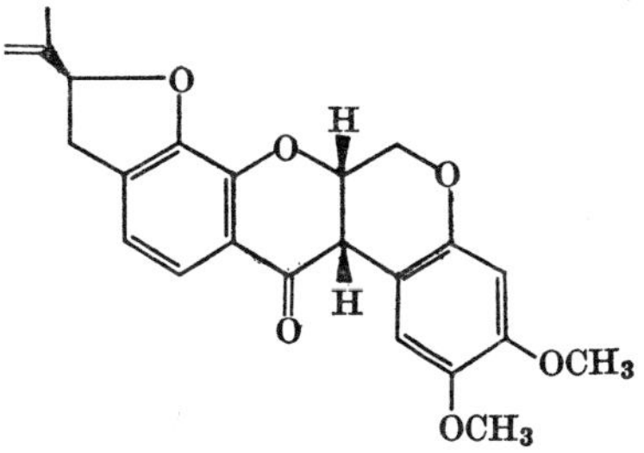

Rotenone (211)

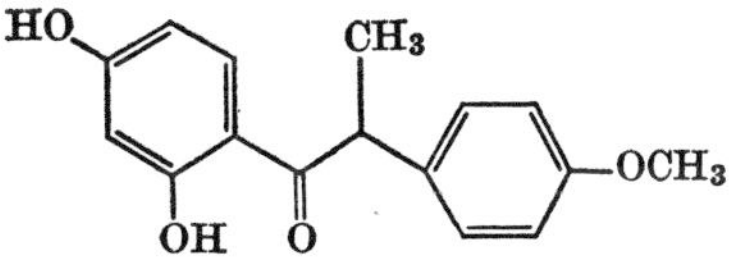

Angolensin (212)

isoflavones, e.g. pisatin (209) (Cruickshank and Perrin, 1961), coumestrol (210), rotenone (211) and angolensin (212) (King, King and Warwick, 1952). It is probable that the biogenesis of all of these involves a 1,2 aryl migration as in isoflavone biosynthesis. Direct evidence for this pathway has been obtained for coumestrol (210) biosynthesis in lucerne (Grisebach and Barz, 1964) and for rotenone (211) biosynthesis in *Derris elliptica* (Crombie and Thomas, 1965).

One of the most interesting recent developments in flavonoid chemistry has been the isolation of a number of structurally related compounds with the basic C_6–C_3–C_6 skeleton shown in (213). Included

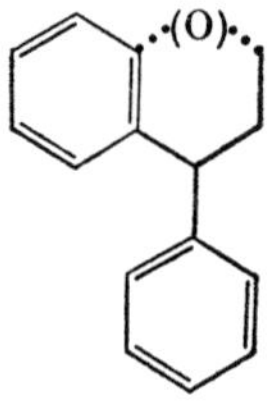

(O)

(213)

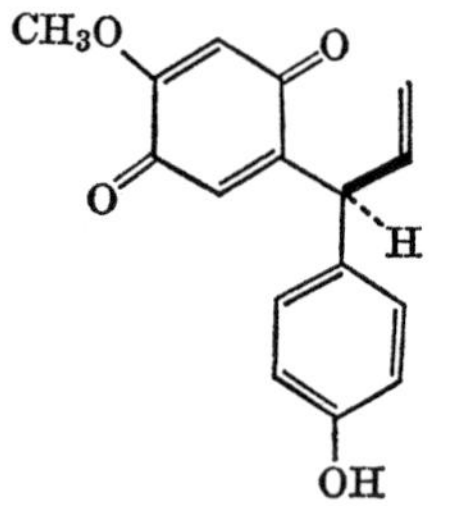

CH_3O O O H OH

(214)

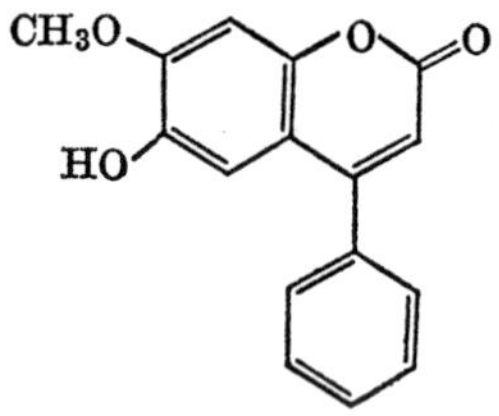

CH_3O O O HO

Dalbergin (215)

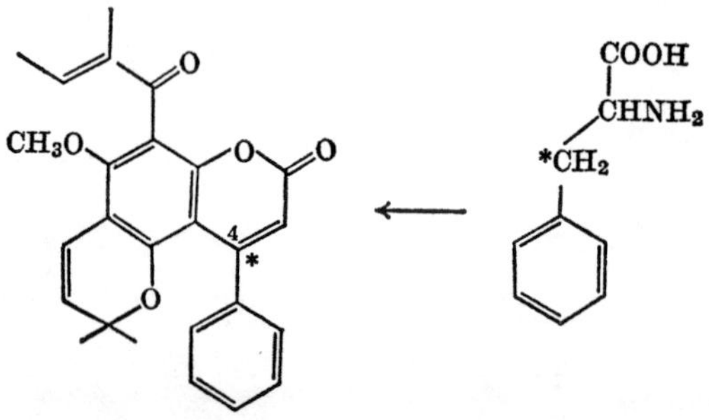

O CH_3O O O 4 * O

Calophyllolide (216)

COOH CHNH$_2$ *CH$_2$

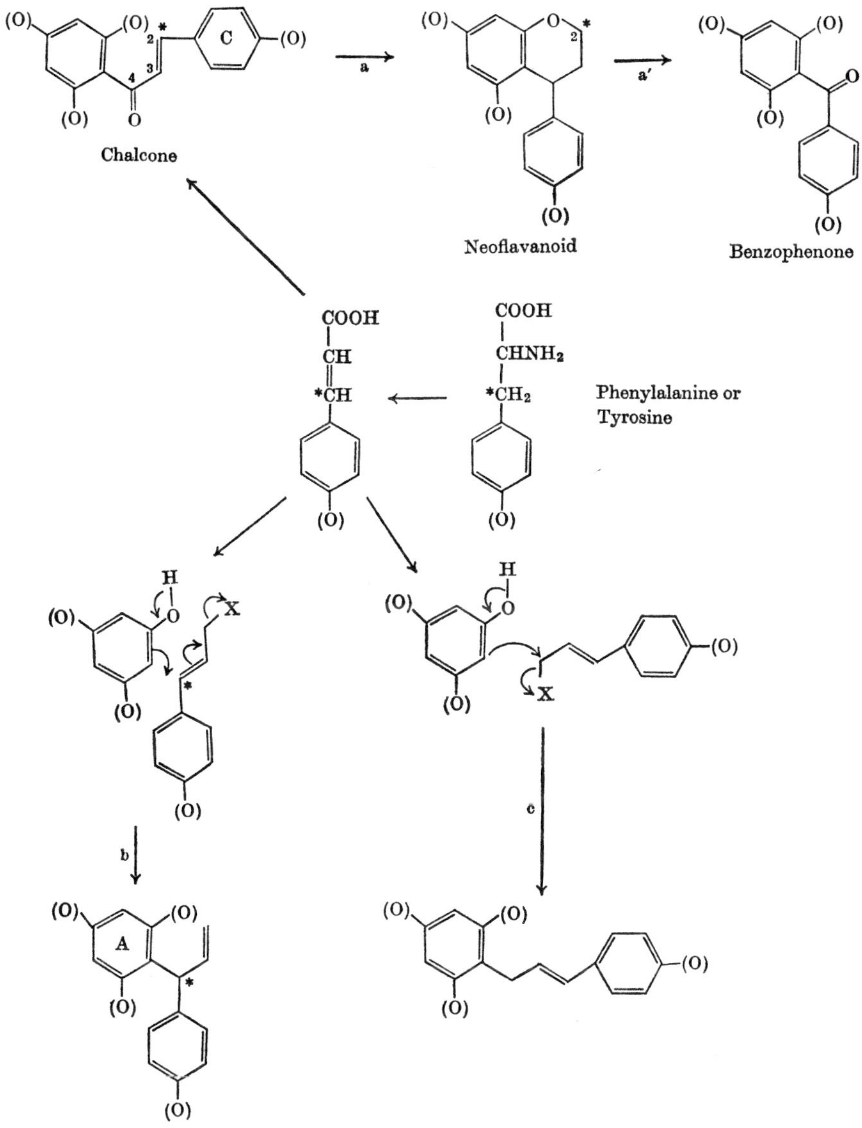

Chalcone

Neoflavanoid

Benzophenone

Phenylalanine or Tyrosine

FIG. 65. Possible pathways of neoflavanoid biosynthesis.

in this class, for which the name *neoflavanoid* has been suggested (Eyton, Ollis, Sutherland, Gottlieb, Taveira Magalhães and Jackman, 1965; Ollis, 1966), are the dalbergiones (e.g. (214)) (Ollis, 1966), the dalbergins (e.g. dalbergin (215)) (Ahluwalia and Seshadri, 1957) and the 4-phenylcoumarins (e.g. calophyllolide (216)) (Polonsky, 1956, 1957, 1958).

Two general schemes for the biosynthesis of the neoflavanoids have been considered (Ollis, 1966). In one, two 1,2 aryl migrations (or a 1,3 migration) in a normal flavonoid precursor were considered (pathway *a*, Fig. 65). This route is an extension of the type of rearrangement involved in isoflavonoid biogenesis. The second mechanism is of quite a different kind and is regarded as proceeding by nucleophilic attack of a phloroglucinol derivative on a cinnamyl substrate with elimination of a suitable leaving group (which could be a phosphate or pyrophosphate anion), as shown in Fig. 65, pathways *b* and *c*. Attack could take place either at the position α to the aromatic ring to give the neoflavanoid skeleton (pathway *b*) or directly at the carbon atom bearing the leaving group X to give a cinnamyl-phenol structure (pathway *c*). In this connection, it is interesting to note the co-occurrence of neoflavanoids such as (*S*)-4′-hydroxy-4-methoxydalbergione (214) and cinnamyl-phenols such as violastyrene (217) in *Dalbergia violacea* (Ollis, 1966). The co-occurrence

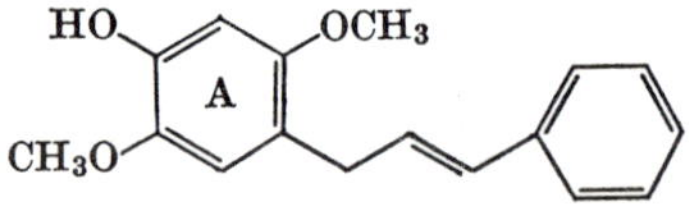

HO, OCH_3, A, CH_3O

Violastyrene (217)

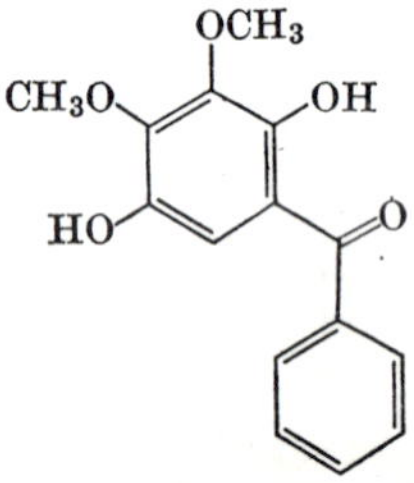

OCH_3, CH_3O, OH, HO, O

Scleroin (218)

of the benzophenone scleroin (218) and neoflavanoids in *Machaerium scleroxylon* (Gottlieb, Fineberg, Salignac de Souza Guimarães, Taveira Magalhães, Ollis and Eyton, 1964; Eyton, Ollis, Fineberg, Gottlieb, Salignac de Souza Guimarães and Taveira Magalhães, 1965) suggests that the former arises by oxidative degradation of a neoflavanoid precursor (Fig. 65, pathway *a′*) (Ollis, 1966).

Recent experiments by Kunesch and Polonsky have provided clear evidence in favour of the alkylation pathway *b* (Fig. 65) for neoflavanoid biosynthesis (Kunesch and Polonsky, 1967). [3-^{14}C]-Phenylalanine was administered to *Calophyllum inophyllum* and the labelled calophyllolide (216) was degraded in an unambiguous manner to show that almost all of the activity was located at C(4). If pathway *a* (Fig. 65) were involved, the activity would be located at C(2) as in normal flavonoid biosynthesis. It is thus apparent that in the neoflavanoids we have an entirely novel type of construction for the C_6–C_3–C_6 system which is quite different mechanistically from the normal pathway of flavonoid biosynthesis. A remarkable series of compounds has recently been isolated from *Mammea americana*; this includes 4-phenyl coumarins such as (219) and also 4-alkylcoumarins

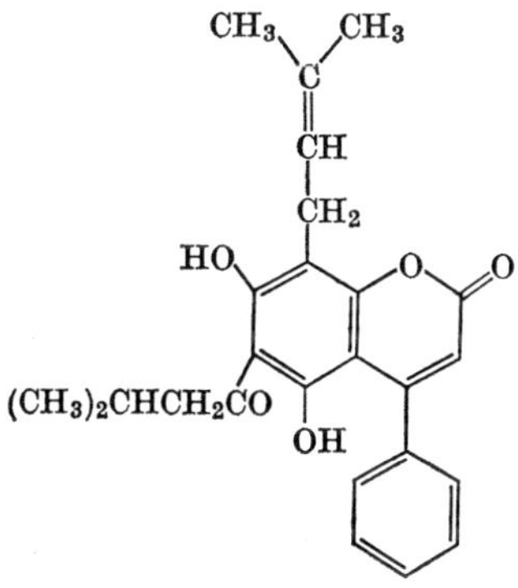

(219)

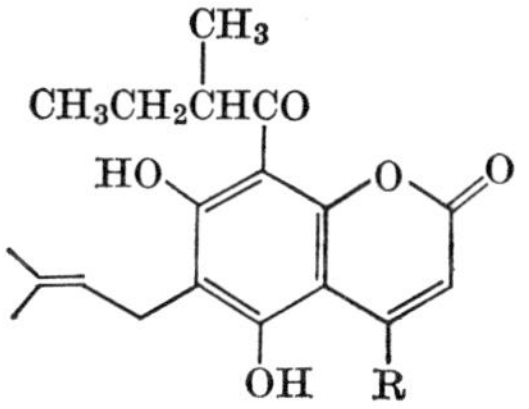

$R = CH_3CH_2CH_2$ (220)

$R = CH_3(CH_2)_4$ (221)

such as (220) and (221) (see Crombie, Games and McCormick, 1967). It would be extremely interesting to know if the alkyl-lactone system in these compounds is derived from straight chain acids in a manner analogous to the biogenesis from phenylalanine of the corresponding system in the neoflavanoids.

Ollis has pointed out (Ollis, 1966) that isoflavonoids and neoflavanoids co-occur in certain *Dalbergia* and *Machaerium* species. This raises an interesting question with regard to the biosynthesis of ring A in these compounds, since the alkylation mechanism (pathway *b*, Fig. 65) for neoflavanoid biosynthesis strongly suggests a phloroglucinol derivative as the substrate for this reaction. Even if an acyclic precursor is considered, the type of coupling observed in the biosynthesis of calophyllolide (216) is mechanistically quite different from the polyketide-type chain extension postulated for flavonoid biosynthesis. In addition, the cinnamyl-phenols (e.g. 217) which co-occur with neoflavanoids, have the C_6–C_3–C_6 carbon skeleton of the chalcones and flavonoids. Ollis has suggested that the cinnamyl phenols could arise by reduction of a chalcone (Ollis, 1966), but the reverse reaction, conversion of a cinnamyl phenol to a chalcone by oxidation at the highly activated position adjacent to ring A (as in (217)), is equally feasible. This suggests the possibility that the cinnamyl-phenols may be biosynthetically related to the flavonoids and isoflavonoids. Militating against this hypothesis is the demonstrable biogenetic similarity between the flavonoids and plant stilbenes (see below). Biosynthesis of the latter could not follow the alkylation pathway but must certainly follow a normal polyketide chain extension pathway. Nevertheless, as was mentioned previously, investigations into the biogenesis of ring A of the flavonoids have revealed no better precursor than acetate, although experiments on these lines have been less energetically pursued than those relating to the more immediate problems of the biogenetic relationships between the various classes of flavonoid. In the light of the results obtained on calophyllolide biosynthesis, there are now strong reasons for examining more closely the biosynthesis of ring A in this series.

2. *Stilbenes*

The original suggestion concerning the now established biogenesis of the flavonoids was made by Birch and Donovan (Birch and Donovan, 1953). The same authors pointed out that by simply changing the mode of cyclisation of the polyketide precursor, the basic structure of the naturally occurring stilbenes (e.g. pinosylvin (222)) could be obtained (Fig. 66). This hypothesis has been confirmed in a number of investigations; the evidence has been summarised by Billek and Schimpl (Billek and Schimpl, 1966).

In the majority of stilbenes, the carboxyl group in the presumed intermediate (223) is lost. However, in a few examples, such as hydrangenol (224) from *Hydrangea* species, it is retained. The biosynthesis of this compound has been investigated by Billek and Kindl and also by Ibrahim and Towers, and has been shown to follow the expected pathway (Billek and Kindl, 1961, 1962; Ibrahim and Towers, 1960, 1962).

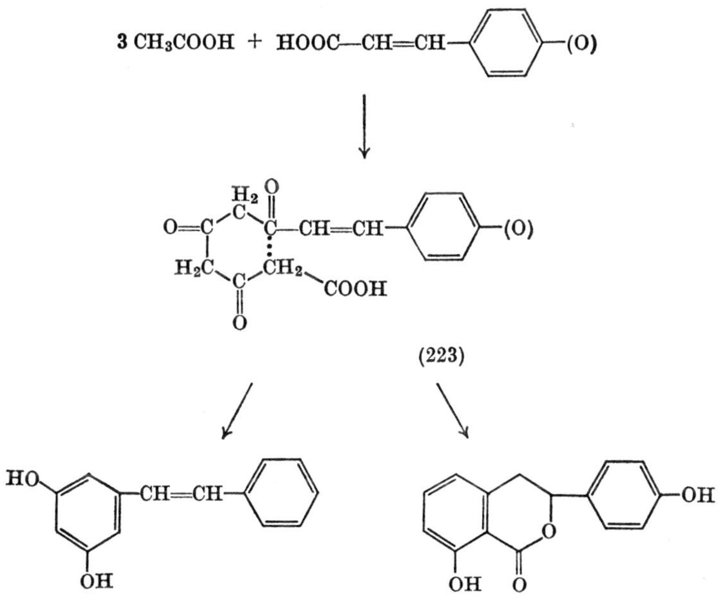

Fig. 66. Stilbene biosynthesis.

3. *Intermediates of Terpene Biosynthesis*

Many secondary metabolites have a mixed biogenesis and it is common to find compounds which are partly derived from intermediates of terpene and steroid biosynthesis. The associated biosynthetic pathway has been described in detail in recent reviews (Richards and Hendrickson, 1964; Clayton, 1965); the early stages, which are relevant to the present discussion, are illustrated in Fig. 67.

The condensation of three molecules of acetyl-CoA as shown leads, after two reduction steps, to mevalonic acid (225). It is noteworthy that the first condensation reaction leading to acetoacetyl-CoA does *not* involve malonyl-CoA, so that the pathways of polyketide and terpene biosynthesis can be seen to diverge even at this early stage. Phosphorylation of mevalonic acid (225), followed by decarboxylation with concomitant elimination of the C(3) phosphate group leads to isopentenyl pyrophosphate (226) which is the 'active isoprene unit' directly involved in the coupling reactions leading to the terpenes, steroids, carotenoids, rubber, etc. Initiation of the coupling process

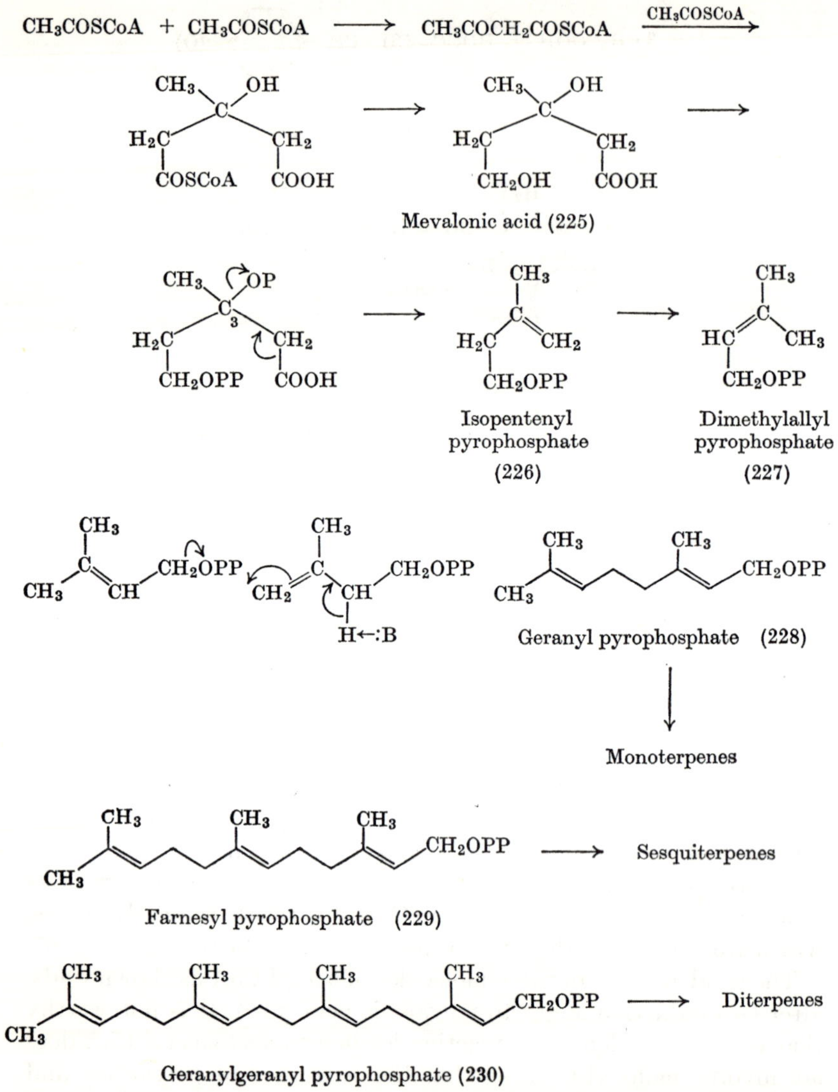

$CH_3COSCoA + CH_3COSCoA \longrightarrow CH_3COCH_2COSCoA \xrightarrow{CH_3COSCoA}$

Mevalonic acid (225)

Isopentenyl pyrophosphate (226)

Dimethylallyl pyrophosphate (227)

Geranyl pyrophosphate (228)

Monoterpenes

Farnesyl pyrophosphate (229) ⟶ Sesquiterpenes

Geranylgeranyl pyrophosphate (230) ⟶ Diterpenes

FIG. 67. Early stages in the pathway of terpene and steroid biosynthesis.

requires isomerisation of isopentenyl pyrophosphate (226) to dimethylallyl pyrophosphate (227). In the following coupling step the excellent properties of the pyrophosphate anion as a leaving group, coupled with the stabilisation of the developing positive charge over the allylic system in dimethylallyl pyrophosphate, facilitates attack by the olefinic double bond of isopentenyl pyrophosphate. Loss of a proton from C(2) of the latter as shown leads to geranyl pyrophosphate (228), the precursor of the monoterpenes. Coupling of geranyl pyrophosphate with another molecule of isopentenyl pyrophosphate gives farnesyl pyrophosphate (229), the precursor of the sesquiterpenes, and further coupling leads to geranylgeranyl pyrophosphate (230), the precursor of the diterpenes, etc.

It is evident that the cations produced by ionisation of any of these allylic pyrophosphate intermediates would be effective alkylating agents towards aromatic substrates. Many natural products have structural features which are clearly obtained by either *C*- or *O*-alkylation in this way, although relatively few have had this biogenesis confirmed experimentally. A very brief selection will be mentioned here.

C-alkylation of a polyketide nucleus is evident in the structures of the cannabis compounds such as cannabidiolic acid (231) (Mechoulam

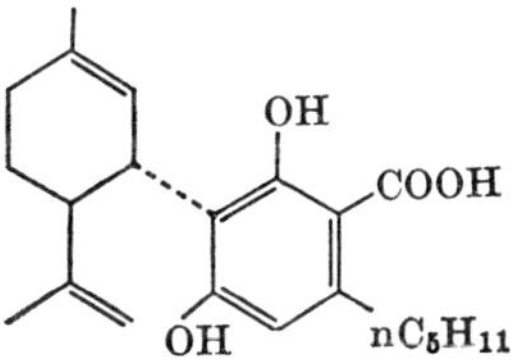

OH COOH OH nC_5H_{11}

Cannabidiolic acid (231)

OH HO nC_5H_{11}

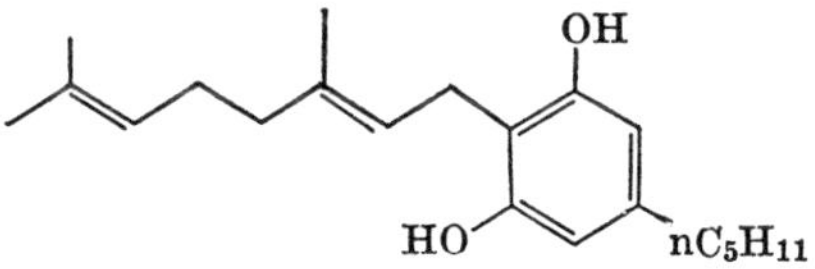

Cannabigerol (232)

and Gaoni, 1965) and cannabigerol (232) (Gaoni and Mechoulam, 1964), and of the hop constituents such as humulone (233) (Stevens, 1967). In the former, alkylation of a polyketide substrate by farnesyl pyrophosphate (229) can be postulated, whereas in humulone biosynthesis the alkylating agent is presumably dimethylallyl pyrophosphate (227). Incorporation experiments with [1-^{14}C]-acetate have supported the latter pathway (Stevens, 1967). The isovaleryl side chain in humulone (233) was only slightly labelled, which suggests that it is probably derived by acylation of the polyketide substrate with isovaleryl CoA, a product of leucine metabolism (cf. section H.5). This biogenesis is supported by the observation that radioactivity is incorporated into hop resins from [^{14}C]-leucine (Mori, 1964). The probable biogenesis of humulone is illustrated in Fig. 68.

O-alkylation, although less common than *C*-alkylation, is evidently

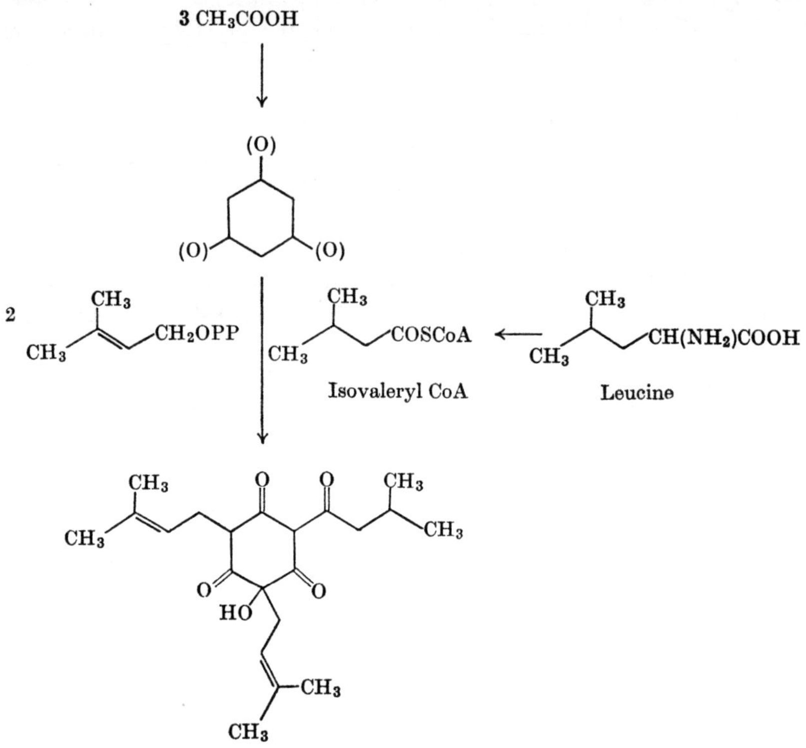

3 CH_3COOH

2 $(CH_3)_2C{=}CHCH_2OPP$

Isovaleryl CoA

Leucine

Humulone (233)

FIG. 68. Biogenesis of humulone.

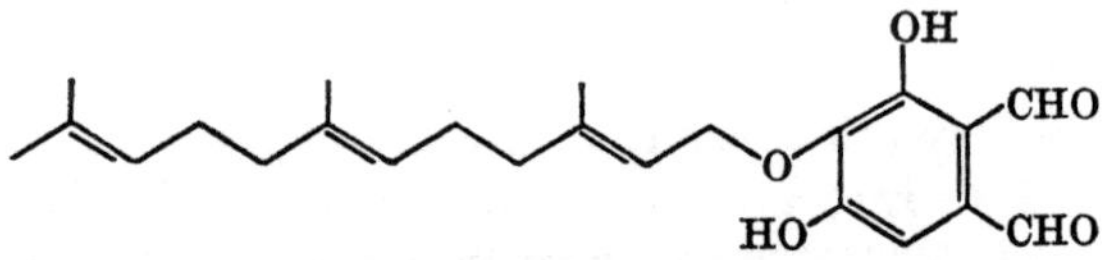

Asperugin B (234)

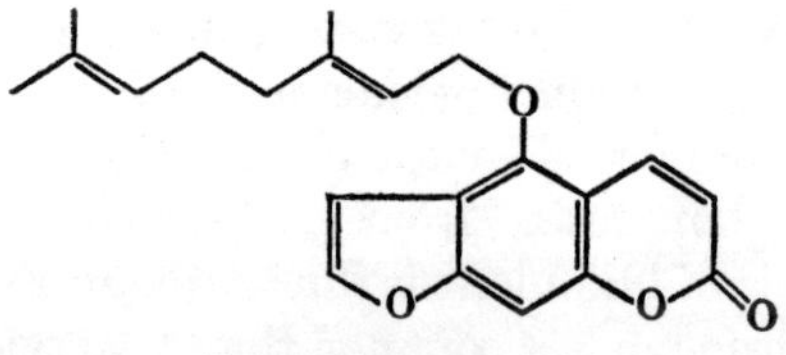

Bergaptin (235)

involved in the biosynthesis of metabolites such as asperugin B (234) (Ballantine, Hassall, Jones and Jones, 1967) and bergaptin (235) (Späth and Kainrath, 1937). The feasibility of these processes has been supported by model studies with dimethylallyl and related phosphate esters (Miller and Wood, 1965). For example, phenol, when warmed with 3,3-dimethylallyl diphenyl phosphate (236), gave 2,2-dimethylchroman (237) (Fig. 69). Orcinol (20), when treated with farnesyl diphenyl phosphate (238), gave the compound (239).

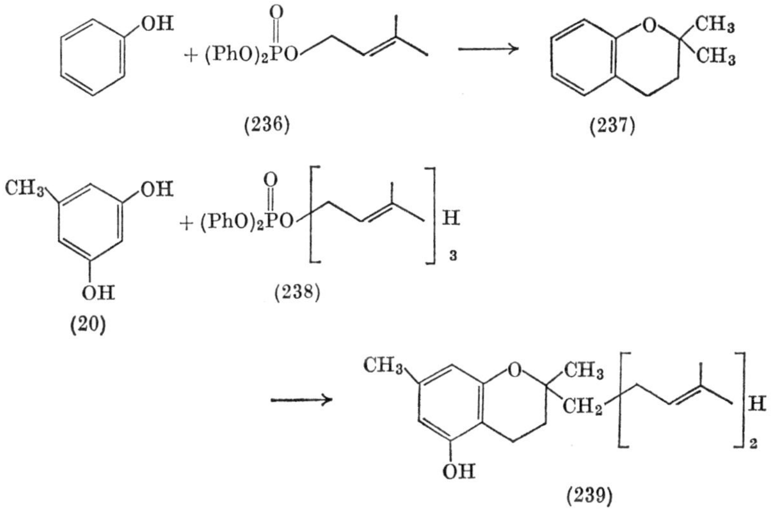

OH + $(PhO)_2PO$ (236) → (237) CH_3 CH_3

CH_3 OH OH (20) + $(PhO)_2PO$ [ ] H 3 (238)

→ CH_3 OH CH_3 CH_2 [ ] H 2 (239)

FIG. 69. Alkylation of phenols by allylic phosphate esters.

Richards and Hendrickson have given a detailed theoretical discussion of the biosynthetic implications of alkylation by intermediates of terpene biosynthesis (Richards and Hendrickson, 1964).

4. *Terpenoid Benzoquinones*

An inspection of the structures of the terpenoid quinones such as the ubiquinones (240), plastoquinone (241), α-tocopherolquinone (242) and the related tocotrienols (243) and tocopherols (244) (Fig. 70) might suggest a biogenesis from acetate for the quinone nucleus similar to that found for the fungal benzoquinones (section H.6). It is quite clear, however, that a completely different pathway is involved, since acetate is not incorporated, whereas intermediates of the shikimic acid pathway and related compounds are found to be effective precursors.

Ubiquinone ($n = 6$–10)
(240)

Plastoquinone (241)

α-Tocopherolquinone (242)

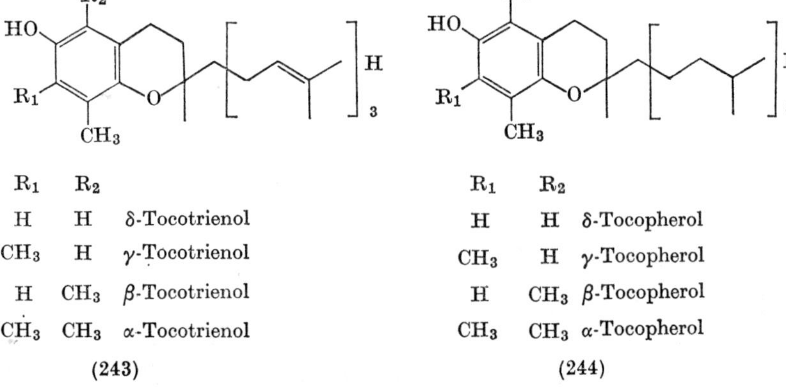

R_1	R_2	
H	H	δ-Tocotrienol
CH_3	H	γ-Tocotrienol
H	CH_3	β-Tocotrienol
CH_3	CH_3	α-Tocotrienol

(243)

R_1	R_2	
H	H	δ-Tocopherol
CH_3	H	γ-Tocopherol
H	CH_3	β-Tocopherol
CH_3	CH_3	α-Tocopherol

(244)

FIG. 70. Terpenoid benzoquinones and related phenols.

Of the various terpenoid quinones, the ubiquinones (240) have been studied most intensively. It has been shown that *p*-hydroxybenzoic acid is incorporated into the ring system in micro-organisms and in higher plants (Miller, 1965; Ramsey, Zwitkowits, Bentley and Olsen, 1966; Whistance, Threlfall and Goodwin, 1966; Threlfall, 1967).

Incorporation studies have shown that the *p*-hydroxybenzoic acid is probably derived from phenylalanine and/or tyrosine (Bentley and Lavate, 1966, Friis, Daves and Folkers, 1966). [^{14}C]-Shikimate is also incorporated both in micro-organisms (Cox and Gibson, 1964) and in higher plants (Whistance *et al.*, 1966).

The expected biogenesis of the terpenoid side chain from mevalonate has been demonstrated in a number of investigations (Threlfall, 1967).

Extensive investigations by Folkers and his collaborators have led to the provisional scheme for ubiquinone biosynthesis in *Rhodospirillum rubrum* shown in Fig. 71 (Friis, Daves and Folkers, 1966). One of

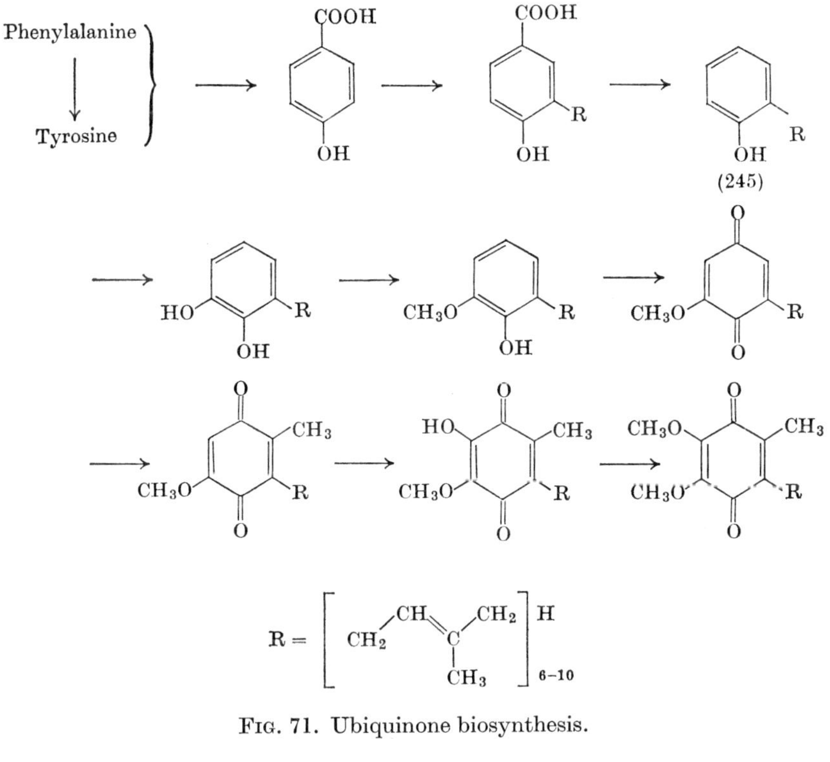

FIG. 71. Ubiquinone biosynthesis.

the key features of this scheme is the early appearance of terpenoid phenols (245). These were postulated as intermediates after isolation of 2-decaprenylphenol ((245, $R = (-CH_2CH{=}C(CH_3)CH_2)_{10}H$) from *R. rubrum* cultures (Olsen, Smith, Daves, Moore, Folkers, Parson and

Rudney, 1965). It would appear probable that the various ubiquinones are constructed by insertion of the appropriate terpenoid side chain at this early stage in the biosynthesis.

The role of *p*-hydroxybenzoic acid as a precursor of the ubiquinones is well established, but the biogenesis of the aromatic system of the other terpenoid quinones and phenols of Fig. 70 is less well documented. Whistance *et al.* found that [^{14}C]-*p*-hydroxybenzoic acid was incorporated only into ubiquinone in a number of plant species; the other terpenoid quinones were not labelled. However, with [^{14}C]-shikimate as substrate, all the quinones examined were found to be labelled (Whistance *et al.*, 1966). Similar results were obtained by Powls and Hemming in *Euglena gracilis* (Powls and Hemming, 1966). Evidence has recently been obtained to show that the tyrosine nucleus is incorporated intact, together with the β-carbon atom of the side chain, in the biosynthesis of plastoquinone (241), α- and γ-tocopherol (244) and α-tocopherolquinone (242). In each of these compounds, the methyl group *meta* to the terpenoid side chain is probably derived from the β-carbon atom of tyrosine (Whistance and Threlfall, 1967). The remaining *C*- and *O*-methyl groups are derived from methionine as expected (Threlfall, Whistance and Goodwin, 1968).

The biosynthesis of the terpenoid quinones has been expertly summarised by Threlfall (Threlfall, 1967).

5. *Naphthaquinones*

Although the terpenoid nature of the side chain in the K vitamins (246) and (247) is apparent, evidence regarding the biogenesis of the naphthaquinone nucleus has been lacking until recently. Cox and Gibson investigated the biogenesis of vitamin K_2 in *Escherichia coli* and found that shikimic acid was selectively incorporated into the benzene ring of the naphthaquinone moiety (Cox and Gibson, 1966). [1,2-^{14}C]-acetate was also incorporated and the activity was distributed nearly equally between the terpenoid side chain and the naphthaquinone system, with most of the activity of the latter located at C(2) and C(3). Addition of phenylpyruvate, *p*-hydroxy phenylpyruvate and 3,4-dihydroxybenzoate to cultures of *Aerobacter aerogenes* did not suppress the conversion of shikimate into vitamin K_2, which tends to suggest that a pre-aromatic intermediate of the shikimic acid pathway is a precursor. The incorporation of shikimic acid into vitamin K_1 (phylloquinone) (247) in higher plants has also been demonstrated (Whistance *et al.*, 1966). Incorporation experiments with *Mycobacterium phlei* have shown that the ring methyl group is derived from methionine (Jaureguiberry, Lenfant, Das and Lederer, 1966).

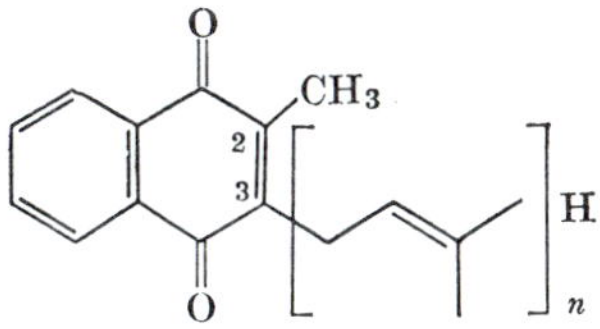

Vitamin K_2 (Menaquinone) $n = 4, 6–9$)

(246)

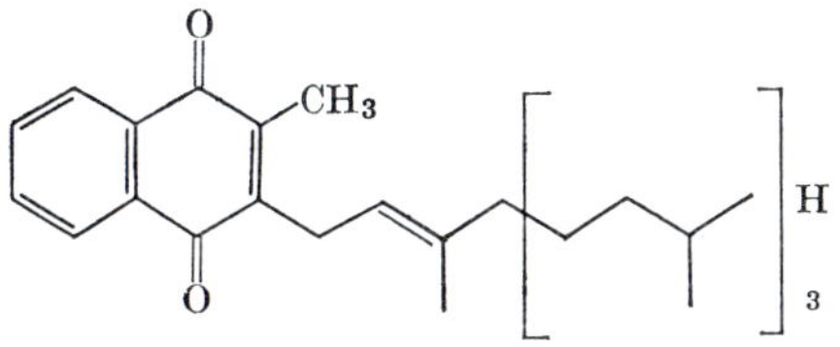

Vitamin K_1 (Phylloquinone) (247)

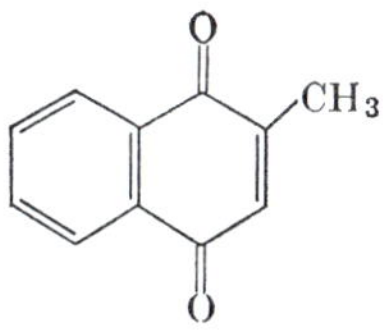

(248)

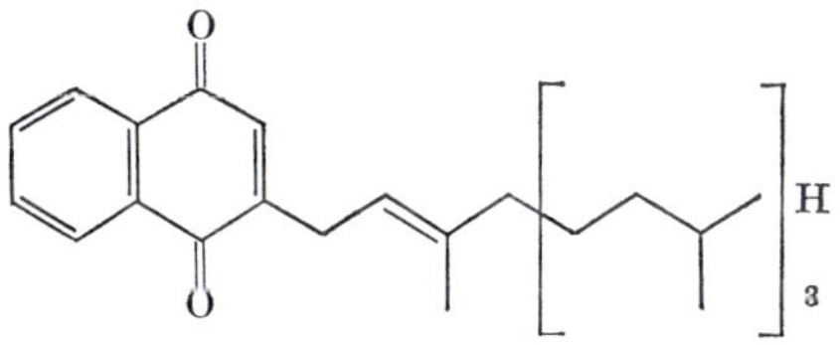

Desmethylphylloquinone (249)

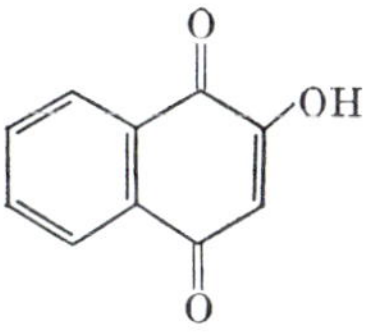

Lawsone (250)

Evidence for the insertion of the terpenoid side chain into the naphthaquinone nucleus *after* introduction of the ring methyl group was obtained by Martius and Esser, who fed [Me-^{14}C]-2-methyl-1,4-naphthaquinone (248) to rats and chicks and observed incorporation of radioactivity into vitamin K_2 ((246), $n = 4$) (Martius and Esser, 1959). This sequence is the reverse of that found in ubiquinone biosynthesis. On the other hand, a desmethylnaphthaquinone (probably desmethylphylloquinone (249)) has recently been isolated from spinach chloroplasts and various isoprenologues of desmethyl vitamin K_2 have been found in bacterial species (Pennock, 1967), an indication that prenylation can also take place after methylation. It is quite possible that the order of the two substitution steps might vary in different organisms. The probable mechanistic course of the alkylation reaction suggests that the substrate is probably the reduced (hydroquinone) form of the naphthaquinone in all of these pathways.

Zenk and Leistner have demonstrated a similar biogenesis from shikimate for the benzene ring of lawsone (2-hydroxy-1,4-naphthaquinone) (250) in *Impatiens balsamina*. These workers were also able to show that the carboxyl group of shikimic acid was incorporated to provide one or (less likely) both of the keto groups of the quinone ring (Zenk and Leistner, 1967). This result again indicates the utilisation of a pre-aromatic intermediate of the shikimic acid pathway, a conclusion which is partially supported by the lack of incorporation of [^{14}C]-benzoic acid.

In comparing the biogenesis of javanicin (page 98), vitamin K_2 and lawsone, it can be seen that there exists a dichotomy similar to that found in the benzoquinone series, where two separate pathways

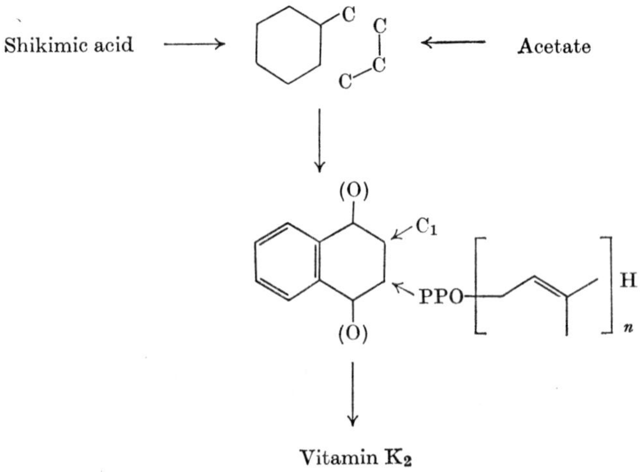

FIG. 72. Biogenesis of vitamin K_2.

to the ring system, from acetate and from shikimic acid, are found. The biogenesis of vitamin K_2 is summarised in Fig. 72. (The equivalence or non-equivalence of the quinone carbonyl positions in the final product is still undecided.)

6. *Anthraquinones*

Evidence has recently been presented by two groups to show that the biosynthesis of anthraquinones in higher plants follows a completely different course from the one established in micro-organisms (section H.1), but which is closely related to the naphthaquinone pathway discussed in the previous section.

Leistner and Zenk were the first to report on this problem. These workers found that whereas [^{14}C]-phenylalanine was not converted into alizarin (251) and purpurin (252) in *Rubia tinctorum*, [^{14}C]-

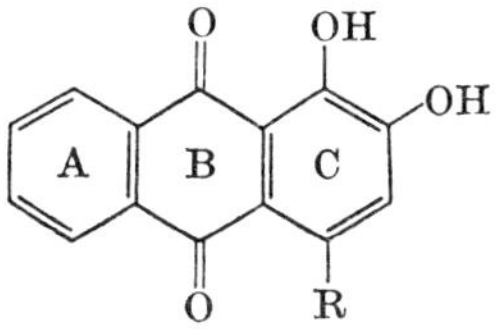

R = H, Alizarin (251)

R = OH, Purpurin (252)

shikimate was incorporated to the relatively high extent of 0·774 per cent (Leistner and Zenk, 1967a, b). Degradation of the labelled anthraquinones showed that shikimate provided ring A specifically and that label from the carboxyl group was incorporated into the quinone carbonyl positions. [2-^{14}C]-acetate was also incorporated into the quinone carbonyl positions and into ring C. The construction of the A/B ring system was therefore similar to that of lawsone (250) and raised the possibility that biosynthesis of the anthraquinone system might be developed by construction of ring C on a preformed naphthaquinone nucleus. Leistner and Zenk were able to show conclusively that this was in fact the pathway in *R. tinctorum* by demonstrating the incorporation of 1,4-naphthaquinone (253) (Fig. 73) into alizarin without prior degradation of the molecule (Leistner and Zenk, 1968a).

Investigations by Thomson and his collaborators have neatly complemented this work by showing that ring C in the *Rubia* anthraquinones is derived from mevalonic acid (Burnett and Thomson, 1967a). This biogenesis was suggested by the co-occurrence of C_{15} anthraquinones, isoprenylnaphthaquinones and naphthapyrones such as (254), (255) and (256) respectively, in teak and in species of the family Bignoniaceae. The laboratory conversion of (255) into (254) supported the possibility of a similar transformation *in vivo* (Burnett and Thomson, 1967b). Burnett and Thomson have provided clear evidence for this pathway by showing that [2-^{14}C]-mevalonate was

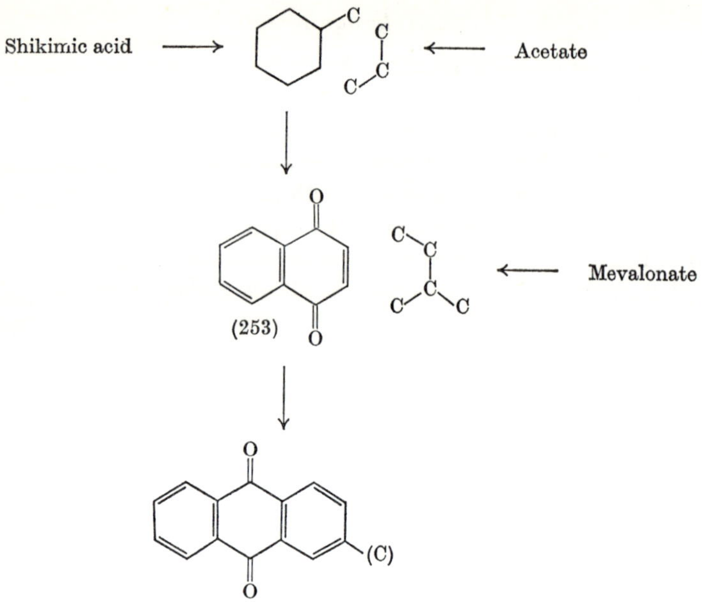

Shikimic acid → ← Acetate

← Mevalonate

(253)

Fig. 73. Pathway of anthraquinone biogenesis in *Rubia* species.

incorporated into the anthraquinone pigments of *R. tinctorum* (Burnett and Thomson, 1967a). By degradation of the labelled pseudopurpurin (257) it was established that half of the activity resided in the carboxyl

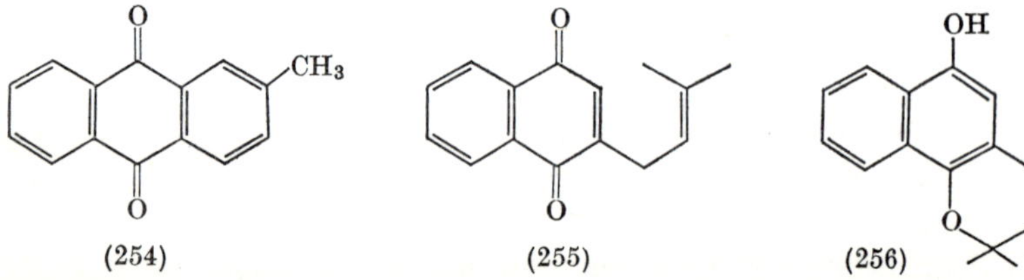

(254) (255) (256)

group and the remainder in ring C (presumably at C(4). This incorporation pattern indicates that C(2) and C(6) of mevalonate (225) become equivalent at some point in the pathway. In confirmation of this

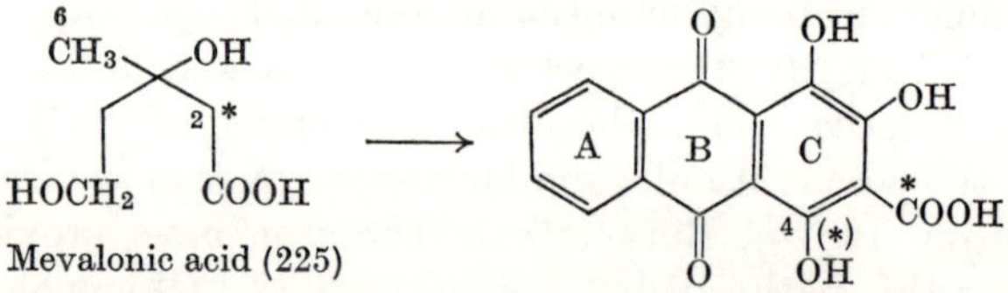

Mevalonic acid (225) → Pseudopurpurin (257)

result, Leistner and Zenk also found that [2-^{14}C]-mevalonate was incorporated into pseudopurpurin, but observed that 85 per cent of the activity was located in the carboxyl group, from which it would appear that C(2) of mevalonic acid (225) largely retained its integrity in their experiment (Leistner and Zenk, 1968b). This discrepancy between the results of the two groups, although unexpected, is not without precedent, since it has been found that in biosynthesis of the monoterpenoid *Skytanthus* alkaloids, for example, C(2) and C(6) of mevalonate may or may not become equivalent depending on the age of the plants used in the feeding experiments (Auda, Juneja, Eisenbraun, Waller, Kays and Appel, 1967).

Once again, the pathway of anthraquinone biosynthesis in higher plants can be seen to differ radically from the polyketide route established for the fungal anthraquinones, and reveals a complexity not hinted at in various hypothetical schemes (Neelakantan and Seshadri, 1960).

The outstanding problem now remaining for both naphthaquinone and anthraquinone biosynthesis is the mode of incorporation of acetate into the ring B positions not furnished by shikimate or mevalonate, as shown in the summary of anthraquinone biogenesis given in Fig. 73.

L. Cyclopropane Derivatives

The cyclopropane ring system is a not uncommon structural feature in the terpene series, where it probably arises by neutralisation of a cationic centre through attack by an olefinic double bond (as in

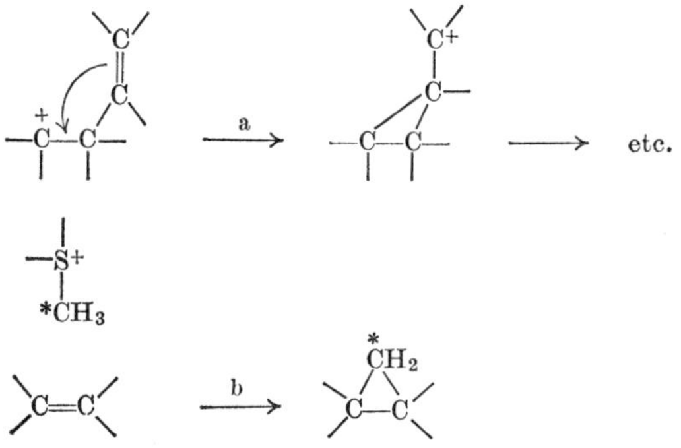

FIG. 74. Formation of the cyclopropane system *in vivo*.

Fig. 74, pathway *a*, for example). However, it has now been established that isolated cyclopropane and cyclopropene systems in a number of non-terpenoid compounds arise by direct insertion into a double bond of a methylene group derived from the methyl group of methionine (Fig. 74, pathway *b*). The first unambiguous indication of such a pathway came from the work of Hofmann and his collaborators who observed the incorporation of activity from [$^{14}CH_3$]-methionine and [^{14}C]-formate specifically into the cyclopropane ring of lactobacillic acid (258) in *Lactobacillus arabinosus* (Liu and Hofmann, 1962). It was shown that the substrate for this reaction was vaccenic acid (259) which, in labelled form, was efficiently converted into lactobacillic acid by the bacterium.

$$\underset{\text{Vaccenic acid (259)}}{CH_3(CH_2)_5CH\overset{cis}{=}CH(CH_2)_9COOH} \longrightarrow \underset{\text{Lactobacillic acid (258)}}{CH_3(CH_2)_5\overset{CH_2}{CH-CH}(CH_2)_9COOH}$$

These results have been extended by other workers, in bacterial systems (Zalkin, Law and Goldfine, 1963; Chung and Law, 1964) and more recently in higher plants (Hooper and Law, 1965). Hooper and Law investigated the cyclopropene acids of *Hibiscus syriacus* seed oil

$$CH_3(CH_2)_7\overset{CH_2}{C=C}(CH_2)_nCOOH$$

$n = 7$, Sterculic acid (260)

$n = 6$, Malvalic acid (261)

which contains sterculic acid (260) (3 per cent) and malvalic acid (261) (17 per cent) together with small amounts of the corresponding saturated acids. Incorporation experiments with [^{14}C]-methionine resulted in labelling of the acids and the activity in the cyclopropene acids was shown to be located in the methylene bridge.

It is evident that the biogenetic route to cyclopropane derivatives is closely allied to the normal process of biological methylation. Jaureguiberry *et al.* distinguish two mechanisms for this process. In mechanism CD_2, only two of the three methyl protons of methionine are transferred to the substrate, as in the biosynthesis of tuberculostearic acid in *Mycobacterium smegmatis* and of ergosterol in *Neurospora crassa*, whereas in mechanism CD_3, all three protons are transferred as in the biosynthesis of sclerotiorin in *Penicillium sclerotiorum*, and of vitamin $K_2(45)H$ and α-smegmamycolic acid in *Mycobacterium smegmatis* (Jaureguiberry, Lenfant, Das and Lederer, 1966). The designations CD_2 and CD_3 signify that these results were obtained by measuring the extent of deuterium incorporation into the substrates from $[C^2H_3]$-methionine. Jaureguiberry *et al.* have suggested that mechanism CD_2 occurs with weakly nucleophilic double bonds

(262)

(264)

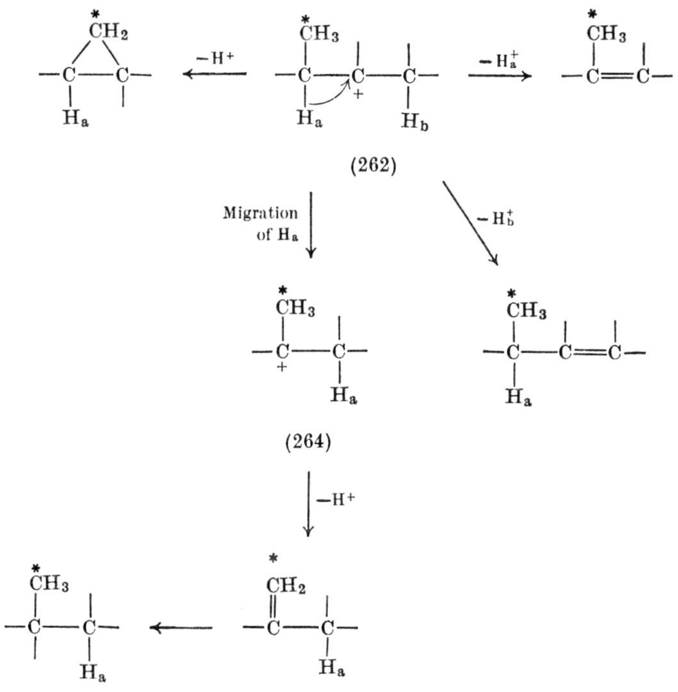

FIG. 75. Possible mechanisms of biological methylation.

and mechanism CD_3 with strongly nucleophilic double bonds. For both mechanisms, as for the formation of the cyclopropane system, the first step in the reaction is postulated to lead to the cation (262), which can then be stabilised in several ways, as shown in Fig. 75. The formation of an intermediate methylene species (263) followed by reduction to give the C-methyl group serves to explain the loss of one of the methionine methyl protons in reactions which proceed by way of the CD_2 mechanism. It is of interest that the pathway to the intermediate (264) involves a hydride ion transfer if hydrogen is present at the methylated position, and experiments in the phytosterol series with ^{3}H-labelled substrates tend to support this mechanism (Akhtar, Hunt and Parvez, 1966; Frantz and Schroepfer, 1967).

The cyclopropene system could arise either by secondary dehydrogenation of an initially formed cyclopropane derivative (Hooper and Law, 1965) or by insertion of a methylene group into an acetylenic substrate (Bu'Lock and Smith, 1964; Smith and Bu'Lock, 1965).

There are a number of laboratory analogies for the formation of cyclopropane derivatives from sulphonium compounds structurally related to *S*-adenosylmethionine. Corey and Chaykovsky, for example, were able to form the cyclopropane derivative (265) from the sulph-

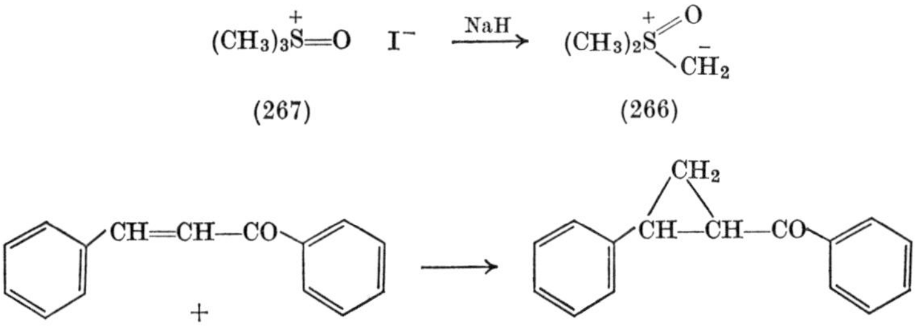

$$(CH_3)_3\overset{+}{S}{=}O \quad I^- \xrightarrow{NaH} (CH_3)_2\overset{+}{S}(=O)\overset{-}{C}H_2$$

(267) (266)

$$C_6H_5{-}CH{=}CH{-}CO{-}C_6H_5 + (266) \longrightarrow C_6H_5{-}\underset{CH_2}{CH{-}CH}{-}CO{-}C_6H_5$$

(266) (265)

FIG. 76. Laboratory synthesis of the cyclopropane system using a sulphonium ylide.

onium ylide (266) as shown in Fig. 76. The ylide was produced by treatment of the sulphonium salt (267) with sodium hydride (Corey and Chaykovsky, 1962). Related examples of this type of reaction have been listed by Nozaki *et al.* (Nozaki, Ito, Tunemoto and Kondo, 1966). The active methylenating species in these reactions is, of

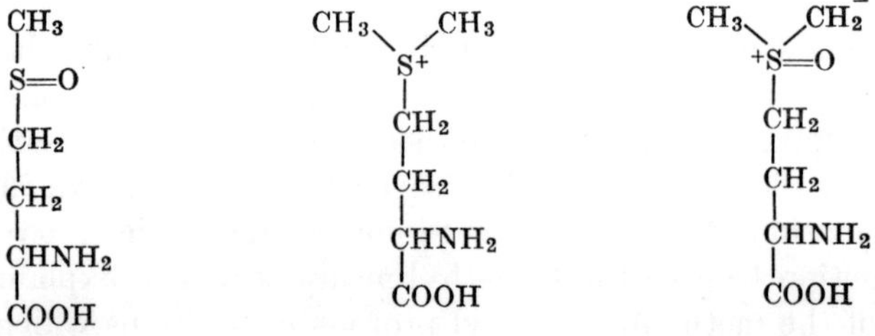

$$CH_3{-}S(=O){-}CH_2{-}CH_2{-}CHNH_2{-}COOH$$

(268)

$$(CH_3)_2\overset{+}{S}{-}CH_2{-}CH_2{-}CHNH_2{-}COOH$$

(269)

$$CH_3(\overset{-}{C}H_2)\overset{+}{S}(=O){-}CH_2{-}CH_2{-}CHNH_2{-}COOH$$

(270)

course, the ylide (as in (266)), which is not strictly comparable with *S*-adenosylmethionine. However, both methionine sulphoxide (268) and salts of the methyl methionine sulphonium cation (269) have been found in higher plants, and transfer of the methyl group of methionine in biosynthesis of the latter has been demonstrated (Splittstoesser and Mazelis, 1967). It is possible therefore that the methylene donor in cyclopropane formation *in vivo* is the sulphonium ylide (270), which could be produced by methylation of methionine sulphoxide followed by proton removal as in the preparation of the synthetic reagent (266).

A review of lipids containing cyclopropane and cyclopropene rings is in preparation (Christie, 1969).

M. Cyclopentane Derivatives

1. 1-*Amino-2-nitrocyclopentanecarboxylic Acid*

This metabolite (271) of *Aspergillus wentii* has been shown to be derived from lysine (272) (Burrows, Mills and Turner, 1965). The same organism is able to convert aspartic acid into β-nitropropionic acid.

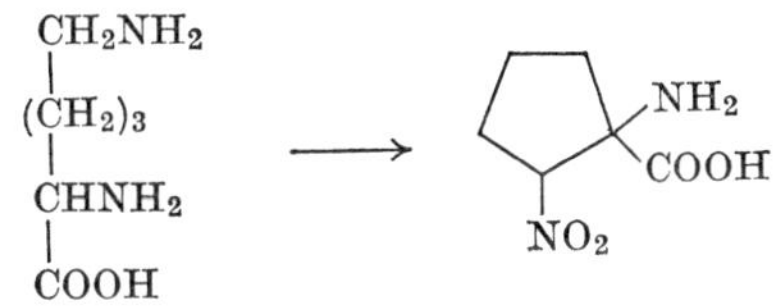

CH_2NH_2
$(CH_2)_3$
$CHNH_2$
COOH

Lysine (272) ⟶ (271)

NH_2 COOH NO_2

2. *Caldariomycin*

The biosynthetic pathway to caldariomycin (273), an unusual chlorinated cyclopentane derivative from *Caldariomyces fumago* (Fig. 77), has been elucidated by Hager and his collaborators.

Shikimic acid was clearly a precursor since it was incorporated with greater efficiency than any other primary metabolite tested (Beckwith, Clark and Hager, 1963). The chlorination steps, which require hydrogen peroxide and inorganic chloride, are catalysed by the enzyme chloroperoxidase. The conversions of 1,3-cyclopentanedione (274) into 2-chloro-1,3-cyclopentanedione (275) and of the latter into 2,2-dichloro-1,3-cyclopentanedione (276) were demonstrated with $K^{36}Cl$ as the chloride source. These reactions showed an interesting

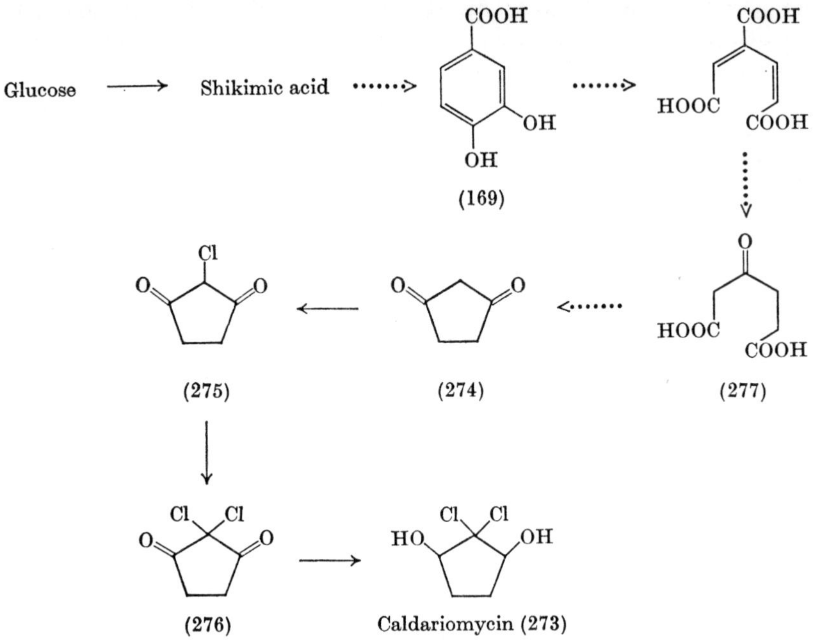

FIG. 77. Caldariomycin biosynthesis.

dependence on the acidity of the medium. At pH 5·6, the enzymatic chlorination of 1,3-cyclopentanedione (274) yielded the monochloro-derivative (275), whereas at pH 3, the dichloro compound (276) was obtained. The final step was demonstrated by the extremely efficient conversion of [^{36}Cl]-2,2-dichloro-1,3-cyclopentanedione (276) into caldariomycin (273) in *C. fumago* cultures (Beckwith and Hager, 1963).

The role of β-ketoadipate (277) as an intermediate was suggested by the fact that it could be derived by ring cleavage of an aromatic precursor such as protocatechuic acid (169) (Dagley, Evans and Ribbons, 1960) which would account for the incorporation of shikimate, and by a laboratory analogy for the cyclisation to 1,3-cyclopentanedione (Boothe, Wilkinson, Kushner and Williams, 1953). Unfortunately, ^{14}C-labelled β-ketoadipate was only incorporated into caldariomycin at a very low rate ($<0{\cdot}00047$ per cent), but this was possibly due to rapid decarboxylation of the precursor under the experimental conditions (pH 3) (Beckwith, Clark and Hager, 1963). Nevertheless, incorporation experiments with specifically labelled glucose gave caldariomycin labelled in a manner consistent with the proposed pathway via shikimate and protocatechuate (Fig. 77).

Chloroperoxidase has recently been obtained in crystalline form (Morris and Hager, 1966) and has been shown to be a glycoprotein of molecular weight 42,000 with ferriprotoporphyrin IX as the prosthetic group. The enzyme appears to be remarkably catholic in its choice of substrates, since it has been reported that if the chloride in the nutrient medium of *C. fumago* cultures is replaced by bromide, the bromo-analogue of caldariomycin is produced (Patterson, Andres and Mitscher, 1967). The organic substrate can also be varied; 16-keto-progesterone (278) for example gave the halogenated product (279) on incubation with added bromide or chloride (Neidleman, Diassi, Junta, Palmere and Pan, 1966).

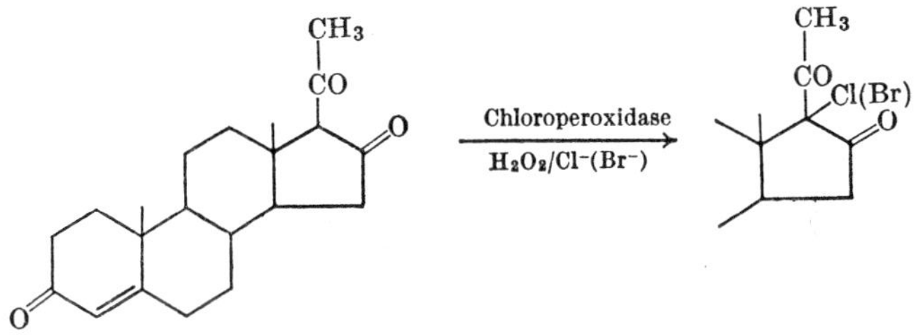

CH_3 CO O O

Chloroperoxidase

$H_2O_2/Cl^-(Br^-)$

CH_3 CO Cl(Br) O

16-Ketoprogesterone (278) (279)

3. *The Prostaglandins*

The prostaglandins comprise an extremely interesting group of closely related compounds which occur in highest concentration in mammalian seminal plasma and vesicular glands. Their physiological activity is evident in the cardiovascular system (lowering of blood pressure) and in a pronounced stimulating effect on smooth muscle. The biological action of the prostaglandins has been the subject of intensive investigation over the last few years and it is clear that they are of fundamental importance in mammalian physiology. Recent reviews summarise current knowledge of the chemistry, biochemistry and physiological activity of these compounds (Bergström and Samuelsson, 1965; Bergström, 1966).

The prostaglandins are C_{20} cyclopentane acids which can be classified into two series. Members of the PGE series have a carbonyl function at C(9), whereas in the PGF series there is a secondary hydroxyl group at this position (Fig. 78). Thirteen prostaglandins have so far been isolated from human seminal plasma. The structural variety of these compounds is illustrated by the six primary prostaglandins shown in Fig. 78 (only $PGF_{3\alpha}$ has so far not been found in human seminal plasma).

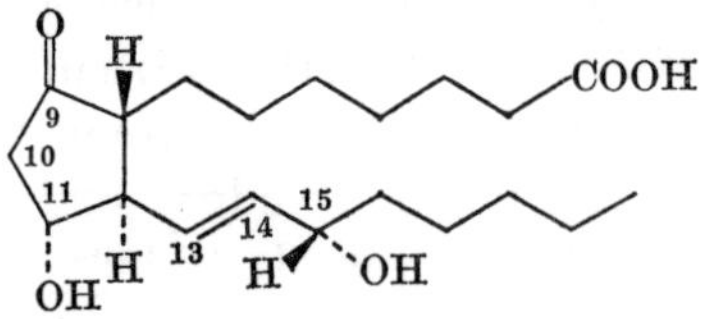

PGE_1 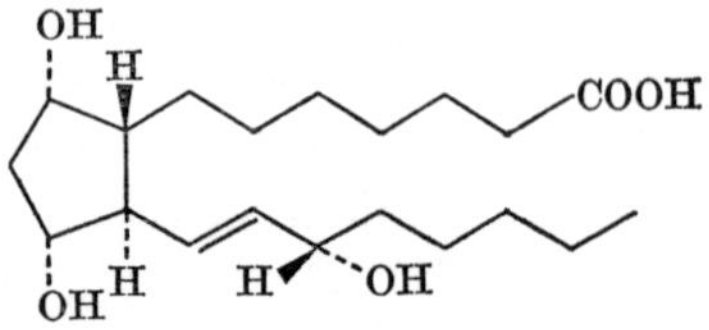 $PGF_{1\alpha}$

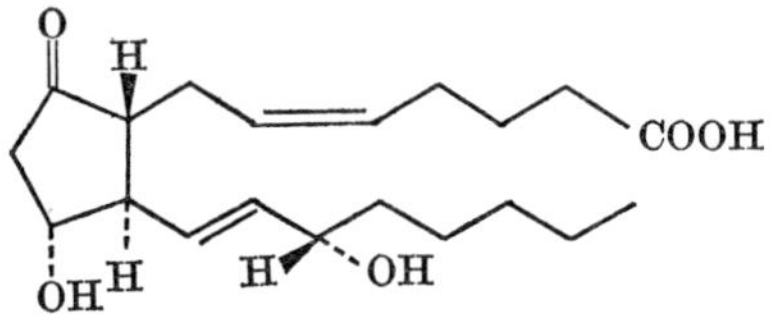

PGE_2 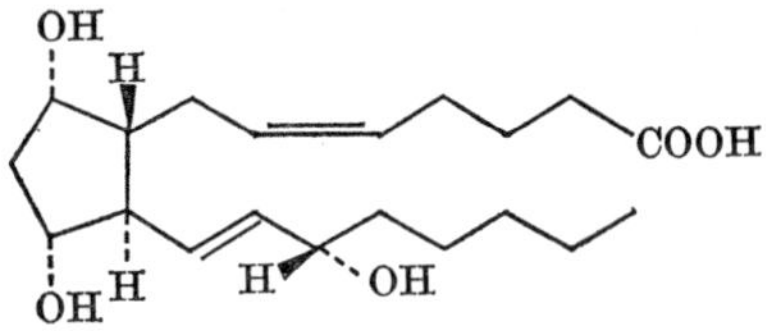 $PGF_{2\alpha}$

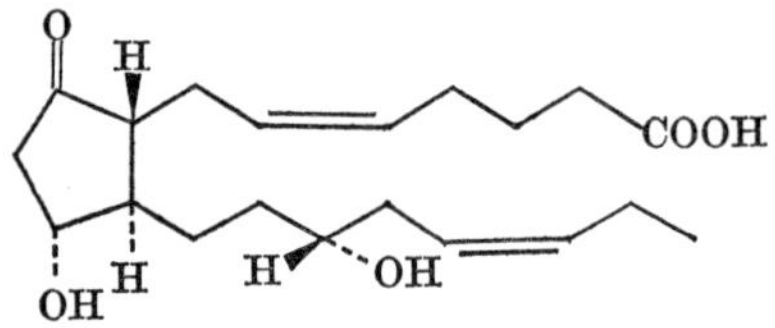

PGE_3 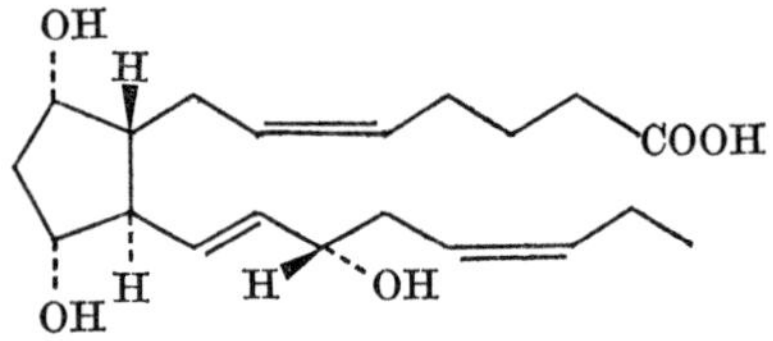 $PGF_{3\alpha}$

FIG. 78. Structures of the primary prostaglandins.

Remarkable progress has been made in the investigations into the biosynthesis of the prostaglandins, mainly due to the efforts of groups at the Karolinska Institute, Stockholm, and at the Unilever Research Laboratories, Vlaardingen, Holland.

The structures of the prostaglandins suggest a biogenesis from C_{20} essential fatty acid precursors, a prediction which has been verified by incorporation studies using homogenates of sheep vesicular glands. Thus [^{3}H]-arachidonic acid (5,8,11,14-eicosatetraenoic acid (280)), synthesised by catalytic reduction of the tetrayne ester (281) with tritium over a Lindlar catalyst, followed by saponification, was efficiently converted (16–20 per cent) into PGE_2 (Fig. 79) (van Dorp, Beerthuis, Nugteren and Vonkeman, 1964; Anggard and Samuelsson,

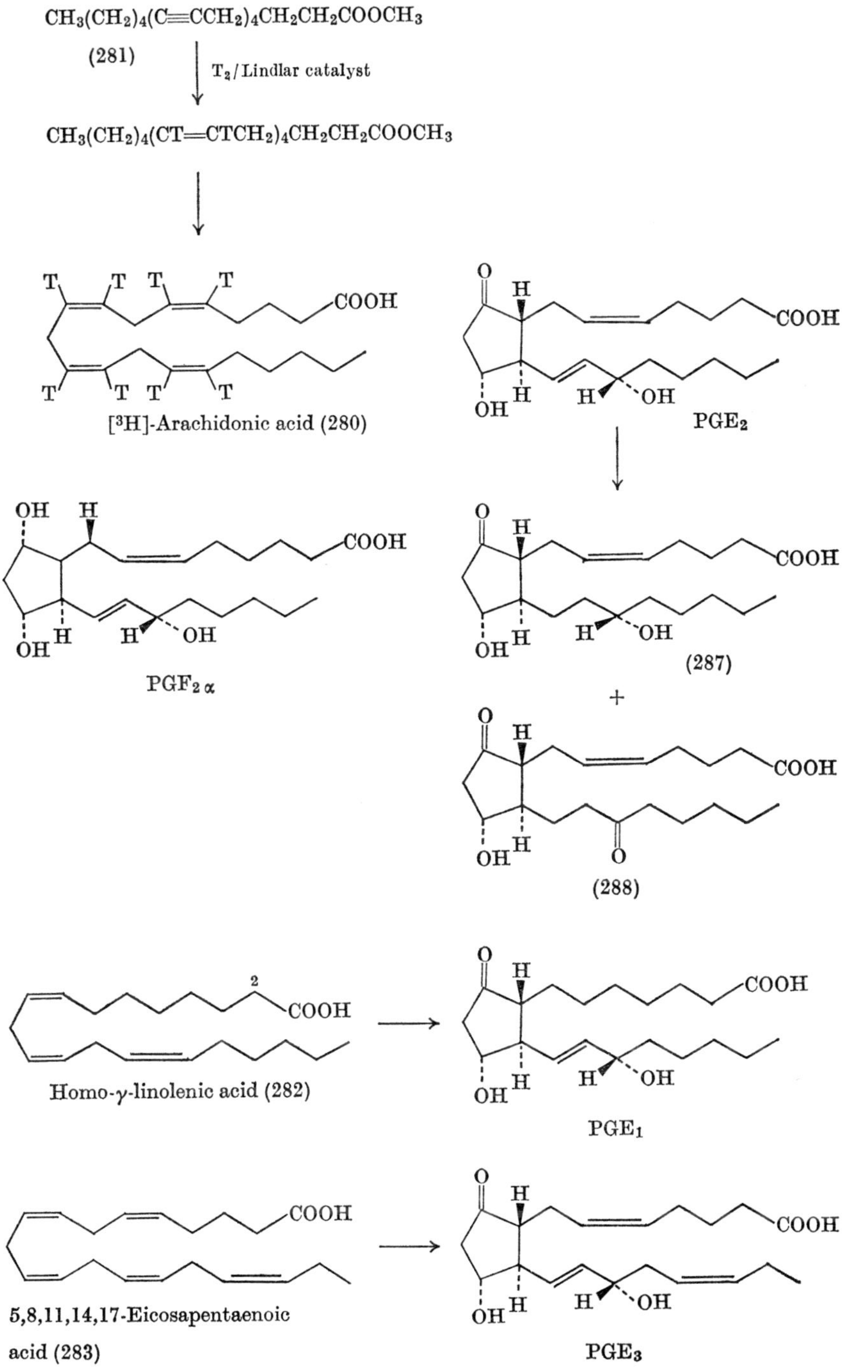

$CH_3(CH_2)_4(C{\equiv}CCH_2)_4CH_2CH_2COOCH_3$

(281)

$\downarrow$ T_2/Lindlar catalyst

$CH_3(CH_2)_4(CT{=}CTCH_2)_4CH_2CH_2COOCH_3$

$\downarrow$

T T T T COOH

T T T T

[^{3}H]-Arachidonic acid (280)

O H COOH

OH H H OH

PGE_2

OH H COOH

OH H H OH

$PGF_{2\alpha}$

O H COOH

OH H H OH

(287)

+

O H COOH

OH H O

(288)

2 COOH

Homo-γ-linolenic acid (282)

$\longrightarrow$

O H COOH

OH H H OH

PGE_1

COOH

5,8,11,14,17-Eicosapentaenoic acid (283)

$\longrightarrow$

O H COOH

OH H H OH

PGE_3

FIG. 79. Biosynthesis of prostaglandins from C_{20} fatty acids.

1965). Similarly, [2-^{14}C]-homo-γ-linolenic acid (8,11,14-eicosatrienoic acid (282)) was converted (30 per cent) into PGE_1 and the formation of PGE_3 from 5,8,11,14,17-eicosapentaenoic acid (283) (unlabelled) was also demonstrated (Fig. 79) (Bergström, Danielsson, Klenberg and Samuelsson, 1964; see also Bergström, 1966).

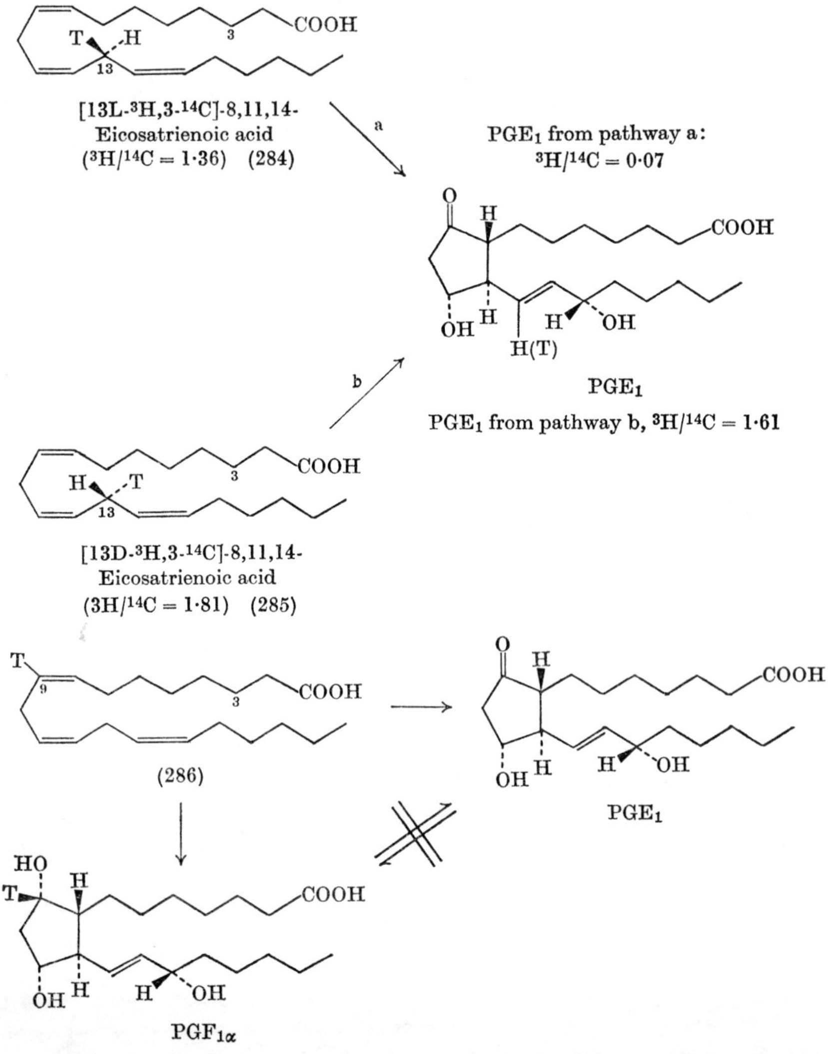

[13L-^{3}H,3-^{14}C]-8,11,14-Eicosatrienoic acid ($^3H/^{14}C = 1{\cdot}36$) (284)

PGE$_1$ from pathway a: $^3H/^{14}C = 0{\cdot}07$

PGE$_1$

PGE$_1$ from pathway b, $^3H/^{14}C = 1{\cdot}61$

[13D-^{3}H,3-^{14}C]-8,11,14-Eicosatrienoic acid ($3H/^{14}C = 1{\cdot}81$) (285)

(286)

PGE$_1$

PGF$_{1\alpha}$

FIG. 80. Experimental investigation of prostaglandin biosynthesis.

A common structural feature of the prostaglandins is the unsaturation at C(13,14) (as in PGE_1, Fig. 78). The stereospecific removal of one of the C(13) methylene protons of the acyclic precursor in the formation of this double bond, was demonstrated by incorporation studies with the specifically labelled acids (284) and (285). The 13L-acid (284), doubly labelled with tritium at C(13) and ^{14}C at C(3) as shown (Fig. 80), was converted into PGE_1 with a change in the $^3H/^{14}C$ ratio from 1·36 to 0·07, whereas with the similarly labelled 13D-isomer (285) the $^3H/^{14}C$ ratio only changed from 1·81 to 1·61 (Hamberg and Samuelsson, 1967).

The possibility that the PGF compounds might be derived *in vivo* from the corresponding PGE acid by reduction of the C(9) carbonyl group has been eliminated by further incorporation studies with doubly-labelled precursors. The [9-^{3}H,3-^{14}C]-labelled acid (286) was converted into PGE_1 (Fig. 80) with loss of tritium as expected. However, the tritium label was retained on conversion of the same acid (286) into $PGF_{1\alpha}$, which would not be possible if this were derived by reduction of PGE_1 (Hamberg and Samuelsson, 1967). It was also found that homogenates of guinea pig lung were able to convert arachidonic acid (280) into PGE_2 and $PGF_{2\alpha}$ (Fig. 79) but that $PGF_{2\alpha}$ added to the same system was not converted into PGE_2 (Anggard and Samuelsson, 1965). A precursor-product relationship in either sense between the PGE and the PGF series can therefore be ruled out.

Incubation of arachidonic acid in the guinea pig lung system gave rise to two further compounds, 11α,15-dihydroxy-9-ketoprost-5-enoic acid (287) and 11α-hydroxy-9,15-diketoprost-5-enoic acid (288). These were shown to be products of the further metabolism of PGE_2 in which reduction of the C(13,14) double bond, and oxidation of the C(15) hydroxyl function had taken place (Fig. 79).

Atmospheric oxygen has been shown to be the source of the oxygen at C(9), C(11) and C(15) in the prostaglandins (Ryhage and Samuelsson, 1965; Nugteren and van Dorp, 1965) and the all-*cis* stereochemistry of the acyclic precursor is evidently important in the cyclisation step. A mechanism for the cyclisation, involving an intermediate endoperoxide as shown in Fig. 81, has been suggested (Hamberg and Samuelsson, 1967). The demonstration that all of the oxygen substituents are derived from atmospheric oxygen, together with the evidence that the two ring oxygen atoms originate in the same molecule of oxygen (Samuelsson, 1965) would appear to rule out suggested

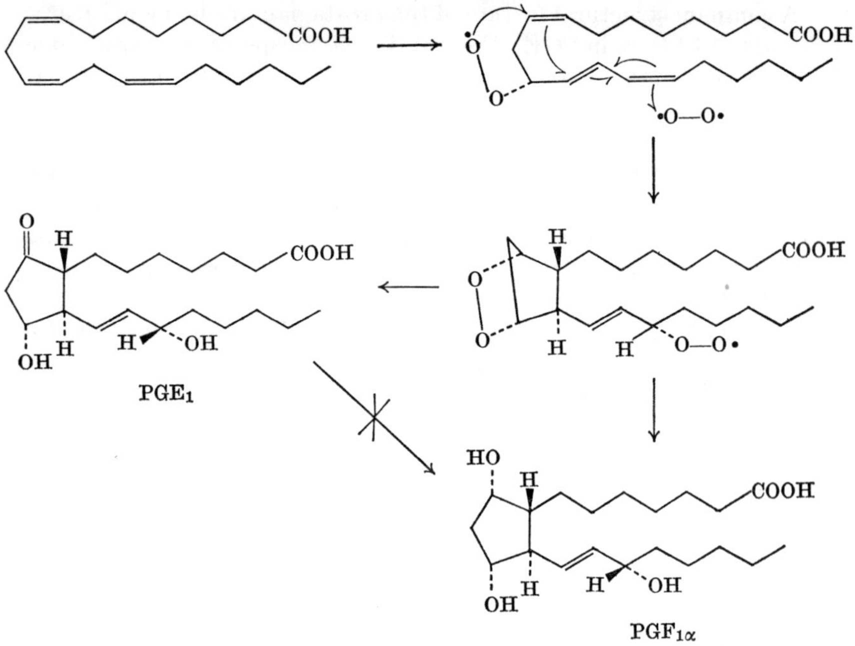

Fig. 81. Possible mechanism for the cyclisation step in prostaglandin biosynthesis.

alternative mechanisms involving bis-epoxide intermediates (Gunstone, 1966).

N. Cyclitols

The cyclitols are a widely distributed group of hydroxylated cyclohexane derivatives which occur in micro-organisms, higher plants, and animals (Posternak, 1965). Typical examples are *myo*-inositol (*meso*-inositol) (289), sequoyitol (290), D-inositol (291), L-leucanthemitol (292) and L(+)-quercitol (293) (Fig. 82).

From the point of view of biogenesis, the most extensively investigated cyclitol is *myo*-inositol (289), which has been shown to arise by direct cyclisation of glucose in parsley, *Sinapis alba*, yeast (*Candida utilis*) and rat testis (Anderson and Wolter, 1966; Kindl, Scholda and Hoffmann-Ostenhof, 1966).

Chen and Charalampous have shown that the intermediate involved in the cyclisation step in yeast is glucose 6-phosphate (294) (Chen and

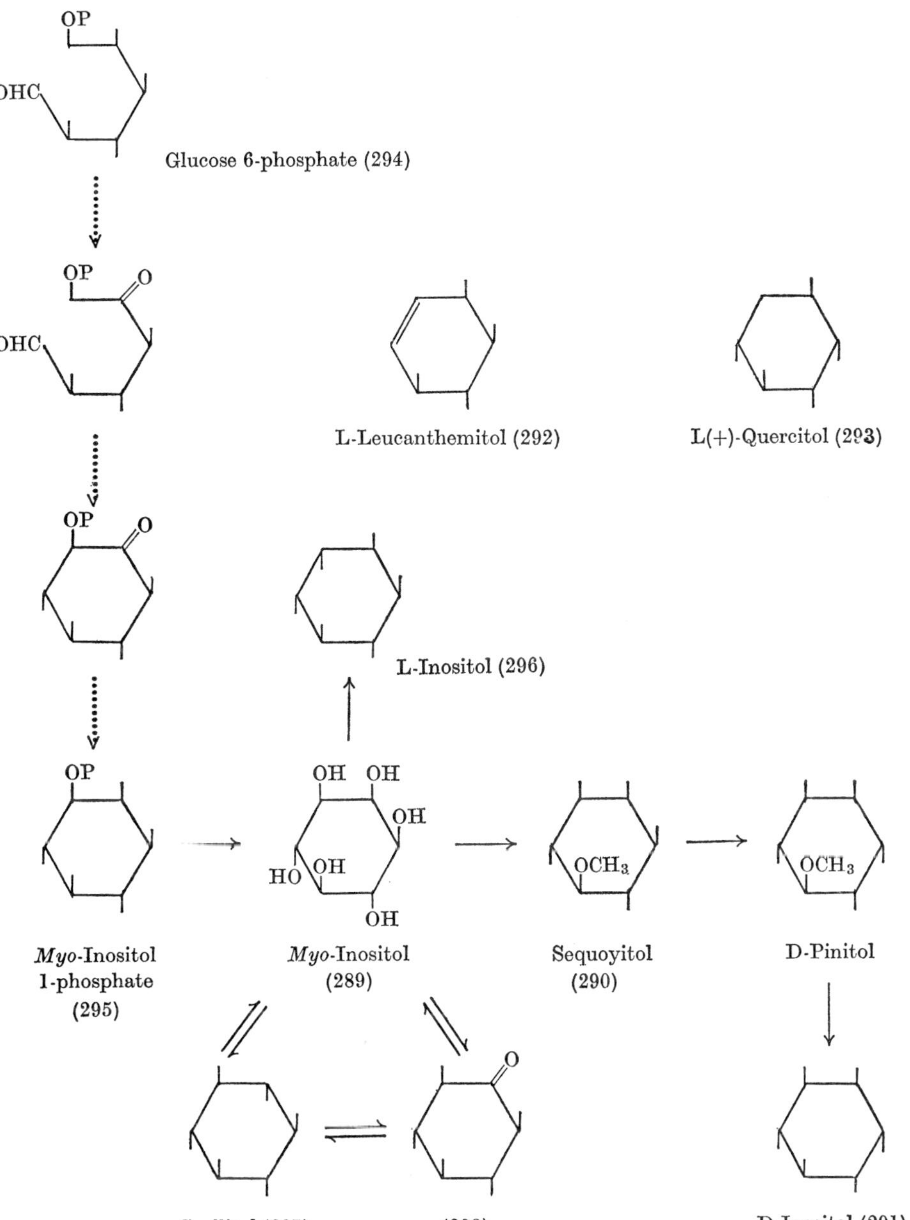

FIG. 82. The biosynthesis of *myo*-inositol and related cyclitols.

Charalampous, 1964a) and that specifically labelled glucose is incorporated without appreciable randomisation of the label (Chen and Charalampous, 1964b), a result which has also been confirmed for other systems (Kindl *et al.*, 1966). The primary product of the cyclisation is *myo*-inositol 1-phosphate (295) (Fig. 82) which has been isolated as the crystalline cyclohexylamine salt from the rat testis system (Eisenberg and Bolden, 1965). Experiments in which tritiated water was added to the incubation mixture have indicated that an aldol rather than an acyloin condensation is involved in the cyclisation step, and on the basis of the available evidence, the pathway of *myo*-inositol biosynthesis from glucose shown in Fig. 82 has been proposed (Kindl *et al.*, 1966).

The biogenetic relationships between the cyclitols of higher plants have been partially clarified. Hoffmann-Ostenhof and his collaborators have demonstrated the sequence from *myo*-inositol (289) to D-inositol (291) (Fig. 82), and have also shown that L-inositol (296) and scyllitol (297) are derived from the same precursor (Kindl *et al.*, 1966; Kindl, Scholda and Hoffmann-Ostenhof, 1967). The conversion of *myo*-inositol (289) into scyllitol (297) in animal tissue has also been reported (Posternak, Schopfer, Kaufmann-Boetsch and Edwards, 1963).

It can be seen from Fig. 82 that the configurational change in each of these conversions is restricted to epimerisation at one position only, which strongly suggests that the isomerisation is brought about by selective oxidation to a keto-compound followed by stereospecific reduction. In support of such a pathway it was shown that in *Calycanthus occidentalis* (Scholda, Billek and Hoffmann-Ostenhof, 1964) and also in animal tissue (Posternak *et al.*, 1963), the ^{14}C-labelled keto-inositol (298) was converted into both *myo*-inositol and its 2-epimer, scyllitol (Fig. 82).

Since the various cyclitols just described could be shown to be derived from *myo*-inositol, it appeared possible that the latter might serve as precursor of other members of the series. Kindl *et al.* showed that a hypothetical scheme could be set up in which the transformation of *myo*-inositol into various cyclitols could be readily explained in terms of sequential epimerisation, methylation and dehydration steps (Kindl *et al.*, 1967). Recent investigations, however, have shown that this simple scheme is not adequate to explain the observed results of incorporation experiments involving L-viburnitol (299) and L(+)-quercitol (293) (Kindl and Hoffmann-Ostenhof, 1967; Kindl *et al.*, 1967). A more direct pathway was suggested by the fact that [^{14}C]-D-glucose was incorporated into these to a significantly greater extent than various ^{14}C-labelled cyclitols in *Chrysanthemum leucanthemum* and in *Quercus robur* respectively. Degradation of L-viburnitol derived from [1-^{14}C]-D-glucose revealed a considerable degree of randomisation of the label (Fig. 83) (Kindl and Hoffmann-Ostenhof, 1967), which suggests that the biosynthetic pathway from D-glucose is complicated by fragmentation-recombination processes similar to those involved in shikimate biosynthesis.

CH_2 CHO
OH

[1-^{14}C]-Glucose

6 1 5 2 4 3

L-Viburnitol (299)

Percentage incorporation of activity into each position of L-Viburnitol from [1-^{14}C]-Glucose.

Position in L-Viburnitol	Percentage incorporation
C(1)	14
C(2)	10
C(3)	59
C(4)	6*
C(5)	4
C(6)	7

* Calculated by difference.

FIG. 83. The biosynthesis of L-viburnitol from glucose.

Myo-inositol appears to be important in higher plants as a source of the glucuronic acid and pentose residues utilised in the biosynthesis of cell-wall polysaccharides (Loewus, 1965; Roberts and Loewus, 1966; Roberts, Shah and Loewus, 1967). *Myo*-inositol is also converted into apiose (300) in higher plants *via* glucuronic acid (301) as shown in Fig. 84 (Roberts, Shah and Loewus, 1967, and references cited therein).

Although it might have been expected that the cyclitols could serve as precursors of aromatic compounds, the conversion of *myo*-inosital

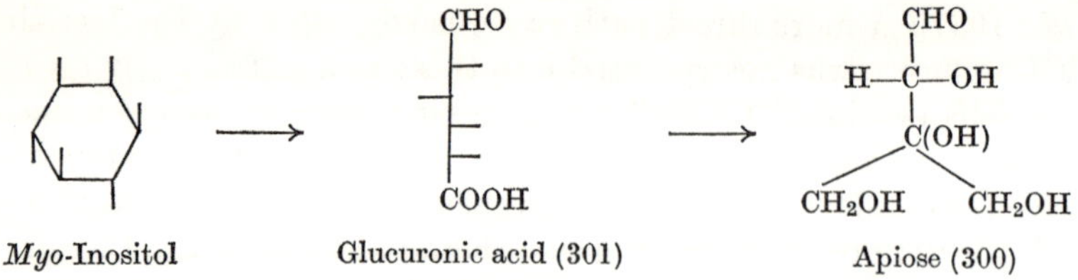

CHO
H—C—OH
C(OH)
CH_2OH CH_2OH

Myo-Inositol → Glucuronic acid (301) → Apiose (300)

FIG. 84. The biosynthesis of apiose from *myo*-inositol.

into tetrahydroxybenzoquinone (302) in *Pseudomonas Beijerinckii* is the only authenticated example of this type of transformation (Weygand, Brucker, Grisebach and Schulze, 1957).

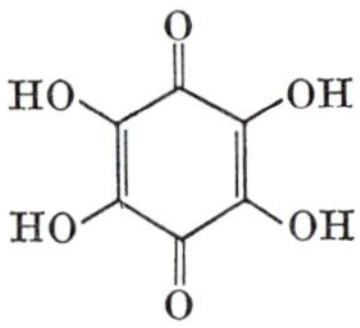

(302)

O. THE NONADRIDES

A small group of fungal metabolites, the 'nonadrides', are characterised by the presence of a nine-membered ring. The three known examples are glaucanic (303) and glauconic (304) acids from *Penicillium glaucum* and *P. purpurogenum*, and byssochlamic acid (305) from *Byssochlamys fulva*. The investigations which led up to the elucidation of these novel structures have been reviewed in detail by Sutherland (1967).

Inspection of the structures of glauconic (304) and byssochlamic (305) acids gave reason to believe that they could arise by different modes of dimerisation of a single C_9 precursor, as shown in Fig. 85. Sutherland and his collaborators, in a remarkably concise investigation (Bloomer, Moppett and Sutherland, 1965, 1968), have identified this precursor as the unsaturated anhydride (306) (Fig. 86). When this compound, labelled at C(7) with ^{14}C, was fed to *P. purpurogenum*, it was incorporated into glauconic acid (304) to the extraordinary extent of 51·5 per cent, which leaves little doubt that it is a true intermediate. Almost all of the activity was located at C(7) and C(16) in (304) as predicted by the postulated dimerisation pathway (Fig. 85).

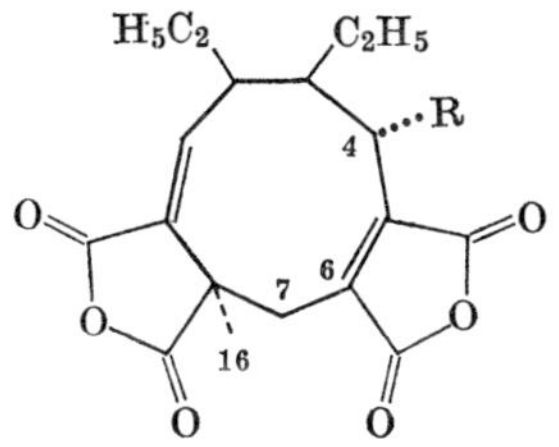

R = H, Glaucanic acid (303)
R = OH, Glauconic acid (304)

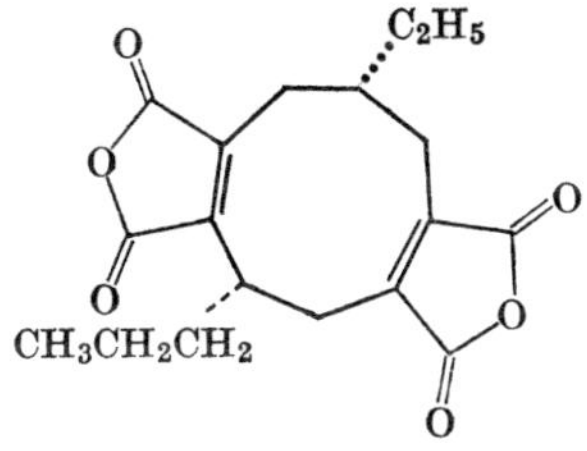

Byssochlamic acid (305)

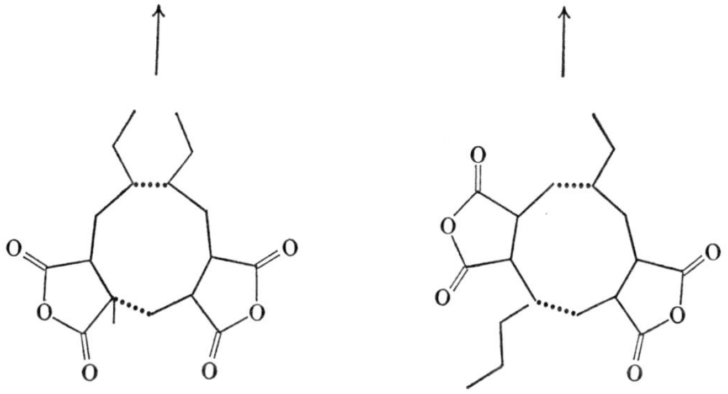

FIG. 85. Biosynthesis of the nonadrides by dimerisation of a C_9 precursor.

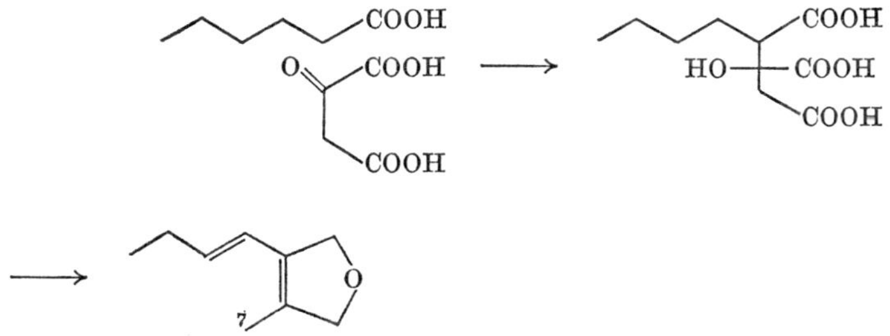

(306)

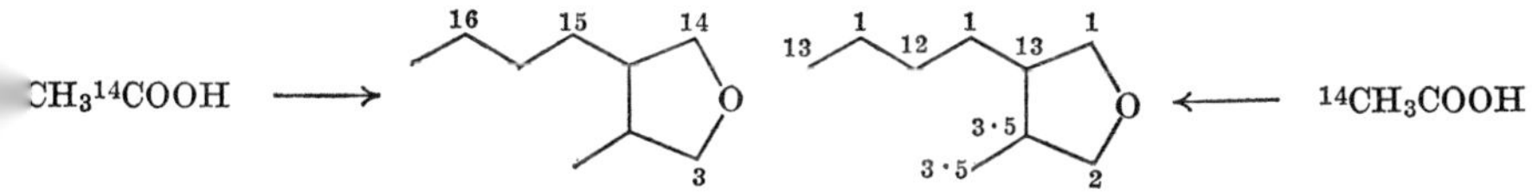

$CH_3{}^{14}COOH$ → (307) (308) ← $^{14}CH_3COOH$

FIG. 86. Biosynthesis of the C_9 precursor of the nonadrides.

The C_9 precursor probably arises by condensation of hexanoate with oxaloacetate (or some other four-carbon acid of the citric acid cycle) followed by decarboxylation, dehydration and dehydrogenation as shown in Fig. 86. The pattern of incorporation of labelled acetate into glauconic acid, expressed in terms of the labelling pattern of the C_9 unit ((307) and (308)), strongly supports this pathway. The six-carbon chain can be seen to be labelled alternately as would be expected if it arises by the normal pathway of fatty acid biosynthesis. The labelling of the remaining three-carbon component reveals some randomisation which is consistent with the proposal that it arises from a four-carbon intermediate of the citric acid cycle. In theory, introduction of

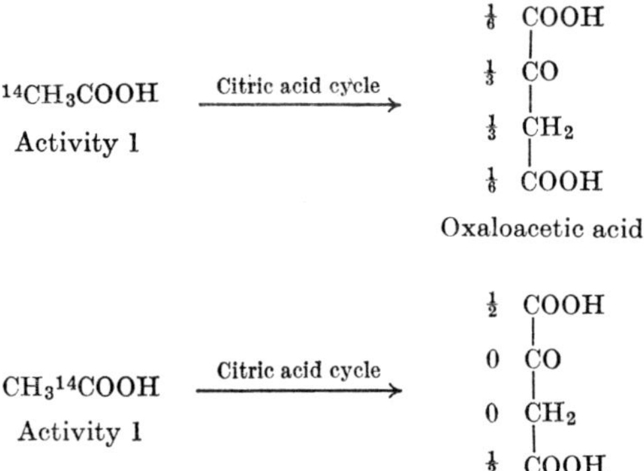

FIG. 87. Distribution of activity in oxaloacetic acid derived from labelled acetate via the citric acid cycle.

labelled acetate into the citric acid cycle should give, after complete equilibration, oxaloacetate labelled as shown in Fig. 87. This theoretical labelling pattern can be seen to resemble closely the observed labelling pattern in the three carbon unit of (307) and (308), assuming that a terminal position of the four-carbon intermediate has been lost. Feeding experiments with [2,3-^{14}C]-succinate (which is converted into oxaloacetate in the citric acid cycle), supported this view. Fifty-five per cent of the total activity of the glauconic acid (304) (7·8 per cent incorporation) was found at C(6) and C(7) as expected.

It has been shown that glaucanic acid (303) is the precursor of glauconic acid (304) in the mould (Sutherland, 1967) and this information throws an interesting light on the stereospecificity of the reactions affecting C(4) in the biosynthetic pathway. The sequence anhydride

(306) → glaucanic acid (303) → glauconic acid (304) involves addition of hydrogen at C(4), followed by removal of one hydrogen atom and substitution by —OH (Fig. 88). Since the tritiated anhydride (307)

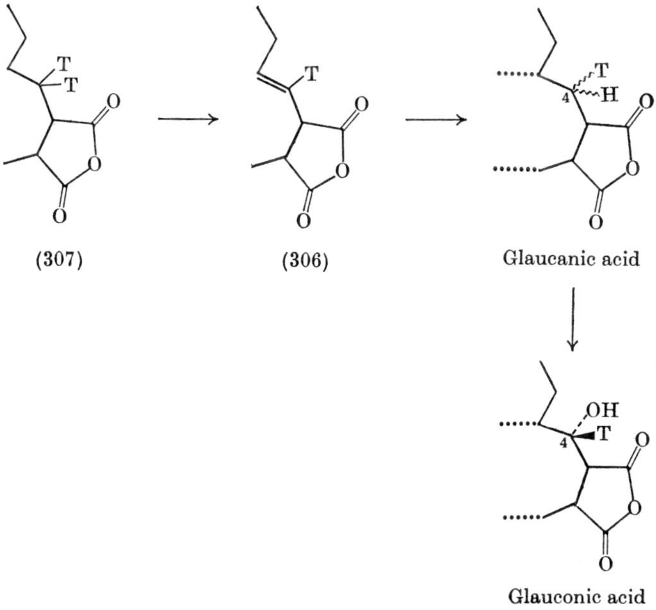

(307)

(306)

Glaucanic acid

Glauconic acid

FIG. 88. The stereochemical implications of the changes at C(4) during glauconic acid biosynthesis.

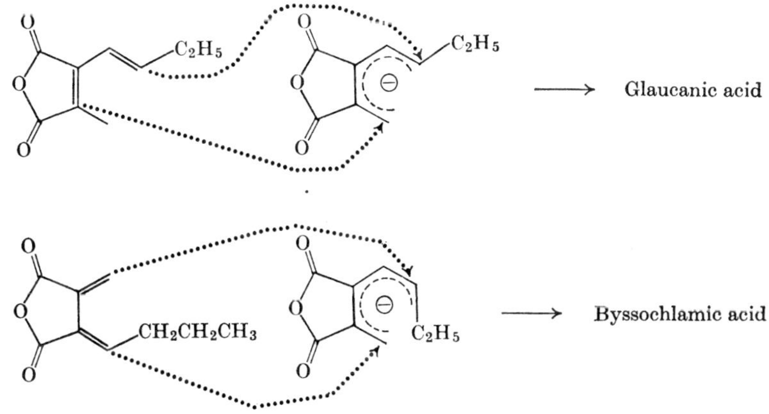

C_2H_5 C_2H_5 → Glaucanic acid

$CH_2CH_2CH_3$ C_2H_5 → Byssochlamic acid

FIG. 89. Possible mechanism of the cyclisation step in nonadride biosynthesis.

was incorporated equally into the two halves of glauconic acid, presumably via the tritiated unsaturated anhydride (306), it follows that as shown in Fig. 88, the hydroxylation step must be stereospecific, the incoming —OH group replacing the hydrogen atom that was added in the prior reduction step and *not* the proton which was originally present at this position in the unsaturated anhydride (306).

Sutherland has suggested that the dimerisation reactions consist of electrocyclic additions between the isomeric forms of the anhydride (306) and the anionic form of the anhydride as shown in Fig. 89 (Sutherland, 1967).

P. Conclusion

The previous discussions will have illustrated the points made in section G concerning the economy of methods by which cyclic compounds are formed in nature. The major pathways provide not one, but many types of cyclic intermediate which are used in the biosynthesis of a wide range of carbocyclic compounds. This is particularly well exemplified by the diversity of natural products, ranging from the aromatic amino acids to the anthraquinones, derived from intermediates of the shikimic acid pathway. The shikimic acid pathway operates in both micro-organisms and higher plants, but not in mammals, although the latter have a fundamental nutritional requirement for the amino acids produced by this pathway. The evolutionary significance of these facts is that a common ancestry for all forms of life dependent on the aromatic amino acids is implied although the ability of mammals to synthesise the amino acids *de novo* has been lost during the course of evolution.

The discussion will also have supported the contention that one of the most fruitful ways of approaching biosynthetic problems is by the application of principles derived from the study of organic reaction mechanisms. For example, reactions related to the aldol condensation such as the Claisen ester, Dieckmann and Michael condensations etc. have been widely used in laboratory syntheses of carbocyclic compounds and for all of these corresponding examples can be found in biosynthetic pathways. As a result it is no longer permissible to propose hypothetical biosynthetic pathways which cannot be rationalised in terms of known reactions or acceptable reaction mechanisms, although this does not mean that all established biological reactions will yield to interpretation in the present state of our knowledge.

Many aspects of the biosynthesis of secondary metabolites have barely been touched upon experimentally. Ways in which metal ions can influence the reactivity of model compounds, for example, have been little investigated in spite of the importance of metal ions as cofactors in many enzymatic reactions. That metal ions can drastically affect the reactivity of organic molecules is, of course, well known. Amino acids, to take just one example, can be induced to enter into aldol-type condensations under very mild conditions if their α-hydrogen atoms are activated by formation of a copper complex (Sato, Okawa and Akabori, 1957; Akabori, Otani, Marshall, Winitz and Greenstein, 1959; Benoiton, Winitz, Colman, Birnbaum and Greenstein, 1959). This type of condensation has been demonstrated in the reversible formation of α-hydroxymethylserine (308) from alanine and formaldehyde (Fig. 90) by an enzyme system from a *Pseudomonas* species (Wilson and Snell, 1962). By analogy with the

$$CH_3CH(NH_2)COOH + HCHO \rightleftharpoons HOCH_2C(CH_3)(NH_2)COOH \quad (308)$$

FIG. 90. The formation of α-hydroxymethylserine from alanine and formaldehyde by *Pseudomonas*.

laboratory reaction, activation of the amino acid by chelation with a metal ion in this equilibrium is very probable. A related carbon–carbon bond forming reaction may well be involved in the cyclisation of lysine to 1 amino 2 nitrocyclopentane carboxylic acid (271) (section M.1).

As more becomes known about the enzymology of the biosynthetic pathways leading to secondary metabolites, an approach to the synthesis of enzyme models will become more feasible and this is an area of research which will undoubtedly continue to expand.

Problems in biosynthesis differ from many chemical problems in that they can be attacked by widely different techniques, ranging from enzymology to radiotracer methodology. The considerable progress which can be made by the joint application of these techniques is vividly illustrated by the recent investigations in the field of cholesterol biosynthesis. The more intractable problems in biosynthesis will no doubt yield to combined operations of this type rather than to solo efforts in which a single approach is adopted.

Acknowledgments

I should like to thank Professor C. H. Hassall, Dr. G. Read and Dr. J. G. Woolley for kindly allowing me to quote results before publication. I should also like to record my thanks to Dr. W. Carruthers and Dr. G. Read for many helpful discussions and for pointing out a number of errors of omission and commission during the preparation of this chapter.

References

Achenbach, H. and Grisebach, H. (1965) *Z. Naturforsch.* **20(B)**, 137.
Acker, T. E., Brenneisen, P. E. and Tanenbaum, S. W. (1966) *J. Amer. Chem. Soc.* **88**, 834.
Adler, E. (1957) *Tech. Pap. Addr. tech. Ass. Pulp Pap. Ind.* **40**, 294.
Ahluwalia, V. K. and Seshadri, T. R. (1957) *J. Chem. Soc.* 970.
Akabori, S., Otani, T. T., Marshall, R., Winitz, M. and Greenstein, J. P. (1959) *Archs. Biochem. Biophys.* **83**, 1.
Akhtar, M., Hunt, P. F. and Parvez, M. A. (1966) *Chem. Commun.* 565.
Allport, D. C. and Bu'Lock, J. D. (1960) *J. Chem. Soc.* 654.
Anderson, L. and Wolter, K. E. (1966) *Ann. Rev. Plant Physiol.* **17**, 209.
Andrew, I. G. and Segal, W. (1964) *J. Chem. Soc.* 607.
Anggard, E. and Samuelsson, B. (1965) *J. Biol. Chem.* **240**, 3518.
Auda, H., Juneja, H. R., Eisenbraun, E. J., Waller, G. R., Kays, W. R. and Appel, H. H. (1967) *J. Amer. Chem. Soc.* **89**, 2476.
Austin, D. J. and Meyers, M. B. (1966) *Chem. Commun.* 125.
Baker, R. B. and Reid, E. E. (1929) *J. Amer. Chem. Soc.* **51**, 1567.
Ballantine, J. A., Hassall, C. H., Jones, B. D. and Jones, G. (1967) *Phytochem.* **6**, 1157.
Banthorpe, D. V. and Turnbull, K. W. (1966) *Chem. Commun.* 177.
Bartels-Keith, J., Johnson, A. W. and Taylor, W. I. (1951) *J. Chem. Soc.* 2354.
Barton, D. H. R. (1964) *Pure Appl. Chem.* **9**, 35.
Barton, D. H. R., DeFlorin, A. M. and Edwards, O. E. (1956) *J. Chem. Soc.* 530.
Barton, D. H. R. and Scott, A. I. (1958) *J. Chem. Soc.* 1767.
Battersby, A. R. (1967) *Oxidative Coupling of Phenols*, edited W. I. Taylor and A. R. Battersby. E. Arnold, London, p. 119.
Barz, W., Patschke, L. and Grisebach, H. (1965) *Chem. Commun.* 400.
Beck, W. S. and Ochoa, S. (1958) *J. Biol. Chem.* **232**, 931.
Beckwith, J. R., Clark, R. and Hager, L. P. (1963) *J. Biol. Chem.* **238**, 3086.
Beckwith, J. R. and Hager, L. P. (1963) *J. Biol. Chem.* **238**, 3091.
Benoiton, L., Winitz, M., Colman, R. F., Birnbaum, S. M. and Greenstein, J. P. (1959) *J. Amer. Chem. Soc.* **81**, 1726.
Bentley, R. (1958) *Biochim. biophys. Acta* **29**, 666.
Bentley, R. (1962) *Ann. Rev. Biochem.* **31**, 589.
Bentley, R. (1963a) *J. Biol. Chem.* **238**, 1889.
Bentley, R. (1963b) *J. Biol. Chem.* **238**, 1895.
Bentley, R. and Lavate, W. V. (1965) *J. Biol. Chem.* **240**, 532.

Bentley, R. and Lavate, W. V. (1966) *Antibiot., Advan. Res. Prod. Clin. Use*, edited M. Herold and Z. Gabriel, Butterworths, London, p. 532.
Bentley, R. and Keil, J. G. (1962) *J. Biol. Chem.* **237**, 867.
Bentley, R. and Zwitkowits, P. M. (1967a) *J. Amer. Chem. Soc.* **89**, 676.
Bentley, R. and Zwitkowits, P. M. (1967b) *J. Amer. Chem. Soc.* **89**, 681.
Bergström, S. (1966) *Recent Prog. Horm. Res.* **22**, 153.
Bergström, S., Danielsson, H., Klenberg, D. and Samuelsson, B. (1964) *J. Biol. Chem.* **239**, PC 4006.
Bergström, S. and Samuelsson, B. (1965) *Ann. Rev. Biochem.* **34**, 101.
Billek, G. and Kindl, H. (1961) *Monatsh.* **92**, 493.
Billek, G. and Kindl, H. (1962) *Monatsh.* **93**, 814.
Billek, G. and Schimpl, A. (1966) *Biosynthesis of Aromatic Compounds*, edited G. Billek, Pergamon Press, Oxford, p. 37.
Birch, A. J. (1957) *Fortschr. Chem. org. Naturstoffe* **14**, 186.
Birch, A. J. (1966) *Biosynthesis of Aromatic Compounds*, edited G. Billek, Pergamon Press, Oxford, p. 3.
Birch, A. J., Cassera, A., Fitton, P., Holker, J. S. E., Smith, H., Thompson, G. A. and Whalley, W. B. (1962) *J. Chem. Soc.* 3583.
Birch, A. J., Cassera, A. and Jones, A. R. (1965) *Chem. Commun.* 167.
Birch, A. J., Cassera, A. and Rickards, R. W. (1961) *Chem. and Ind.* 792.
Birch, A. J. and Donovan, F. W. (1953) *Austral. J. Chem.* **6**, 360.
Birch, A. J., English, R. J., Massy-Westropp, R. A., Slaytor, M. and Smith, H. (1958) *J. Chem. Soc.* 365.
Birch, A. J., Fitton, P., Pride, E., Ryan, A. J., Smith, H. and Whalley, W. B. (1958) *J. Chem. Soc.* 4576.
Birch, A. J., Fryer, R. I. and Smith, H. (1958) *Proc. Chem. Soc.* 343.
Birch, A. J., Massy-Westropp, R. A. and Moye, C. J. (1955a) *Chem. and Ind.* 683.
Birch, A. J., Massy-Westropp, R. A. and Moye, C. J. (1955b) *Austral. J. Chem.* **8**, 539.
Birch, A. J., Massy-Westropp, R. A., Rickards, R. W. and Smith, H. (1958) *J. Chem. Soc.* 360.
Birch, A. J., Musgrave, O. C., Rickards, R. W. and Smith, H. (1959) *J. Chem. Soc.* 3146.
Birch, A. J., Pride, E., Rickards, R. W., Thompson, P. J., Dutcher, J. D., Perlman, D. and Djerassi, C. (1960) *Chem. and Ind.* 1245.
Birch, A. J., Willis, J. L., Hellyer, R. O. and Salahud-Din, M. (1966) *J. Chem. Soc. (C)* 1337.
Birkinshaw, J. H., Raistrick, H., Ross, D. J. and Stickings, C. E. (1952) *Biochem. J.* **50**, 610.
Bloomer, J. L., Moppett, C. E. and Sutherland, J. K. (1965) *Chem. Commun.* 619.
Bloomer, J. L., Moppett, C. E. and Sutherland, J. K. (1968) *J. Chem. Soc. (C)* 588.
Bohlmann, F. (1967) *Fortschr. Chem. org. Naturstoffe* **25**, 1.
Bohlmann, F., Bonnet, H. and Jente, R. (1968) *Chem. Ber.* **101**, 855.
Bohlmann, F. and Jente, R. (1966) *Chem. Ber.* **99**, 995.
Bohlmann, F., Jente, R., Lukas, W., Laser, J. and Schulz, H. (1967) *Chem. Ber.* **100**, 3183.

Bohm, B. A. (1965) *Chem. Rev.* **65**, 435.
Booth, A. N., Murray, C. W., DeEds, F. and Jones, F. T. (1955) *Fedn. Proc. Fedn. Am. Socs. exp. Biol.* **14**, 321.
Boothe, J. H., Wilkinson, R. G., Kushner, S. and Williams, J. H. (1953) *J. Amer. Chem. Soc.* **75**, 1732.
Brown, S. A. (1963) *Phytochem.* **2**, 137.
Brown, S. A. (1964) *Biochemistry of Phenolic Compounds*, edited J. B. Harborne, Academic Press, London and New York, p. 361.
Brown, S. A. (1966) *Biosynthesis of Aromatic Compounds*, edited G. Billek, Pergamon Press, Oxford, p. 15.
Brown, S. A., Towers, G. H. N. and Chen, D. (1964) *Phytochem.* **3**, 469.
Bu'Lock, J. D. (1966) *Comparative Phytochemistry*, edited T. Swain, Academic Press, London and New York, p. 79.
Bu'Lock, J. D. and Smalley, H. M. (1961) *Proc. Chem. Soc.* 209.
Bu'Lock, J. D., Smalley, H. M. and Smith, G. N. (1962) *J. Biol. Chem.* **237**, 1778.
Bu'Lock, J. D. and Smith, G. D. (1964) *Biochem. Biophys. Res. Commun.* **17**, 433.
Bu'Lock, J. D. and Smith, G. N. (1967) *J. Chem. Soc.* (*C*) 332.
Bunton, C. A., Kenner, G. W., Robinson, M. J. T. and Webster, B. R. (1963) *Tetrahedron* **19**, 1001.
Burnett, A. R. and Thomson, R. H. (1967a) *Chem. Commun.* 1125.
Burnett, A. R. and Thomson, R. H. (1967b) *J. Chem. Soc.* (*C*) 2100.
Burrows, B. F., Mills, S. D. and Turner, W. B. (1965) *Chem. Commun.* 75.
Challenger, F., Bywood, R., Thomas, P. and Hayward, B. J. (1957) *Archs. Biochem. Biophys.* **69**, 514.
Chambers, K., Kenner, G. W., Robinson, M. J. T. and Webster, B. R. (1960) *Proc. Chem. Soc.* 291.
Chandra, P., Read, G. and Vining, L. C. (1965) *Can. J. Biochem. Physiol.* **44**, 403.
Chen, I. W. and Charalampous, F. C. (1964a) *J. Biol. Chem.* **239**, 1905.
Chen, I. W. and Charalampous, F. C. (1964b) *Biochem. Biophys. Res. Commun.* **17**, 521.
Christie, W. W. (1969) *Topics in Lipid Chemistry*, edited by F. D. Gunstone, Vol. 1. Logos Press, London.
Chung, A. E. and Law, J. H. (1964) *Biochemistry, N.Y.* **3**, 967.
Clayton, R. B. (1965) *Quart. Rev. Chem. Soc.* **19**, 168, 201.
Collie, J. N. (1907) *J. Chem. Soc.* 1806.
Comer, F. W., Money, T. and Scott, A. I. (1967) *Chem. Commun.* 231.
Conn, E. E. (1964) *Biochemistry of Phenolic Compounds*, edited J. B. Harborne, Academic Press, London and New York, p. 399.
Conn, E. E. and Swain, T. (1961) *Chem. and Ind.* 592.
Coon, M. J., Robinson, W. G. and Bachhawat, B. K. (1956) *Amino Acid Metabolism*, edited W. D. McElroy and B. Glass, Johns Hopkins Press, Baltimore, p. 431.
Corey, E. J. and Chakovsky, M. (1962) *J. Amer. Chem. Soc.* **84**, 867.
Cox, G. B. and Gibson, F. (1964) *Biochim. Biophys. Acta* **93**, 204.
Cox, G. B. and Gibson, F. (1966) *Biochem. J.* **100**, 1.

Crombie, L. and Thomas, M. B. (1965) *Chem. Commun.* 155.
Crombie, L., Games, D. and Knight, M. H. (1966) *Chem. Commun.* 355.
Crombie, L., Games, D. and McCormick, A. (1967) *J. Chem. Soc.* (*C*) 2545.
Cruickshank, I. A. M. and Perrin, D. R. (1961) *Aust. J. Biol. Sci.* **14**, 336.
Curtis, R. F., Hassall, C. H., Jones, D. W. and Williams, T. W. (1960) *J. Chem. Soc.* 4838.
Curtis, R. F., Harries, P. C., Hassall, C. H., Levi, J. D. and Phillips, D. M. (1966) *J. Chem. Soc.* (*C*) 168.
Curtis, R. F., Hassall, C. H. and Pike, R. K. (1968) *J. Chem. Soc.* (*C*) 1807.
Dacre, J. C. and Williams, R. T. (1962) *Biochem. J.* **84**, 81P.
Dagley, S., Evans, W. C. and Ribbons, D. W. (1960) *Nature* **188**, 560.
Daly, J. W. and Witkop, B. (1963) *Angew. Chem. Intern. Ed.* **2**, 421.
Davidson, T. A. and Scott, A. I. (1960) *Proc. Chem. Soc.* 390.
Davis, B. D. (1951) *J. Biol. Chem.* **191**, 315.
Davis, B. D. (1955) *Adv. Enzymol.* **16**, 247.
Day, A. C., Nabney, J. and Scott, A. I. (1961) *J. Chem. Soc.* 4067.
Del Campillo-Campbell, A., Dekker, E. E. and Coon, M. J. (1959) *Biochim. Biophys. Acta* **31**, 290.
Edwards, J. D. and Cashaw, J. L. (1957) *J. Amer. Chem. Soc.* **79**, 2283.
Edwards, K. G. and Stoker, J. R. (1967) *Phytochem.* **6**, 655.
Eisenberg, F. and Bolden, A. H. (1965) *Biochem. Biophys. Res. Commun.* **21**, 100.
El-Basyouni, S. Z., Chen, D., Ibrahim, R. K., Neish, A. C. and Towers, G. H. N. (1964) *Phytochem.* **3**, 485.
Evans, W. C. (1963) *J. Gen. Microbiol.* **32**, 177.
Eykman, J. F. (1885) *Rec. Trav. chim.* **4**, 32.
Eyton, W. B., Ollis, W. D., Fineberg, M., Gottlieb, O. R., Salignac de Souza Guimarães, I. and Taveira Magalhães, M. (1965) *Tetrahedron* **21**, 2697.
Eyton, W. B., Ollis, W. D., Sutherland, I. O., Gottlieb, O. R., Taveira Magalhães, M. and Jackman, L. M. (1965) *Tetrahedron* **21**, 2683.
Fairbrother, J. R. F., Jones, E. R. H. and Thaller, V. (1967) *J. Chem. Soc.* (*C*) 1035.
Floss, H-G. and Mothes, U. (1966) *Phytochem.* **5**, 161.
Franck, B., Hüper, F., Gröger, D. and Erge, D. (1966) *Angew. Chem. Intern. Ed.* **5**, 728.
Frantz, I. D. and Schroepfer, G. J. (1967) *Ann. Rev. Biochem.* **36**, 691.
Freudenberg, K. (1965) *Science, N.Y.* **148**, 595.
Friis, P., Daves, G. D. and Folkers, K. (1966) *J. Amer. Chem. Soc.* **88**, 4754.
Gaoni, Y. and Mechoulam, R. (1964) *Proc. Chem. Soc.* 82.
Gastambide-Odier, M., Delaumeny, J. M. and Küntzel, H. (1966) *Biochim. Biophys. Acta* **125**, 33.
Gatenbeck, S. (1958) *Acta Chem. Scand.* **12**, 1985.
Gatenbeck, S. (1960a) *Acta Chem. Scand.* **14**, 102.
Gatenbeck, S. (1960b) *Acta Chem. Scand.* **14**, 296.
Gatenbeck, S. (1960c) *Svensk. kem. Tidskr.* **72**, 188.
Gatenbeck, S. (1962) *Biochem. Biophys. Res. Commun.* **6**, 422.
Gatenbeck, S. and Bentley, R. (1965) *Biochem. J.* **94**, 478.
Gatenbeck, S. and Hermodsson, S. (1965) *Acta Chem. Scand.* **19**, 65.

Gatenbeck, S. and Mosbach, K. (1959) *Acta Chem. Scand.* **13**, 1561.
Gatenbeck, S. and Mosbach, K. (1963) *Biochem. Biophys. Res. Commun.* **11**, 166.
Geissman, T. A. (editor) (1962) *The Chemistry of Flavonoid Compounds*, Pergamon Press, Oxford.
Gibson, F. (1964) *Biochem. J.* **90**, 256.
Gibson, M. I. and Gibson, F. (1964) *Biochem. J.* **90**, 248.
Goodwin, T. W. (1966) *Comparative Phytochemistry*, edited T. Swain, Academic Press, London and New York, p. 121.
Gordon, P. G., Penttila, A. and Fales, H. M. (1968) *J. Amer. Chem. Soc.* **90**, 1376.
Gottlieb, O. R., Fineberg, M., Salignac de Souza Guimarães, I., Taveira Magalhães, M., Ollis, W. D. and Eyton, W. B. (1964) *Anais. Acad. bras. Cienc.* **36**, 33.
Grisebach, H. (1965a) *Chemistry and Biochemistry of Plant Pigments*, edited T. W. Goodwin, Academic Press, London and New York, p. 279.
Grisebach, H. (1965b) *Biosynthetic Pathways in Higher Plants*, edited J. B. Pridham and T. Swain, Academic Press, London and New York, p. 159.
Grisebach, H. and Barz, W. (1964) *Z. Naturforsch.* **19B**, 569.
Grisebach, H., Barz, W., Hahlbrock, K., Kellner, S. and Patschke, L. (1966) *Biosynthesis of Aromatic Compounds*, edited G. Billek, Pergamon Press, Oxford, p. 25.
Grisebach, H. and Hofheinz, W. (1964) *J. Roy. Inst. Chem.* 332.
Grisebach, H. and Kellner, S. (1965) *Z. Naturforsch.* **20B**, 446.
Grisebach, H. and Vollmer, K. (1964) *Z. Naturforsch.* **19B**, 781.
Gröger, D., Erge, D. and Floss, H-G. (1965) *Z. Naturforsch.* **20B**, 856.
Gross, S. R. (1958) *J. Biol. Chem.* **233**, 1146.
Grove, J. F. (1961) *Quart. Rev. Chem. Soc.* **15**, 56.
Gunstone, F. D. (1966) *Chem. and Ind.* 1551.
Hadfield, J. R., Holker, J. S. E. and Stanway, D. N. (1967) *J. Chem. Soc.* (*C*) 752.
Hamberg, M. and Samuelsson, B. (1967) *J. Biol. Chem.* **242**, 5336.
Harkin, J. M. (1967) *Oxidative Coupling of Phenols*, edited W. I. Taylor and A. R. Battersby, E. Arnold, London, p. 243.
Harris, T. M. and Carney, R. L. (1966) *J. Amer. Chem. Soc.* **88**, 2053.
Haslam, E., Haworth, R. D. and Knowles, P. F. (1961) *J. Chem. Soc.* 1854.
Hassall, C. H. (1957) *Org. Reactions* **9**, 73.
Hassall, C. H. (1965) *Biogenesis of Antibiotic Substances*, Academic Press, London and New York, p. 51.
Hassall, C. H. and Scott, A. I. (1961) *Recent Developments in the Chemistry of Natural Phenolic Compounds*, edited W. D. Ollis, Pergamon Press, Oxford, p. 119.
Hassall, C. H. and Lewis, J. R. (1961) *J. Chem. Soc.* 2312.
Hellyer, R. O. and Pinhey, J. T. (1966) *J. Chem. Soc.* (*C*) 1496.
Holker, J. S. E., Staunton, J. and Whalley, W. B. (1964) *J. Chem. Soc.* 16.
Hooper, N. K. and Law, J. H. (1965) *Biochem. Biophys. Res. Commun.* **18**, 426.
Huxley, T. H. (1870) *Nature* **2**, 401.
Ibrahim, R. K. and Towers, G. H. N. (1960) *Canad. J. Biochem. Physiol.* **38**, 627.

Ibrahim, R. K. and Towers, G. H. N. (1962) *Canad. J. Biochem. Physiol.* **40**, 449.

Ingraham, L. L. (1962) *Biochemical Mechanisms*, Wiley, New York and London, p. 100.

Isherwood, F. A. (1965) *Biosynthetic Pathways in Higher Plants*, edited J. P. Pridham, Academic Press, London and New York, p. 133.

Jaureguiberry, G., Lenfant, M., Das, B. C. and Lederer, E. (1966) *Tetrahedron, Suppl.* **8**, Pt. 1, 27.

Kalan, E. B., Davis, B. D., Srinivasan, P. R. and Sprinson, D. B. (1956) *J. Biol. Chem.* **223**, 907.

Kaneda, T. (1966) *Biochim. Biophys. Acta* **125**, 43.

Karrer, W. (1958) *Constitution und Vorkommen der organischen Pflanzenstoffe*, Birkhäuser Verlag, Basel and Stuttgart.

Kindl, H. and Hoffmann-Ostenhof, O. (1967) *Phytochem.* **6**, 77.

Kindl, H., Scholda, R. and Hoffmann-Ostenhof, O. (1966) *Angew. Chem. Intern. Ed.* **5**, 165.

Kindl, H., Scholda, R. and Hoffmann-Ostenhof, O. (1967) *Phytochem.* **6**, 237.

King, F. E., King, T. J. and Warwick, A. J. (1952) *J. Chem. Soc.* 1920.

Komatsu, E. (1957) *Nippon Nôgei-Kagaku Kaishi* **31**, 905 (*Chem. Abs.* **52**, 16473h).

Kunesch, G. and Polonsky, J. (1967) *Chem. Commun.* 317.

Lane, J. F., Koch, W. T., Leeds, N. S. and Gorin, G. (1952) *J. Amer. Chem. Soc.* **74**, 3211.

Leistner, E. and Zenk, M. H. (1967a) *Tetrahedron Letters* 475.

Leistner, E. and Zenk, M. H. (1967b) *Z. Naturforsch.* **22B**, 865.

Leistner, E. and Zenk, M. H. (1968a) *Tetrahedron Letters* 861.

Leistner, E. and Zenk, M. H. (1968b) *Tetrahedron Letters* 1395.

Lennarz, W. J. (1961) *Biochem. Biophys. Res. Commun.* **6**, 112.

Levin, J. G. and Sprinson, D. B. (1964) *J. Biol. Chem.* **239**, 1142.

Liu, T. Y. and Hofmann, K. (1962) *Biochemistry, N.Y.* **1**, 189.

Loewus, F. (1965) *Fedn. Proc. Fedn. Am. Socs. exp. Biol.* **24**, 855.

Lynen, F. (1967a) *Pure Appl. Chem.* **14**, 137.

Lynen, F. (1967b) *Biochem. J.* **102**, 381.

Lynen, F. and Tada, M. (1961) *Angew. Chem.* **73**, 513.

Martius, C. and Esser, H. O. (1959) *Biochem. Z.* **331**, 1.

McCalla, D. R. and Neish, A. C. (1959) *Canad. J. Biochem. Physiol.* **37**, 537.

McCormick, J. R. D. (1965) *Biogenesis of Antibiotic Substances*, Academic Press, New York and London, p. 73.

McCormick, J. R. D. and Jensen, E. R. (1965) *J. Amer. Chem. Soc.* **87**, 1794.

Mechoulam, R. and Gaoni, Y. (1965) *Tetrahedron Letters* 1223.

Meinwald, J., Koch, K. F., Rogers, J. E. and Eisner, T. (1966) *J. Amer. Chem. Soc.* **88**, 1590.

Mentzner, C. (1966) *Actualités de Phytochemie Fondamentale* (2), Masson et Cie, Paris.

Mentzner, C. and Fatianoff, O. (1964) *Actualités de Phytochemie Fondamentale* (1), Masson et Cie, Paris.

Miller, J. E. (1965) *Biochem. Biophys. Res. Commun.* **19**, 335.

Miller, P. A., McCormick, J. R. D. and Doerschuk, A. P. (1956) *Science, N.Y.* **123**, 1030.
Miller, J. A. and Wood, H. C. S. (1965) *Chem. Commun.* 39.
Mitscher, L. A., Martin, J. H., Miller, P. A., Shu, P. and Bohonos, N. (1966) *J. Amer. Chem. Soc.* **88**, 3647.
Money, T., Qureshi, I. H., Webster, G. B. and Scott, A. I. (1965) *J. Amer. Chem. Soc.* **87**, 3004.
Mori, Y. (1961) *Jozo Kagaku Kenkyu Hokoku* **6**, 28 (*Chem. Abs.* **60**, 7150).
Morris, D. R. and Hager, L. P. (1966) *J. Biol. Chem.* **241**, 1763.
Mosbach, K. (1960) *Acta Chem. Scand.* **14**, 457.
Mosbach, K. (1964) *Biochem. Biophys. Res. Commun.* **17**, 363.
Mosbach, K. (1967) *Acta Chem. Scand.* **21**, 2331.
Mosbach, K. and Ljungcrantz, I. (1964) *Biochim. Biophys. Acta* **86**, 203.
Moustafa, E. and Wong, E. (1967) *Phytochem.* **6**, 625.
Neelakantan, S. and Seshadri, T. R. (1960) *J. Sci. Ind. Res., India* **19A**, 71.
Neidleman, S. L., Diassi, P. A., Junta, B., Palmere, R. M. and Pan, S. C. (1966) *Tetrahedron Letters* 5337.
Neish, A. C. (1960) *Ann. Rev. Pl. Physiol.* **11**, 55.
Neish, A. C. (1964) *Biochemistry of Phenolic Compounds*, edited J. B. Harborne, Academic Press, London and New York, p. 295.
Nozaki, H., Ito, H., Tunemoto, D. and Kondo, K. (1966) *Tetrahedron* **22**, 441.
Nugteren, D. H. and van Dorp, D. A. (1965) *Biochim. Biophys. Acta* **98**, 654.
Ollis, W. D. (1966) *Experientia* **22**, 777.
Ollis, W. D., Sutherland, I. O., Codner, R. C., Gordon, J. J. and Miller, G. A. (1960) *Proc. Chem. Soc.* 349.
Olsen, R. K., Smith, J. L., Daves, G. D., Moore, H. W., Folkers, K., Parson, W. W. and Rudney, H. (1965) *J. Amer. Chem. Soc.* **87**, 2298.
Patschke, L., Barz, W. and Grisebach, H. (1964) *Z. Naturforsch.* **19B**, 1110.
Patschke, L., Hess, D. and Grisebach, H. (1964) *Z. Naturforsch.* **19B**, 1114.
Patterson, E. L., Andres, W. W. and Mitscher, L. A. (1967) *Appl. Microbiol.* **15**, 528.
Pennock, J. F. (1967) *Terpenoids in Plants*, edited J. B. Pridham, Academic Press, London and New York, p. 129.
Penttila, A. and Fales, H. M. (1966a) *J. Amer. Chem. Soc.* **88**, 2327.
Penttila, A. and Fales, H. M. (1966b) *Chem. Commun.* 656.
Penttila, A., Kapadia, G. J. and Fales, H. M. (1965) *J. Amer. Chem. Soc.* **87**, 4402.
Pettersson, G. (1963a) *Acta Chem. Scand.* **17**, 1771.
Pettersson, G. (1963b) *Acta Chem. Scand.* **17**, 1323.
Pettersson, G. (1964a) *Acta Chem. Scand.* **18**, 1202.
Pettersson, G. (1964b) *Acta Chem. Scand.* **18**, 1839.
Pettersson, G. (1964c) *Acta Chem. Scand.* **18**, 1428.
Pettersson, G. (1965a) *Acta Chem. Scand.* **19**, 35.
Pettersson, G. (1965b) *Acta Chem. Scand.* **19**, 1724.
Pettersson, G. (1965c) *Acta Chem. Scand.* **19**, 1827.
Polonsky, J. (1956) *Compt. rend.* **242**, 2961.
Polonsky, J. (1957) *Bull. Soc. chim. France* 1079.
Polonsky, J. (1958) *Bull. Soc. chim. France* 929.

Posternak, T. (1965) *The Cyclitols*, Holden-Day, San Francisco.
Posternak, T., Schopfer, W. H., Kaufman-Boetsch, B. and Edwards, S. (1963) *Helv. Chim. Acta* **46**, 2676.
Powls, R. and Hemming, F. W. (1966) *Phytochem.* **5**, 1249.
Pridham, J. B. (1967) *Terpenoids in Plants*, Academic Press, London and New York.
Ramsey, V. G., Zwitkowits, P. M., Bentley, R. and Olsen, R. E. (1966) *J. Amer. Chem. Soc.* **88**, 1553.
Read, G., Vining, L. C. and Haskins, R. H. (1962) *Canad. J. Chem.* **40**, 2357.
Read, G. and Vining, L. C. (1968) *Chem. Commun.* 935.
Rhoads, S. J. (1963) *Molecular Rearrangements*, edited P. de Mayo, Interscience, New York and London, p. 655.
Rhodes, A., Boothroyd, B., McGonagle, M. P. and Somerfield, G. A. (1961) *Biochem. J.* **81**, 28.
Rhodes, A., McGonagle, M. P. and Somerfield, G. A. (1962) *Chem. and Ind.* 611.
Richards, J. H. and Hendrickson, J. B. (1964) *The Biosynthesis of Steroids, Terpenes and Acetogenins*, Benjamin, New York and Amsterdam.
Rittenberg, D. and Bloch, K. (1944) *J. Biol. Chem.* **154**, 311.
Roberts, R. M. and Loewus, F. (1966) *Pl. Physiol., Lancaster* **41**, 1489.
Roberts, R. M., Shah, R. H. and Loewus, F. (1967) *Pl. Physiol., Lancaster* **42**, 659.
Ruddat, M., Heftmann, E. and Lang, A. (1965) *Archs. Biochem. Biophys.* **110**, 496.
Rudman, D. and Meister, A. (1953) *J. Biol. Chem.* **200**, 591.
Ryhage, R. and Samuelsson, B. (1965) *Biochem. Biophys. Res. Commun.* **19**, 279.
Samuelsson, B. (1965) *J. Amer. Chem. Soc.* **87**, 3011.
Sankawa, U., Taguchi, H., Ogihara, Y. and Shibata, S. (1966) *Tetrahedron Letters* 2883.
Sato, M., Okawa, K. and Akabori, S. *Bull. Chem. Soc. Japan* **30**, 937.
Scheline, R. R., Williams, R. T. and Witt, J. G. (1960) *Nature* **188**, 849.
Scholda, R., Billek, G. and Hoffmann-Ostenhof, O. (1964) *Z. physiol. Chem.* **339**, 28.
Scott, A. I. (1965) *Quart. Rev. Chem. Soc.* **19**, 1.
Scott, A. I. (1967) *Oxidative Coupling of Phenols*, edited W. I. Taylor and A. R. Battersby, E. Arnold, London, p. 95.
Scott, A. I. and Yalpani, M. (1967) *Chem. Commun.* 945.
Shibata, S. (1967) *Chemistry in Britain* 110.
Smith, G. N. and Bu'Lock, J. D. (1965) *Chem. and Ind.* 1840.
Snell, J. F., Birch, A. J. and Thomson, P. L. (1960) *J. Amer. Chem. Soc.* **82**, 2402.
Späth, E. and Kainrath, P. (1937) *Chem. Ber.* **70**, 2272.
Splittstoesser, W. E. and Mazelis, M. (1967) *Phytochem.* **6**, **39**.
Sprinson, D. B. (1960) *Adv. Carbohyd. Chem.* **15**, 235.
Srinivasan, P. R., Katagiri, M. and Sprinson, D. B. (1955) *J. Amer. Chem. Soc.* **77**, 4943.
Srinivasan, P. R., Katagiri, M. and Sprinson, D. B. (1959) *J. Biol. Chem.* **234**, 713.

Srinivasan, P. R., Shigeura, H. T., Sprecher, M., Sprinson, D. B. and Davis, B. D. (1956) *J. Biol. Chem.* **220**, 477.

Srinivasan, P. R., Sprinson, D. B., Kala, E. B. and Davis, B. D. (1956) *J. Biol. Chem.* **223**, 913.

Srinivasan, P. R. and Sprinson, D. B. (1959) *J. Biol. Chem.* **234**, 716.

Srinivasan, P. R., Rothschild, J. and Sprinson, D. B. (1963) *J. Biol. Chem.* **238**, 3176.

Stevens, R. (1967) *Chem. Rev.* **67**, 19.

Sutherland, J. K. (1967) *Fortschr. Chem. org. Naturstoffe* **25**, 131.

Swain, T. (1965) *Biosynthetic Pathways in Higher Plants*, edited J. B. Pridham and T. Swain, Academic Press, London and New York, p. 9.

Taguchi, H., Sankawa, U. and Shibata, S. (1966) *Tetrahedron Letters* 5211.

Tanabe, M. and Detre, G. (1966) *J. Amer. Chem. Soc.* **88**, 4515.

Tanenbaum, S. W. and Bassett, E. W. (1959) *J. Biol. Chem.* **234**, 1861.

Tanenbaum, S. W. and Bassett, E. W. (1960) *Biochim. Biophys. Acta* **40**, 535.

Thomas, R. (1961) *Biochem. J.* **78**, 748.

Threlfall, D. R. (1967) *Terpenoids in Plants*, edited J. B. Pridham, Academic Press, London and New York, p. 191.

Threlfall, D. R., Whistance, G. R. and Goodwin, T. W. (1968) *Biochem. J.* **106**, 107.

Turner, A. B. (1964) *Quart. Rev. Chem. Soc.* **18**, 347.

van Dorp, D. A., Beerthuis, R. K., Nugteren, D. H. and Vonkeman, H. (1964) *Biochim. Biophys. Acta* **90**, 204.

Weiss, U., Gilvarg, C., Mingioli, E. S. and Davis, B. D. (1954) *Science, N.Y.* **119**, 774.

Westerfield, W. W. and Lowe, C. (1942) *J. Biol. Chem.* **145**, 463.

Weygand, F., Brucker, W., Grisebach, H. and Schulze, E. (1957) *Z. Naturforsch.* **12B**, 222.

Whistance, G. R., Threlfall, D. R. and Goodwin, T. W. (1966) *Biochem. Biophys. Res. Commun.* **23**, 849.

Whistance, G. R. and Threlfall, D. R. (1967) *Biochem. Biophys. Res. Commun.* **28**, 295.

Wilson, E. M. and Snell, E. E. (1962) *J. Biol. Chem.* **237**, 3171.

Wong, E. (1965) *Biochim. Biophys. Acta* **111**, 358.

Wong, E. (1966a) *Phytochem.* **5**, 463.

Wong, E. (1966b) *Chem. and Ind.* 598.

Woolley, J. G. (1968) Private Communication.

Zalkin, H., Law, J. H. and Goldfine, H. (1963) *J. Biol. Chem.* **238**, 1242.

Zenk, M. H. (1966) *Biosynthesis of Aromatic Compounds*, edited G. Billek, Pergamon Press, Oxford, p. 45.

Zenk, M. H. (1964a) *Z. Naturforsch.* **19B**, 83.

Zenk, M. H. (1964b) *Z. Naturforsch.* **19B**, 856.

Zenk, M. H. and Leistner, E. (1967) *Z. Naturforsch.* **22B**, 460.

Zenk, M. H. and Müller, G. (1964) *Z. Naturforsch.* **19B**, 398.

ADDENDUM

Since the foregoing chapter was completed there have been reported a number of investigations which amplify or shed new light on several of the topics discussed. The relevant results are briefly noted in the following addendum.

Direct evidence in support of the proposed biosynthesis of the ergochromes by way of oxidative fission of an anthraquinone precursor (emodin) has been obtained (Gröger, D., Erge, D., Franck, B., Ohnsorge, U., Flasch, H. and Hüper, F. (1968) *Chem. Ber.* **101,** 1970). Full details of the further studies of Hassall and his collaborators relating to sulochrin biosynthesis have been published (Curtis, R. F., Hassall, C. H. and Pike, R. K. (1968) *J. Chem. Soc.* (C) 1807). The results obtained are regarded as providing support for the hypothesis that sulochrin is formed from two biogenetically discrete precursors. However, in a preliminary communication, Gatenbeck has reported that emodin undergoes oxidative fission to give sulochrin in the manner discussed on page 97 (Gatenbeck, S. 1968) (Abstracts, *5th. Intern. Symp. Chem. Nat. Prods.*, London, page 114). The conclusions of Hassall *et al.* and of Gatenbeck are mutually incompatible unless the accepted view of emodin biosynthesis is to be modified.

Circumstantial evidence implicating methyl triacetic lactone in stipitatic acid biosynthesis has been obtained (Marx, G. S. and Tanenbaum, S. W. (1968) *J. Amer. Chem. Soc.* **90,** 5302). Formation of the tropolone ring in the fungal metabolite sepedonin follows the pattern observed in stipitatic acid biosynthesis (page 104) (McInnes, A. G., Smith, D. G., Vining, L. C. and Wright, J. L. C. (1968) *Chem. Commun.* 1669).

New, chlorinated cyclopentane acids from *Sporormia affinis* are suggested to be biogenetically related to terrein (McGahren, W. J., van den Hende, J. H. and Mitscher, L. A. (1969) *J. Amer. Chem. Soc.* **91,** 157).

Portentol, a metabolite of the lichen *Roccella portentosa*, is possibly the first example of a polypropionate polyketide containing a six-membered carbocyclic ring constructed entirely of propionate residues (Aberhart, D. J., Overton, K. H. and Huneck, S. (1969) *Chem. Commun.* 162).

Tetra-acetic acid has been synthesised and its acid and base catalysed cyclisation studied (Howarth, T. T., Murphy, G. P. and Harris, T. M. (1969) *J. Amer. Chem. Soc.* **91,** 517).

Studies with blocked mutants have provided further support for the proposed sequence in tetracycline biosynthesis (McCormick, J. R. D., Jenson, E. R., Arnold, N. H., Corey, H. S., Joachim, U. H., Johnson, S., Miller, P. A. and Sjolander, N. O. (1968) *J. Amer. Chem. Soc.* **90,** 7127; McCormick, J. R. D. and Jensen, E. R. (1969) *J. Amer. Chem. Soc.* **91,** 206).

The biogenesis of phytoquinones has been the subject of further studies which confirm and extend previous results (Whistance, G. R. and Threlfall, D. R. (1968) *Biochem. J.* **109,** 577).

Two groups have provided necessary confirmation of the shikimate origin of the bacterial menaquinones (Campbell, I. M., Coscia, C. J., Kelsey, M. and Bentley, R. (1967) *Biochem. Biophys. Res. Commun.* **28,** 25; Leistner, E., Schmitt, J. H. and Zenk, M. H. (1967) *Biochem. Biophys. Res. Commun.* **28,** 845).

Biosynthesis of the naphthquinone juglone is found to follow the expected course from shikimate and acetate with randomisation of the carboxyl carbon of the former between the quinone carbonyl positions (cf. page 166) (Leistner, E. and Zenk, M. H. (1968) *Z. Naturforsch.* **23B,** 259).

A full report on the biosynthesis of the *Rubia* anthraquinones has appeared (Burnett, A. R. and Thomson, R. H. (1968) *J. Chem. Soc.* (C) 2437). However, biosynthesis of the emodin-type anthraquinone, chrysophanol, in higher plants, has been shown to resemble the polyketide pathway established in micro-organisms (Leistner, E. and Zenk, M. H. (1969) *Chem. Commun.* 210).

Preliminary notice has appeared of investigations which indicate that the cyclopropene system of sterculic acid arises by dehydrogenation of a cyclopropane precursor, rather than by insertion of a methylene group into an acetylenic substrate (cf. page 172) (James, A. T. (1968) *Chemistry in Britain*, **4,** 484).

A route to the fungal cyclopentane acid brefeldin-A from palmitic acid by internal cyclisation has been demonstrated. A similar route is suggested for biosynthesis of certain other carbocyclic polyketides (Bu'Lock, J. D. and Clay, P. T. (1969) *Chem. Commun.* 237).

Further evidence for the aldol condensation pathway in *myo*-inositol biosynthesis has been presented (Barnett, J. E. G. and Corina, D. L. (1968) *Biochem. J.* **108,** 125).

Isoflavones may arise by dehydrogenation of isoflavanones. If this is the obligatory route to the former, the aryl migration step may belong to the class of rearrangement typified by the methylmalonate-succinate interconversion (page 107) (Grisebach, H. and Zily, H. (1968) *Z. Naturforsch.* **23B,** 494). The thallic acetate induced rearrangement of chalcones may have some relevance to the rearrangement step in isoflavone biosynthesis (Ollis, W. D., Ormand, K. L. and Sutherland, I. O. (1968) *Chem. Commun.* 1237).

The discovery of new representatives of the neoflavanoid class has provided further chemo-taxonomic evidence relating to the biosynthesis of these compounds (Ollis, W. D. and Gottlieb, O. R. (1968) *Chem. Commun.* 1396).

Rubratoxins A and B from *Penicillium rubrum* are new representatives of the nonadride group (Moss, M. O., Wood, A. B. and Robinson, F. V. (1969) *Tetrahedron Letters* 367).

3

BICYCLO[3,3,1]NONANES AND RELATED COMPOUNDS

G. L. Buchanan

Reader in Organic Chemistry, University of Glasgow, Glasgow, W.2, Scotland

A. Introduction

The bicyclo[3,3,1]nonane ring system I, shown here both in plan and in perspective, has been known for about seventy years, and work on this, and related bicyclic systems, has uncovered and illustrated many significant chemical principles; yet it has never been adequately reviewed.

With a growing body of interesting results emerging from studies on natural and synthetic bicyclononanes, it seemed to the author that the time had come to draw together the threads of this subject, and to place on record the main features of the chemistry of the system. This review will therefore be concerned with the synthesis, properties and reactions of the bicyclo[3,3,1]nonane system. However, to avoid producing a sterile catalogue of facts, the author has deliberately digressed from time to time into related bicyclic systems—e.g. bicyclo[3,2,1]octane II, to observe the influence of ring-size on the

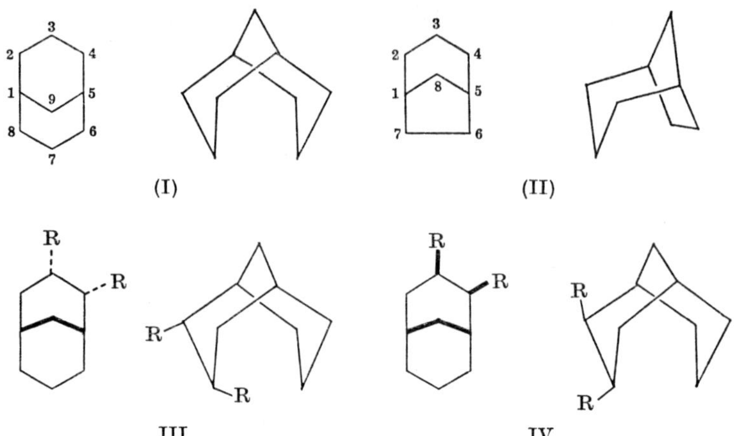

chemistry of the system and, it is hoped, to add a little perspective to the discussion.

The reader is reminded that by convention, the numbering of a bridged bicyclic molecule starts at one bridgehead and proceeds to the other by the longest route, as shown above. The stereochemistry of substituents has been variously described by the prefixes *endo*- or α- as in III, as *exo*- or β- as in IV, or simply as axial and equatorial.

Formulae are indicated by Roman and references by Arabic numerals; each section has been enumerated separately.

B. Synthesis

It is convenient to subdivide all syntheses of bicyclononanes into two types—those which give rise to 9-keto-derivatives and those which lead to a tetrahedral C(9).

1. *Carbonyl-bridged Derivatives*

Compounds of this type are usually prepared by the cyclisation of 1,5-diones of type I (page 201), and these in turn are available by the Michael reaction (R = aryl, H or alkyl). However, formation of the bicyclic compound is not always smooth or uncomplicated.

Aryl ketones are generally cyclised under acid conditions [38], and give rise to the fewest complications. The initial aldol product, a tertiary benzylic alcohol, is readily dehydrated, but affords a $\beta\gamma$-

enone in obedience to Bredt's rule. Consequently, the enone II is readily prepared from cyclohexanone, and the homologous bicyclo-(4,3,1)decenones and bicyclo(5,3,1)undecenones can be obtained from its homologues[22]. However, the related bicyclo(3,2,1)octenones III are distinctly unstable in an acid environment. They can be isolated only under special reaction conditions[22, 26] and readily rearrange to cycloheptene carboxylic acids (see section D.2).

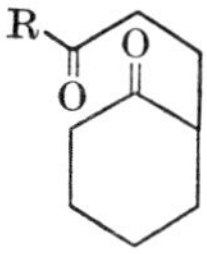

I 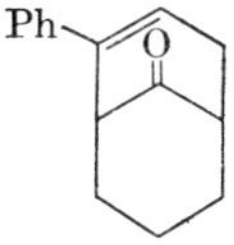 II 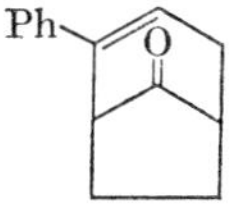 III 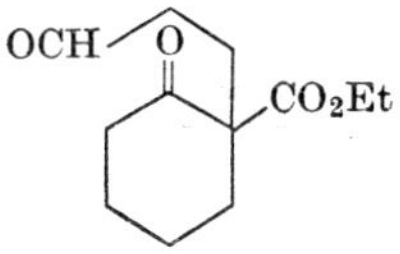 IV

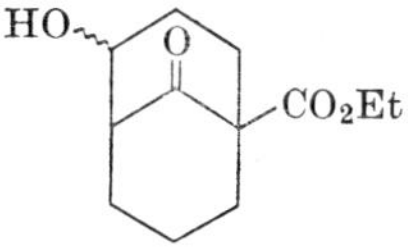

V 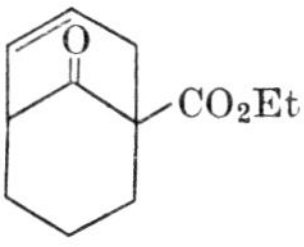 VI 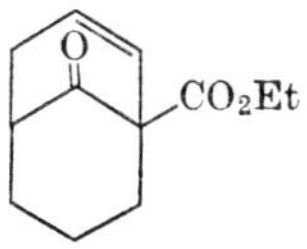 VII

When acrolein is used as the initial enone, the intermediate Michael-product (IV) can be converted into the mixture of stereo-isomeric ketols V, or dehydrated to the enone VI according to the reaction conditions[40]. The dehydration, by concentrated sulphuric acid, is messy, but in this particular case no by-product is isolated, apart from the isomer VII. On prolonged contact with acid the isomers VI and VII are generated in roughly equal amounts, and this isomerism is probably general to the series[33]. The addition of a

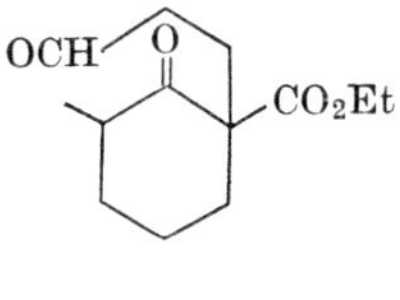

VIII 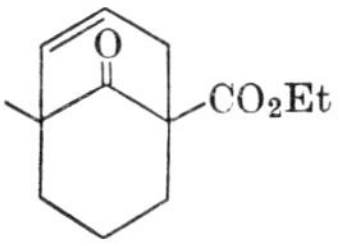 IX 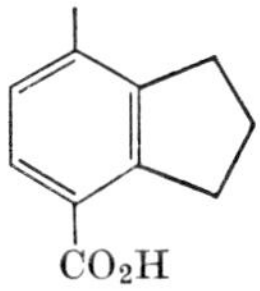 X 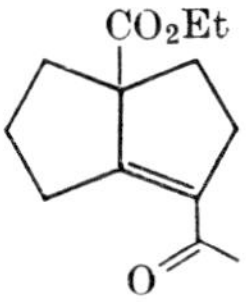 XI

seemingly insignificant methyl function produces two by-products, X and XI, from the ring closure of VIII to IX, and it is interesting to observe in passing, that whilst the alicyclic by-product XI is an ester, the aromatic by-product X is an *acid*. Thus the overall process leading to X probably includes preferential hydrolysis of the initially-formed ester of X, via the derived acylium ion—i.e. by the $A_{AC}1$

mechanism[73]. The by-products themselves presumably arise from IX via the carbonium ions XII and XIII formed by protonation

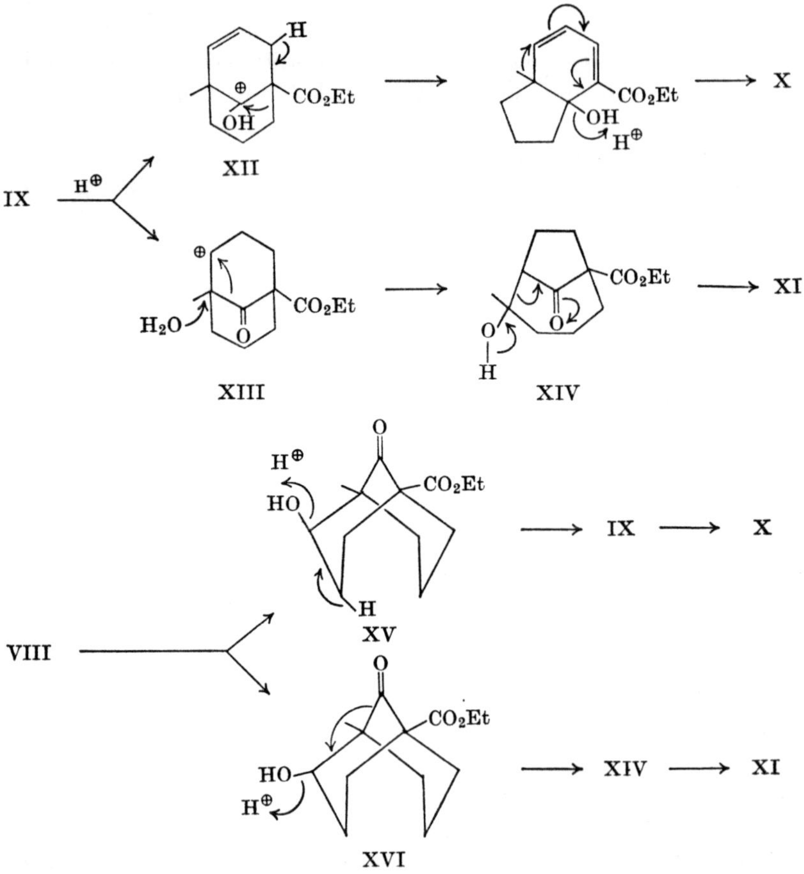

at the two most likely sites[107]. The formation of X from XII requires only the rearrangement indicated, and aromatisation of a dihydrobenzene. The formation of XI from XIII requires a retro-aldol ring opening of XIV followed by ring closure. Alternatively, the olefin IX, and hence X, may be derived specifically from the intermediate axial alcohol XV, and the intermediate XIV from the equatorial epimer XVI by a concerted acyl-shift process. Curiously, the homologous ketol mixture XVII yields *none* of the enone XVIII under the same conditions; only the rearrangement products XIX and XX are isolated[23]. Repeated failure to observe similar rearrangement in the

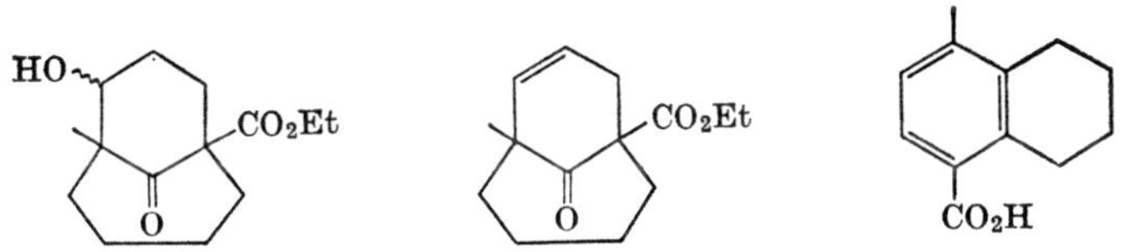

XVII XVIII XIX

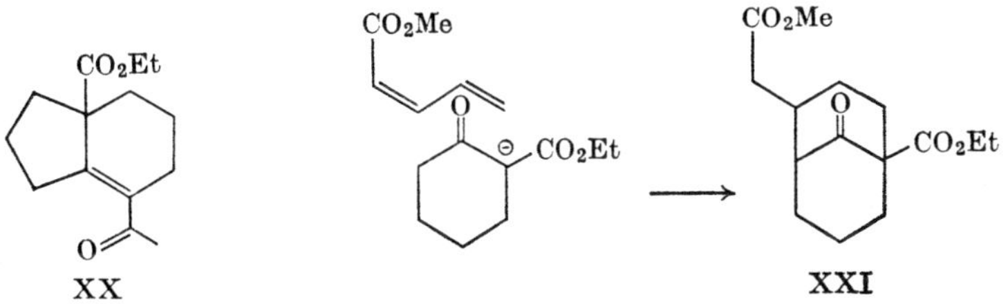

XX XXI

conversion of IV into VI[33] remains unexplained, and the apparently decisive role of the bridgehead methyl group is intriguing.

Substituted bicyclic products, e.g. XXI, can be obtained as shown above, by an elegant one-step double Michael reaction on methyl pentadienoate[46].

Alternatively, substituted $\alpha\beta$-enones, such as chalkones, have been used in place of acrolein in the general synthesis described above. The product is a mixture of 4(*endo* and *exo*)-phenyl-diastereomers, but a degree of *endo*-stereoselectivity has been claimed[77]. The use of an enamine intermediate in place of an enolate in the Michael step has interesting consequences. When cinnamaldehyde is used as the enone, the product is rich in the 2-(*exo*)-phenyl diastereomer[5], i.e. the enolate and enamine condensations appear to be stereochemically complementary.

A more remarkable feature of the enamine syntheses is the fate of the amino-function. Thus the reaction of XXII with acrolein gives[13] the bicyclic amino-ketone XXIII, apparently in one step, and the corresponding bicyclo(3,2,1)octanone derivative is obtained from the enamine of cyclopentanone. The authors have offered no comment on the mechanism of this remarkable reaction which clearly conceals a rearrangement and, experimentally, requires strong heating. An indication of the mechanism seems to be emerging from recent work[108, 117] which has demonstrated that the reaction e.g. of XXIV

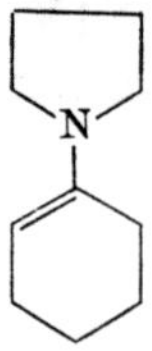

XXII 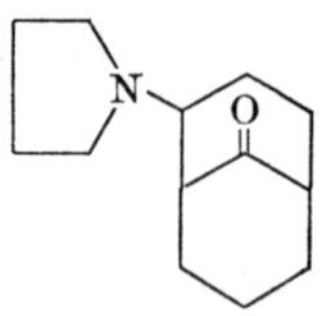 XXIII

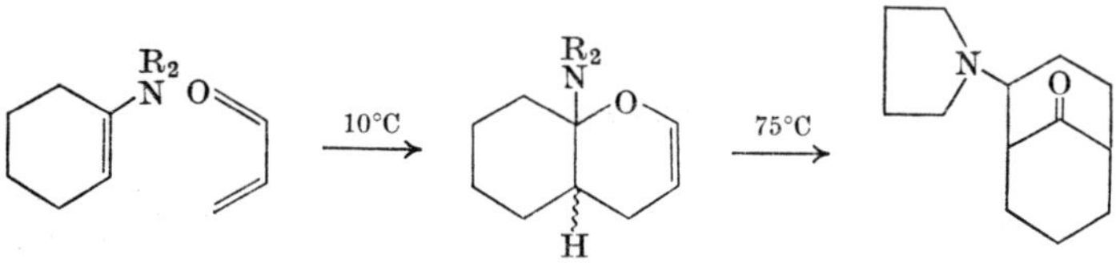

XXIV XXV XXVI

(NR_2 = *N*-phenylpiperazyl) with acrolein at 10° gives an intermediate cycloaddition product XXV which rearranges at higher temperatures to XXVI, or in some instances to an uncyclised product (XXVII →

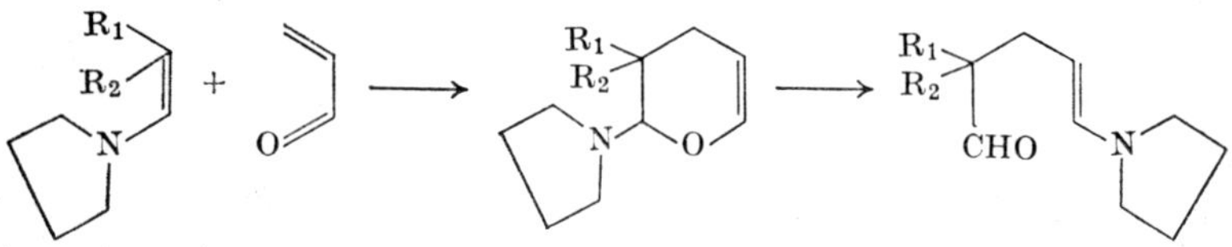

XXVII XXVIII

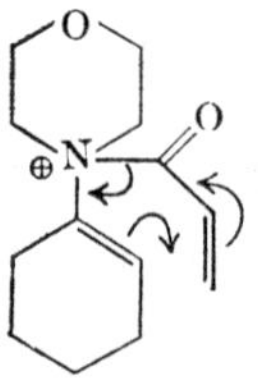

XXIX 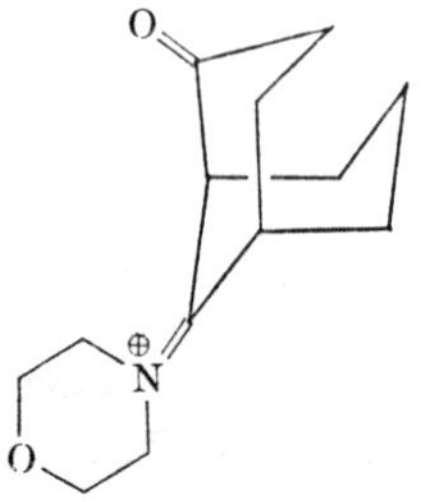 XXX

XXVIII). The intimate details of the amine migration process are still unknown, but it has been suggested that the migration cannot be intramolecular since this would lead to a preferred configuration at C(2). In fact, the amino-ketone XXVI is a mixture of C(2) epimers whose composition is close to that expected at equilibrium [5].

Enamines also react with acrylyl chloride forming initially the salt XXIX which rearranges intramolecularly via a ketone or an acylium intermediate to XXX. From this, the bicyclic β-diketone can be readily obtained[78].

Alicyclic 1,5-diketones of type XXXI are unexceptional in their behaviour, affording, for example, the bridged bicycle XXXII under acid conditions, unless one or both of the rings are 5-membered, when the resulting bicyclo[3,2,1]octenone fragments[26] (see section D.2).

The behaviour of aliphatic-alicyclic 1,5-diketones such as XXXIII is more complex since it can, and does, cyclise by two alternative pathways to bridged (XXXIV) or fused (XXXV) bicyclic products.

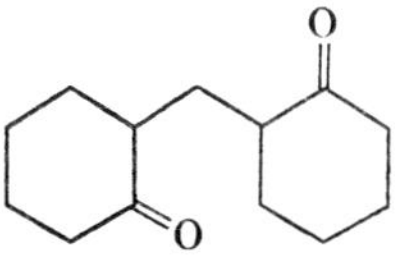

XXXI 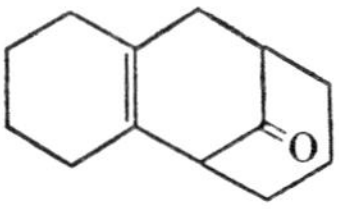 XXXII 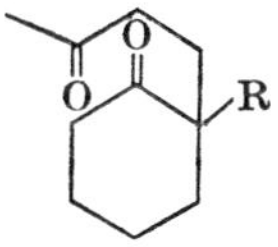 XXXIII

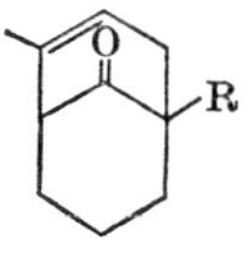

XXXIV 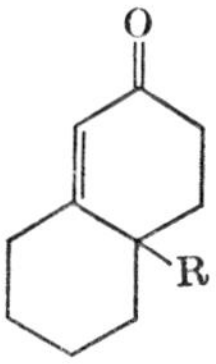 XXXV 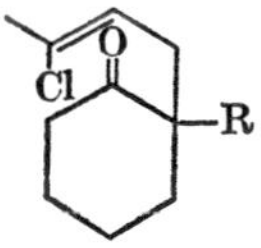 XXXVI

In practice, the synthesis of bridged bicyclic compounds is frequently carried out by a related process, known as the Wichterle reaction[98, 138], in which the vinyl chloride XXXVI is cyclised to XXXIV by concentrated sulphuric acid via the intermediate diketone XXXIII[98, 137]. On the other hand ring closure of the diketones XXXIII may be carried out either by concentrated sulphuric acid or by dilute acid or base, and in an analysis of the factors which control the direction of cyclisation it is important to consider the reaction conditions as well as the effect of change of ring-size.

The historic investigations of Prelog[113] showed that when alcoholic base is employed, the dione XXXVII (R = H or CO_2Et) is invariably cyclised, with simultaneous hydrolysis and decarboxylation, to a conjugated enone, XXXVIII ($n \leqslant 8$) or XXXIX ($n \geqslant 8$) according to the size of the ring involved. The application of this process to

alc. base, $n = 6$–8, $R = H$ or CO_2Et: XXXVII → XXXVIII

H_2SO_4, $n = 6$ or 7: XL → XXXVIII

alc. base, $n \geqslant 8$: XXXVII → XXXIX

XLI → XL

H_2SO_4: XLI → XLII ($n = 6$ or 7); XLI → XLIII ($n \geqslant 8$)

XXXVII XXXVIII XL XLII XXXIX XLI XLIII

n = total number of carbon atoms in ring.

5- or 6-membered rings is known as the Robinson annelation reaction[80]. Similarly, concentrated sulphuric acid converts XL ($n = 6$ or 7) into XXXVIII, but the keto-ester XLI yields XLII ($n = 6$ or 7) or XLIII ($n \geqslant 8$) according to ring size; i.e. *only bridged products* are formed.

Thus at first sight, the determining factor appears to be the presence or absence of a bulky substituent at the bridge junction, and several authors have noted the decisive effect of such a group. Whereas XXXVI (R = H) affords a mixture of XXXIV and XXXV[83, 98, 111], replacement of this proton by an alkyl, aldehyde or ester function specifically favours the bridged bicyclic product XXXIV[9, 47, 83, 98, 111] by making it easier for the butanone side chain to attain the appropriate axial conformation XLIV[83, 98, 115].

However true this may be, it does not provide a completely satisfactory answer; reaction conditions play a significant role. Thus the course of the Robinson annelation reaction is *not* altered by the retention of an 'angular' ester group[80] and even in the absence of an 'angular' substituent the bridged enone XLVI can be obtained from XLV, by the use of appropriate reaction conditions[45]. Finally, each of the three diones XLVII–XLIX affords the corresponding octalone on treatment with alcoholic base but only the bridged bicyclic ketone on treatment with concentrated sulphuric acid[98].

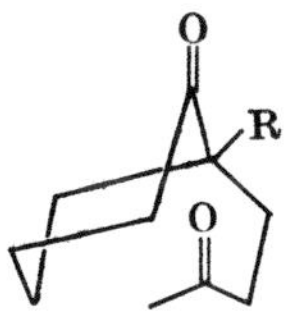

XLIV 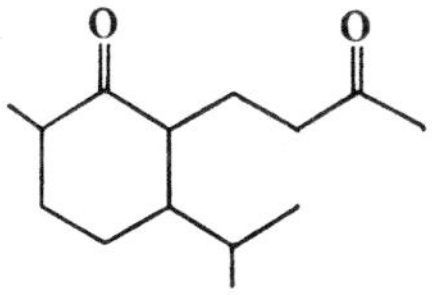 XLV 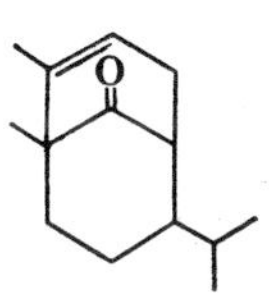 XLVI

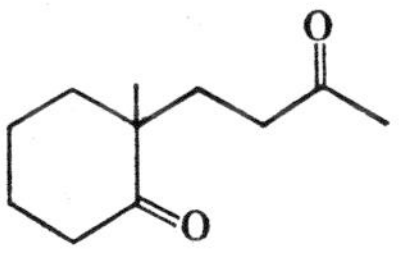

XLVII 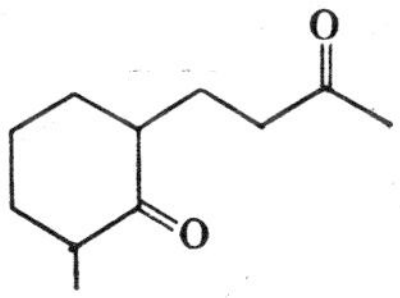 XLVIII 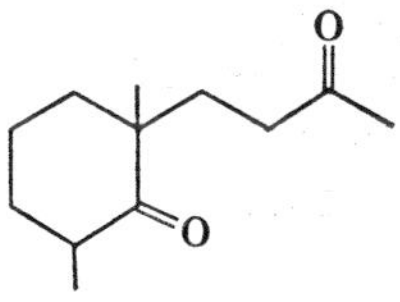 XLIX

It is difficult to avoid the conclusion that in these cyclisations the bridged structure is kinetically favoured, whereas the conjugated enone is the product of thermodynamic control. This hypothesis accounts for the formation of XXXIX rather than XXXVIII in the case of large ring ketones. It also accounts for the observation that the intermediate ketols isolated in the Robinson annelation reaction

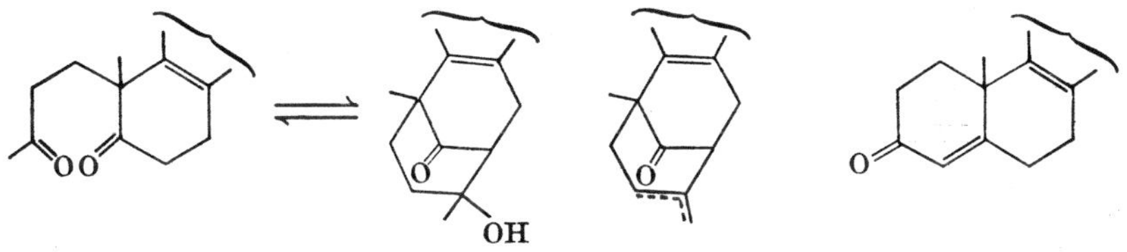

L LI LII

have frequently the bridged structure (L) and can be transformed into either LI or LII by choice of conditions [82].

It is interesting to enquire whether the same equilibrium holds for the cyclopentanone series LIII, for the action of concentrated sulphuric acid on this dione [93], or indeed on the chlorovinyl ketone [47, 112] LIV, uniformly gives the hydro-indanone LV. However, recently it has been shown that here too, under special, strongly acid conditions the bridged structure (LVI) must be formed transiently, for the mixture of cycloheptene carboxylic acids (LVII), which is isolated in good yield, is most plausibly derived from such an intermediate by an established fragmentation reaction [26].

A similar series of cyclisations has been effected on the analogous

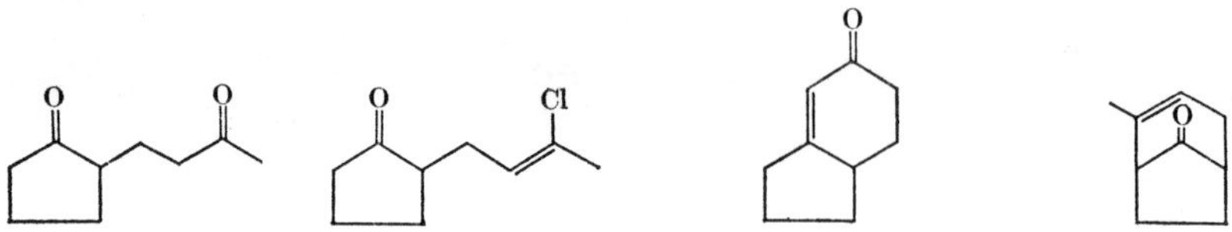

LIII LIV LV LVI

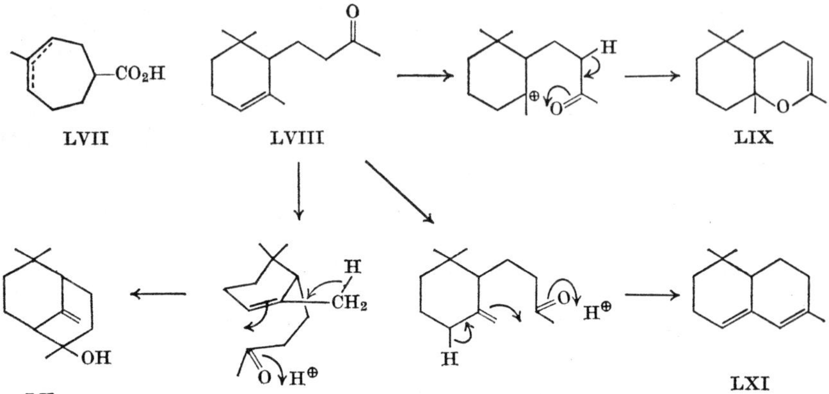

LVII LVIII LIX

LX LXI

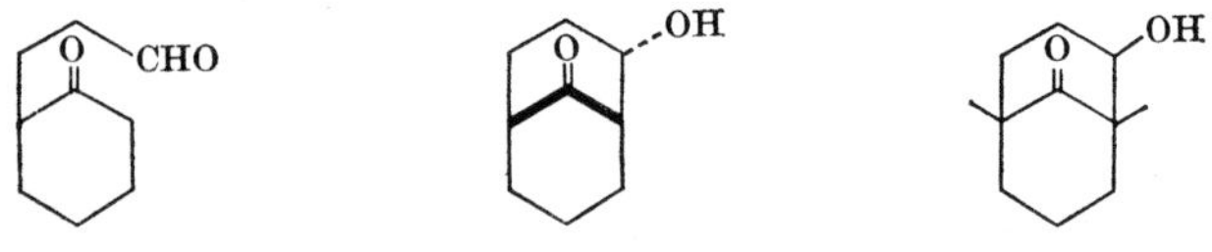

LXII LXIII LXIV

unsaturated ketone, β-ionone (LVIII), giving rise to both bridged and fused bicyclic products LIX, LX and LXI[27, 130]. The products can be rationalised in terms of the mechanism shown above, in which protonation of the olefin leads to LIX, and protonation of the carbonyl function gives LX and LXI.

It has been claimed[43] that when the keto-aldehyde LXII is cyclised by means of dilute acid, the mixture of ketols so formed comprises mainly (99 per cent) the *endo*-isomer LXIII. This result is interesting on two counts. First, it is at variance with the report[102] that at equilibrium, the dimethyl analogue LXIV affords a 2:1 mixture of *endo*- and *exo*-epimers. This implies either that the bridgehead methyls are able to exert an unexpected influence on the stereochemistry of the reaction or that LXIII is being formed by a special, non-equilibrating mechanism. Second, it complements nicely a new and elegant stereoselective synthesis of the *exo*-epimer LXVI from the enol-lactone LXV. Reduction of the latter by lithium tri(t-butoxy)-

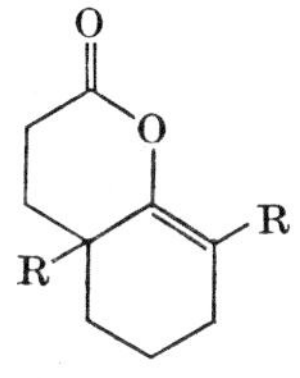

LXV 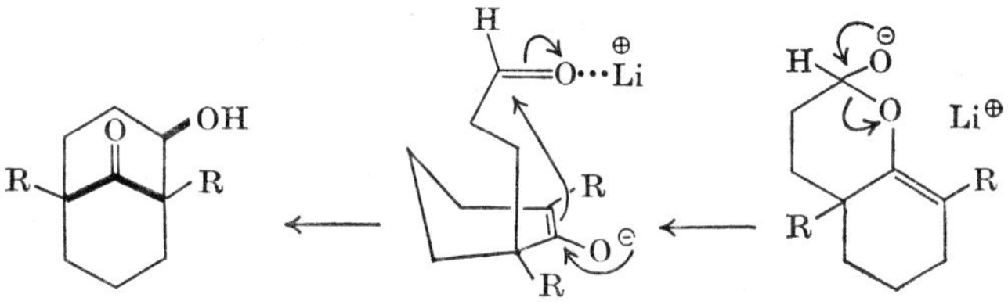 LXVI LXVIII LXVII

aluminium hydride proceeds via LXVII, and stereoselectivity is believed to result from the conformation (LXVIII) of the intermediate, in which the gegen-ion Li^+, or a corresponding aluminate complex, holds the developing aldehyde group in a conformation which can only lead to the axial ketol LXVI[102]. When R = H, the cyclisation step is not fast enough to preclude hydride reduction of the intermediate aldehyde, and the main product is the hemi-acetal LXIX or its dehydration product[62, 102]. However, when R ≠ H, the method has been shown to yield the axial alcohol in a highly specific manner. Curiously, similar reduction of the isomeric enol-lactone LXX yields only the keto-aldehyde LXII.

Enol-lactones of type LXV have also been transformed into bridged bicyclic 1,3-diketones by photolysis. Thus LXXI affords

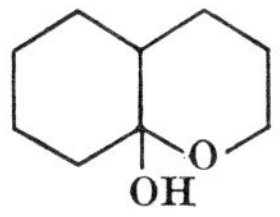

LXIX 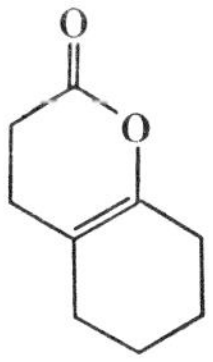 LXX 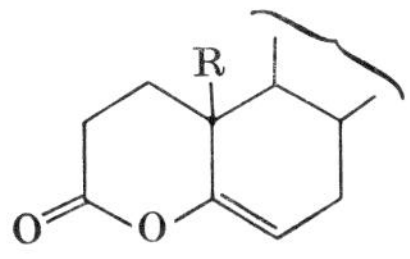 LXXI

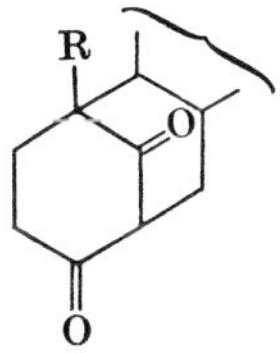

LXXII 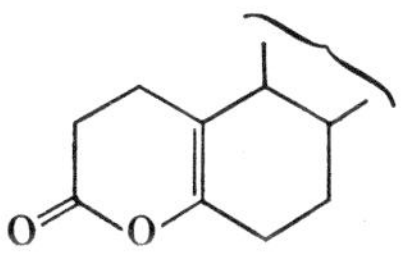 LXXIII 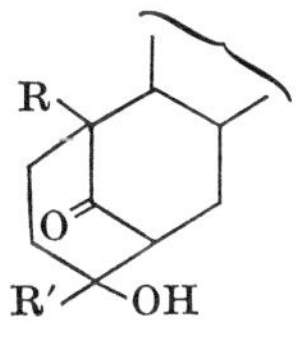 LXXIV

LXXII, but the isomeric lactone LXXIII gives rise to spiro products when similarly treated[140]. Bridged bicyclic ketols are also formed in the reaction of LXXI with Grignard reagents, and although these have been formulated as LXXIV, the stereochemistry of the products has not been discussed[69, 70]. The analogous bicyclo[3,2,1]octanolone system is formed from a γ-enol-lactone[48] and it would be surprising if the transformations brought about by complex hydride and by Grignard reagents did not share a common mechanism. More recently, it has been observed[58] that thermal cyclisation of the acid chloride LXXVI affords mainly the 2-*exo*-chloro-ketone LXXVII, whilst heating in the presence of catalytic amounts of $AlCl_3$ gives mainly its epimer LXXV.

Amongst the many miscellaneous syntheses of carbonyl bridged products, two deserve mention. The intramolecular ring expansion,

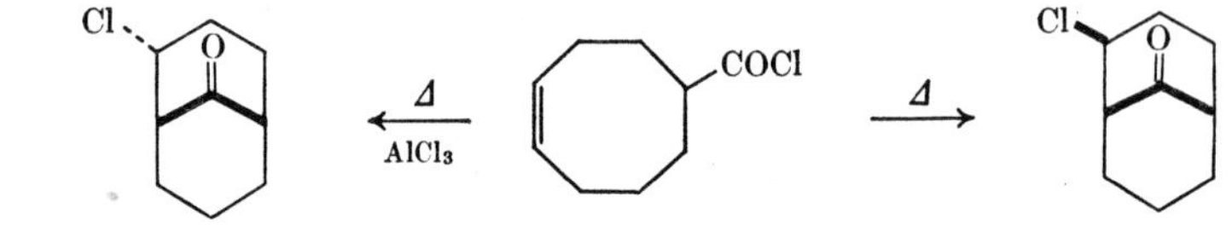

LXXV LXXVI LXXVII

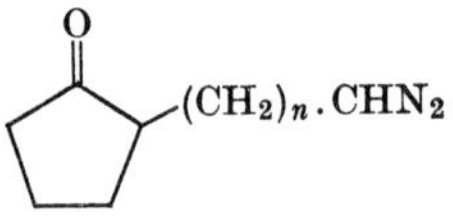

LXXVIII 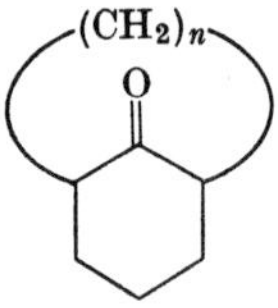 LXXIX 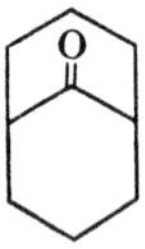 LXXX

LXXVIII → LXXIX, is highly successful if $n = 2$, less so if $n = 3$ and fails, giving only olefin or solvolysis products, if $n = 1$. A very similar set of results was obtained when the cyclohexanone analogue of LXXVIII was investigated, but the cycloheptanone series gave no bicyclic products[8]. The parent ketone LXXX is preparable directly from cycloocta-1,5-diene, by the action of tetracarbonyl nickel[60].

2. *Methylene-bridged Derivatives*

Bicyclic compounds of this type were the first to be described. As early as 1894, Knoevenagel obtained[84] a bis-enone by condensing acetyl-acetone with benzaldehyde and cyclising the resulting product with acid. He formulated it as either Ia or Ib and provided some supporting evidence, but real proof is still lacking. Another early synthesis was described by Rabe in 1908[114]. The addition of ethyl acetoacetate to 3-methylcyclohexenone gave the expected Michael product, and this readily afforded the bridgehead alcohol II, which

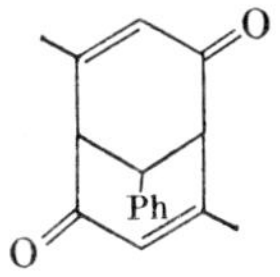

Ia 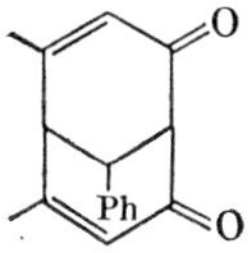 Ib 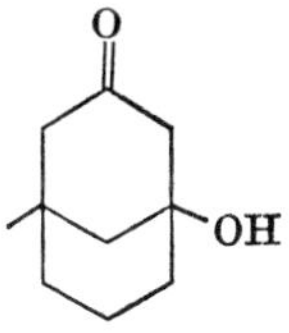 II

resisted all his efforts to bring about dehydration. A more dramatic construction of a bicyclononane III from methyl vinylketone and malonic ester has recently been described[118], but pride of place must

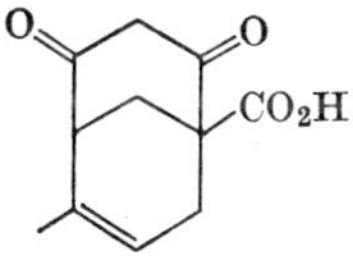

III 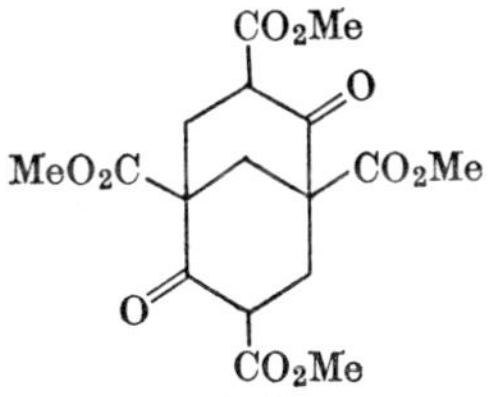 IV 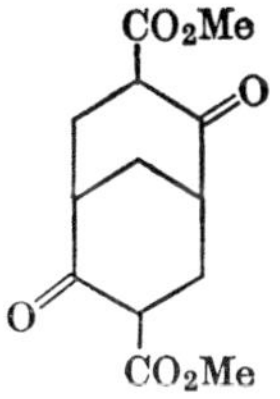 V

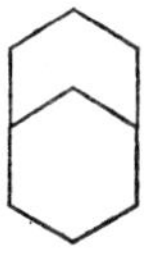

VI 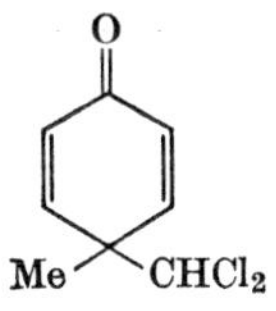 VII 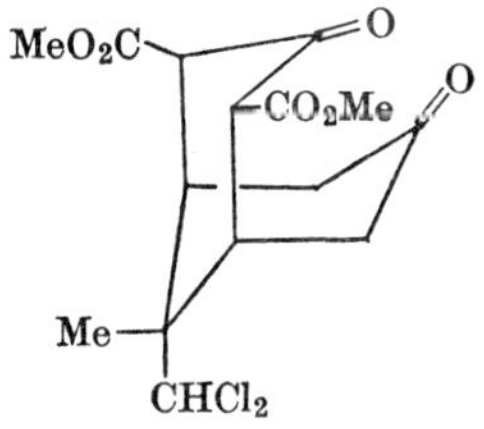 VIII

go to Meerwein[105] who obtained IV in a one step synthesis from malonic ester and formaldehyde. In a fascinating paper[105] written in 1922, he argued out its structure (without any spectroscopic aid), prepared from it the bicyclononan-2,6-dione and thence the parent hydrocarbon VI, whose structure has since been verified[30]; and in a prophetic footnote he commented on what is now termed the conformation of the ring system. He also defined selective hydrolytic conditions which led via the 1,5-dicarboxylic acid to the bis β-ketoester V. Although later generations have improved on the preparation of IV[85] and V[119], the synthetic routes pioneered by Meerwein and by Rabe were destined to play major roles in the chemistry of bicyclononanes.

A bridge-substituted compound VIII has been synthesised by a double Michael reaction on the 'abnormal' Reimer-Tiemann product VII[120].

Attempts have also been made to rearrange the more accessible bicyclo[2,2,2]octyl derivatives into bicyclo[3,3,1]nonanes. The carbonium ion X, when generated from the amine IX ($R = NH_2$)—i.e. under kinetic control conditions—gave mainly XI, via process *a*, accompanied by lesser amounts of XII and XIII, the former being derived by process *b* and the latter by the sequence of shifts shown below. However, the bicyclo[3,2,2]nonane system appears to be thermodynamically less stable than its [3,3,1]-isomer, for the same

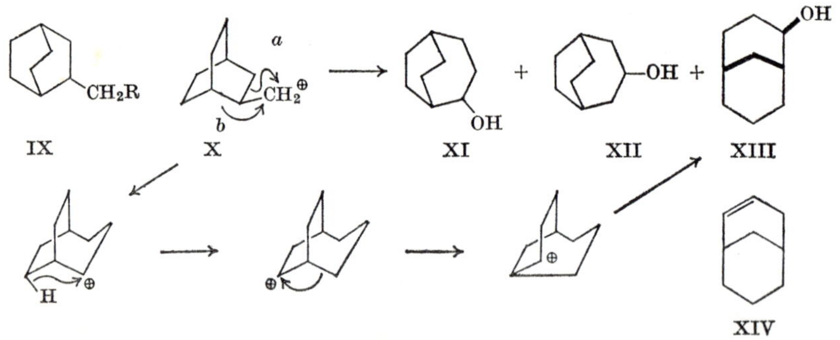

IX X XI XII XIII XIV

carbonium ion X, generated from the alcohols IX (R = OH) or XI by the action of strong acid—i.e. under conditions approaching equilibrium control—yielded only the olefin XIV[116].

The formation of a bridged compound XVIII from any of the three alcohols XV, XVI, or XVII by reaction with polyphosphoric acid is less readily understood. It is remarkable that hydrofluorenes are not detected amongst the products formed in any of these dehydrations, which must produce *inter alia* the carbonium ion XIX. The same process applied to the cyclopentanol analogues gives rise to XX, and in both series the structures of the products have been rigorously proved by synthesis[10, 12]. The dimethyl analogue XXII is interesting since it affords the hydrofluorene XXI when treated with polyphosphoric acid but the bridged product when treated with aluminium chloride[72]. This synthetic approach to bicyclo[3,3,1]nonanes has been employed in a key step (XXIV → XXV) of a recent synthesis of lycopodine[132].

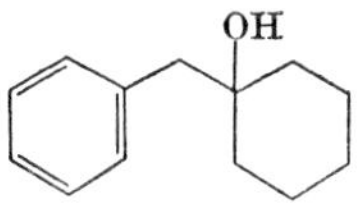

OH OH OH

XV 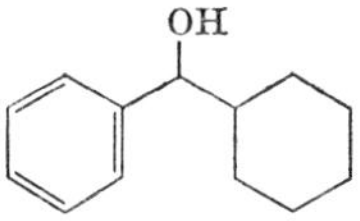 XVI 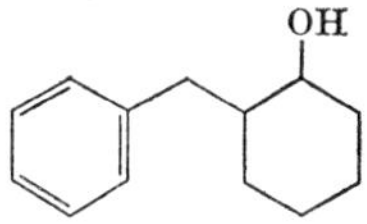 XVII

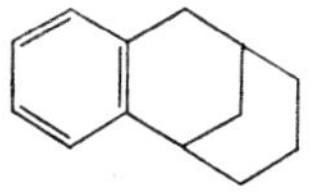

⊕

XVIII 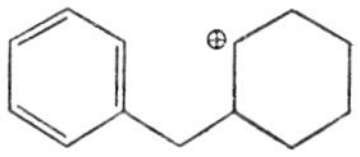 XIX 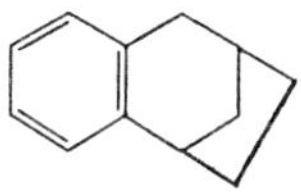 XX

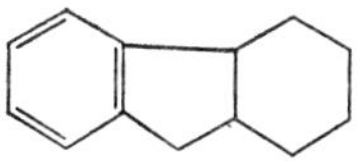

HO

XXI 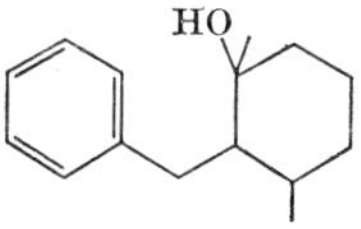 XXII 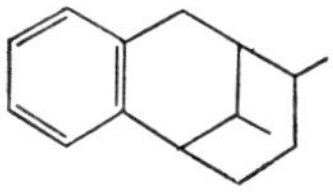 XXIII

H

OMe O N ⊕ H O N H

O N H OMe OMe

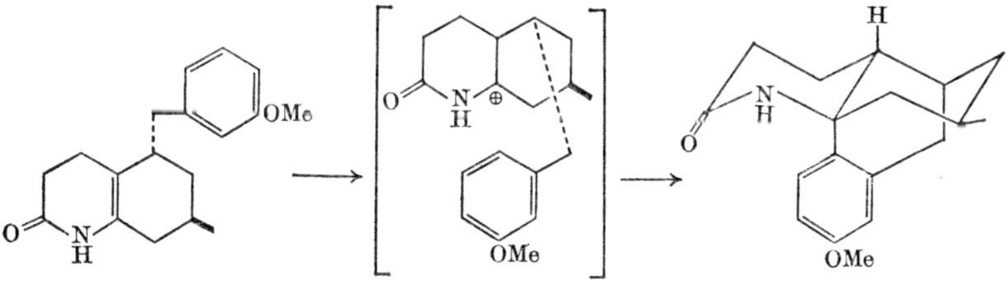

XXIV XXV

The search for synthetic routes to the gibberellins has also uncovered some interesting results. In exploratory work related to this objective, it has been shown that acid catalysed cyclisation of the cyclohexene carboxylic acids XXVI ($n = 1$ or 2) furnishes good yields of the bicyclic diketones XXVII ($n = 1$ or 2), but is less efficient[92] when $n = 3$. The corresponding reduced keto-acid XXVIII (R = H) gives a mixture of two possible products, XXIX (R = H) and XXX.

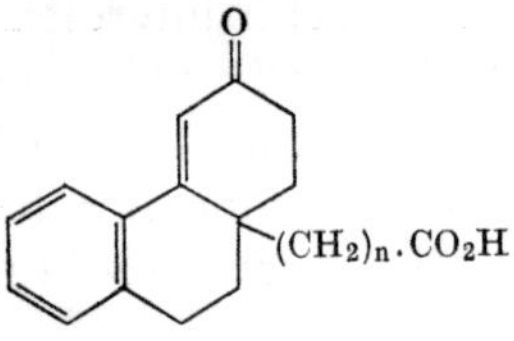

XXVI 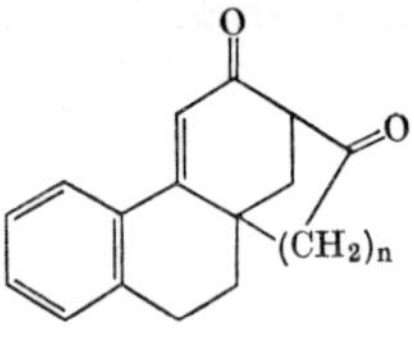 XXVII 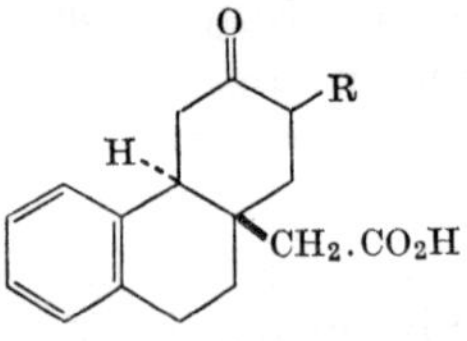 XXVIII

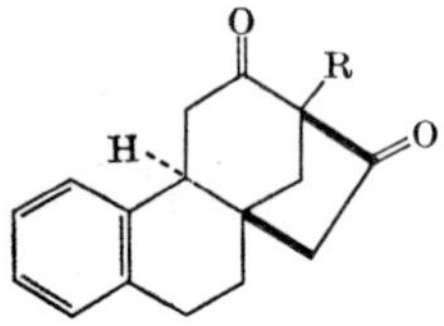

XXIX 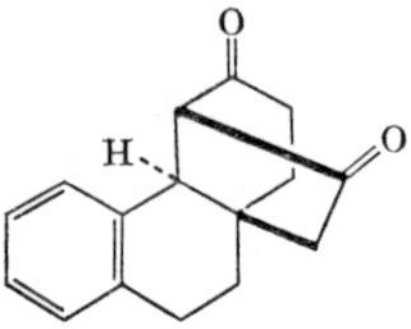 XXX 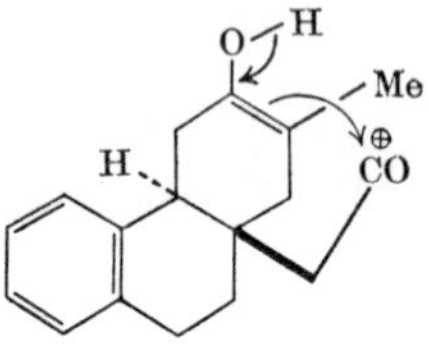 XXXI

However, the presence of a methyl group, e.g. XXVIII (R = Me), favours the enol XXXI and directs the ring-closure exclusively towards XXIX (R = Me)[14, 92].

Bicyclo[3,3,1]nonane derivatives are also to be found amongst the rearrangement products of caryophyllene (XXXII). On treatment with acid, it gives an alcohol caryolan-1-ol (XXXIII), formerly called β-caryophyllene alcohol, and an isomeric olefin clovene (XXXIV), whose synthesis has recently been described[14, 51]. Although it is possible to draw a 'paper' mechanism for the transformation, XXXIII → XXXIV, clovene is *not* derived from caryolan-1-ol; both are derived from XXXII by competing processes. Under different conditions the tertiary alcohol XXXIII *can* be dehydrated, with rearrangement, to a mixture of three olefins: iso-clovene (XXXV), pseudo-clovene A (XXXVI) and an isomer named pseudo-clovene B whose structure is still unknown[61]. It can be seen from models that the conversion of caryophyllene to clovene or caryolan-1-ol is controlled by the conformation of the large ring. If cyclisation of the *exo*-methylene double bond takes place from the *underside* (cf. XXXVII) of the 9-membered ring, one of the cyclobutane bonds is situated antiperiplanar, and can therefore migrate as shown. Loss of a proton from the newly created 5-membered ring completes the formation of clovene XXXIV. On the other hand, if the bridge is formed *above* the 9-membered ring (cf. XXXVIII), no migration is possible, and the ensuing carbonium ion is hydrated to XXXIII.

The generation of iso-clovene and pseudo-clovene A from XXXIII can be explained[61] on the assumption that the more drastic, anhydrous conditions promote a fragmentation (cf. XXXIX) leading to the diene XL. Protonation of this intermediate at C(2) could lead via XLII to XXXVI, but protonation at C(6) would give rise to XLI which can be plausibly rearranged to XXXV as indicated. This mechanism is preferable to that previously suggested[31, 32] for two reasons; it rationalises the simultaneous formation of XXXV and XXXVI, as well as their stereochemistry, and in doing so, it involves a more plausible, i.e. less strained, intermediate (XL).

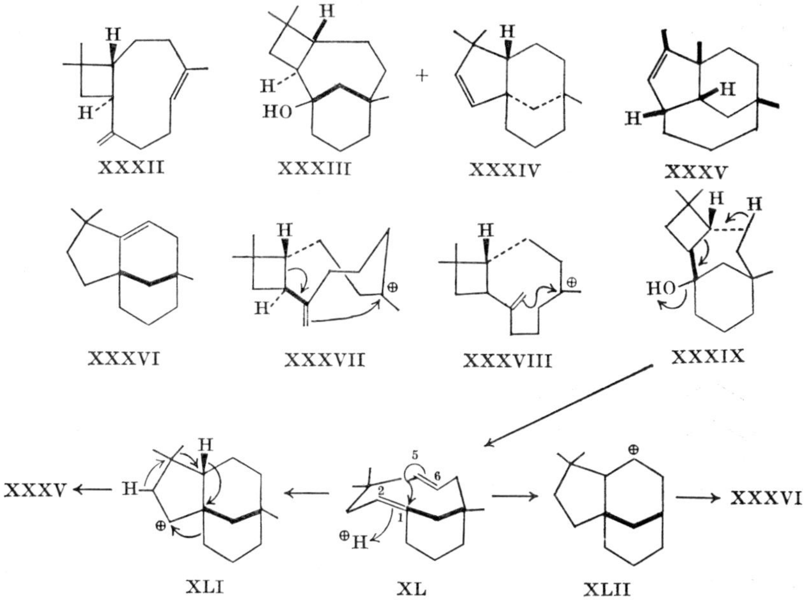

C. Physical Properties

Bicyclo[3,3,1]nonane forms plastic, volatile crystals which have been described as having a camphane-like odour and appearance[105]. It melts at 145–6°, sublimes readily, boils at 170° and is volatile in steam.

The first intuitive three-dimensional representation of the molecule by Meerwein in 1922[105] showed it in the chair–chair conformation I, but later discussions favoured the boat–chair conformation II as a means of relieving the interaction of the endo-hydrogens at C(3) and C(7)[56]. Ironically a recent study has vindicated Meerwein. The infra-red spectrum of III shows[54] unusual C—H bending and stretching

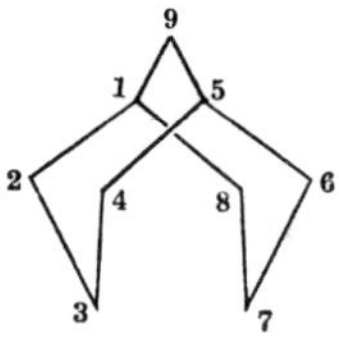

I

II

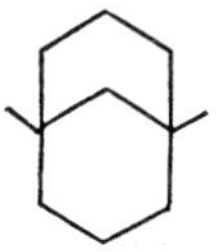

III 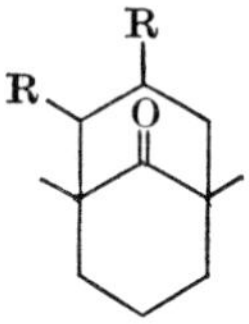 IV 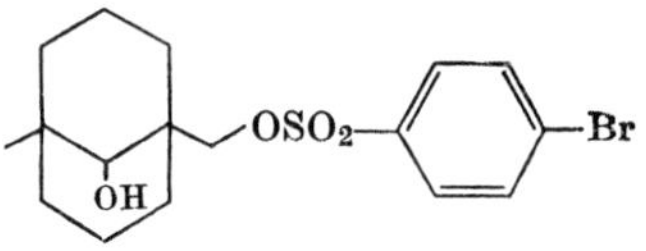 V

absorptions at about 1490 and 2990 cm^{-1} which have been attributed specifically to C(3)–C(7) hydrogen interactions for the following reasons. These bands are also observed in the spectrum of the ketone IV (R = H) and so an alternative assignment to C(3)–C(9) hydrogen interaction may be discounted. Their intensities are halved in the deuterated ketone IV (R = D) and they are totally absent in the bicyclonon-2-ene and in adamantane. They are seen in both solution and solid state spectra and hence the X-ray evidence, which shows that V adopts the chair–chair conformation in the crystal, can be extrapolated to establish the conformation in solution[21]. The molecule relieves C(3)–C(7) hydrogen interaction by flattening both 'wings'. Thus the distance between C(3) and C(7) is found to be 3·06 Å rather than 2·52 Å as measured in the ideal twin chair structure, and the C(2)–C(1)–C(8) angle is 113°, i.e. slightly greater than tetrahedral. There is no evidence of skewing but rather of 'splaying', by bending C(3) and C(7) upwards so that the plane described by the atoms C(2)C(3)C(4) is at an angle of 18° to, rather than parallel to, that of atoms C(1)C(9)C(5). The resulting globular-shaped molecule is able to rotate freely between a solid transition point and the melting point, thus giving rise to the phenomenon of 'plastic' crystals. Bicyclo[3,3,1]nonane also forms mixed crystals with adamantane, consistent with its double chair conformation[88].

Even in the presence of a bulky axial substituent at C(2) the bicyclo-[3,3,1]nonan-9-one VI appears to prefer the double chair conformation despite the absence of C(3)–C(9) hydrogen interactions in the boat form (VII)[58, 136]. It must therefore be assumed that the major factor in determining the conformation is the preference of the cyclohexane

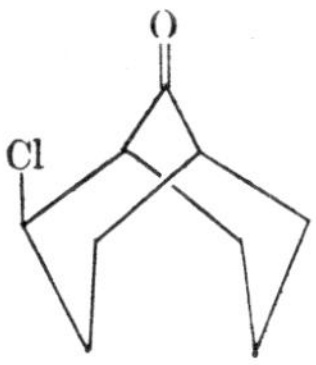

VI 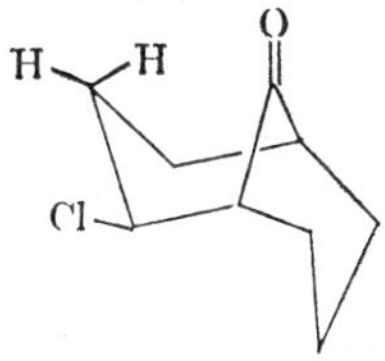 VII

ring for the chair rather than the boat form.

The transannular strain in the double chair conformation has been estimated[4] as approximately 3 kcal. mole^{-1}.

An *exo*-substituent at C(3) has no influence on the preferred conformation, but an *endo*-C(3) substituent such as OH, Br, CO_2H or Me, forces the ring into the boat form[4, 34, 94, 95]. Even the natural bicyclononanolide, swietenine VIII, which is gem-dimethylated at C(4), but has one ring 'flattened out' by a double bond, has been shown[94] to prefer the saturated ring in the boat form. Most surprising of all, the

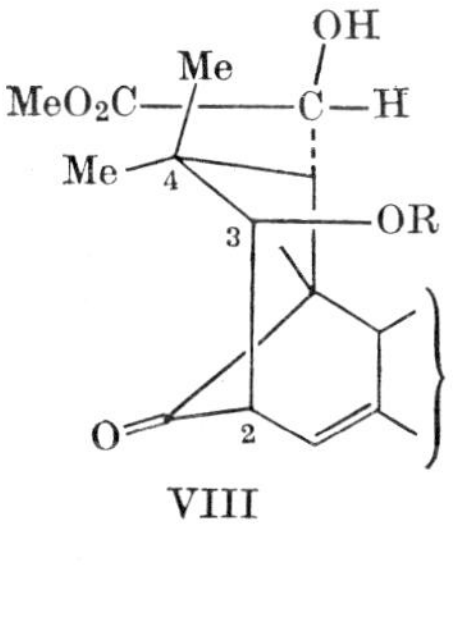

VIII 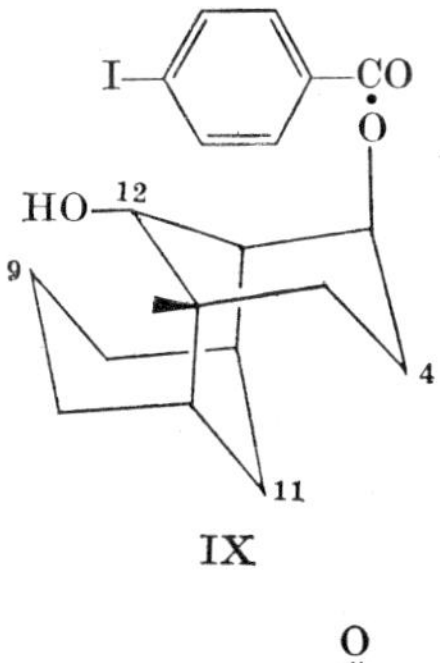 IX

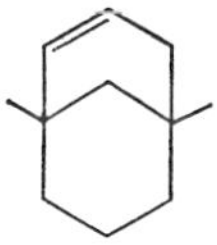

X 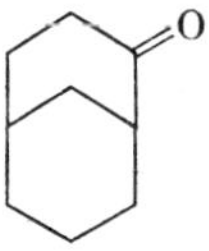 XI 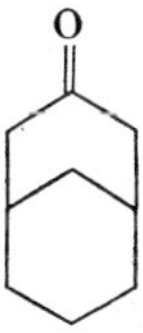 XII

tricyclic alcohol IX adopts the all-chair conformation shown[62, 96] in spite of an *endo*-C(12) hydroxyl. Here again, the molecule adapts itself to congestion by 'splaying' C(4) and C(9) outwards, and bending the C(12)-OH bond upwards. These findings provide a timely reminder to those of us who use molecular models, that nature is less naive than we sometimes imagine.

The bicyclo[3,3,1]nonane skeleton is strainless, and consequently its infrared absorption characteristics resemble those of 6-membered rings. The olefinic C—H stretching and bending and the C=C stretching frequencies observed in compounds such as X[54] are

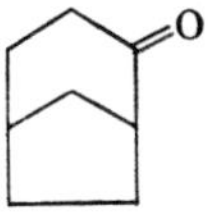

XIII 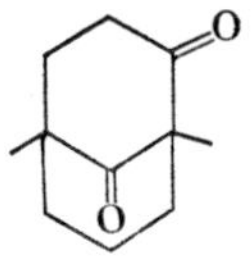 XIV 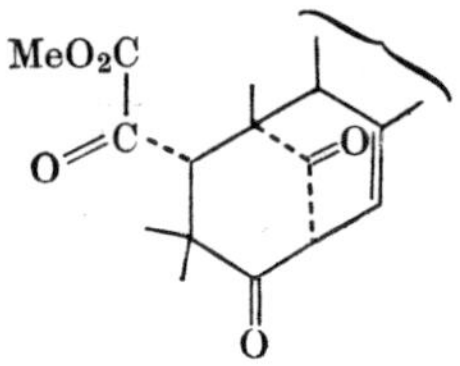 XV

approximately those expected in a *cis*-cyclohexene. Similarly, carbonyl absorptions in XI and XII are unexceptional at 1720 and 1717 cm^{-1} respectively[63, 142] and even the corresponding bicyclo-octane XIII[63] absorbs similarly at 1717 cm^{-1}. However, in carbonyl-bridged bicyclics, the C=O stretching frequency is affected by ring strain as shown in Table 1. It can be seen that as the size of the lower ring decreases, ν_{CO} rises exceptionally. Substituents appear to exert a slight influence, as does an *endo*-cyclic double bond. In some of the bicyclic-enones, the C=O absorption appears as a doublet, an effect which is probably due to Fermi resonance with a suitable overtone rather than to dipole interaction, since the relative intensities of the peaks, and their separation, varies with change in polarity of the solvent (Table 2). On the other hand, the C=O doublet of β-diketones such as XIV is not affected by solvent polarity (Table 3) and is ascribed to dipole interaction. The effect is common amongst 1,3-dicarbonyl compounds such as anhydrides, Meldrum acids, diacyl peroxides and β-diketones[16], and is particularly evident in non-enolic systems such as the β-diketones listed in Table 4. It may also account for the C=O doublets observed in medium-sized-ring β-diketones[55] in which enolisation is stereochemically disfavoured. In the bridged bicyclic diketones, band separation is approximately 30 cm^{-1} and fairly constant, although the factors which affect $\Delta\nu$

TABLE 1

Infrared carbonyl frequencies (cm^{-1}) in carbon tetrachloride solution

A	B	C	D	n	ν_{CO}	Reference
CO_2H	H	H	H	1	1780[a]	15
H	H	H	H	2	1750[b]	65, 104
H	CO_2Et	OH	H	2	1760	25
H	CO_2Et	OAc	H	2	1761	25
H	H	OH	H	2	1742, 1748	90
H	H	H	H	3	1724	66
H	H	H	H	3	1726	65
Me	Me	H	H	3	1714	54
Me	Me	OH	H	3	1713	102
					1716	54
H	CO_2Et	H	H	4	1705	135
H	H	H	H	5	1700	135
H	CO_2Et	H	H	5	1700	135
H	CO_2Et	OH	H	5	1705	135
H	H	H	H	2	1758	65, 91
H	H	Ph	H	2	1758	22
H	CO_2Et	H	H	2	1763	26
H	H	H	H	3	1733, 1725	102, 66
H	H	Ph	H	3	1732, 1722	103
Me	Me	H	H	3	1720	54
H	H	Ph	H	4	1718, 1710	22, 89
H	H	Ph	H	5	1701	89

[a] No solvent. [b] $CHCl_3$.

are not understood. To the reviewer's knowledge, it fails to appear in only one case (XV)[34]. This case is particularly interesting, for it can be seen from molecular models that in this molecule, the C=O

TABLE 2

Variation in ϵ and Δ with solvent polarity[89, 103]

Compound	Hexane		CCl$_4$		CHCl$_3$	
	ν_{CO} (ϵ)	Δ	ν_{CO} (ϵ)	Δ	ν_{CO} (ϵ)	Δ
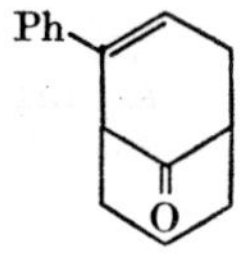 [structure: Ph-substituted bicyclic ketone]	1736 (504) 1724 (160)	12	1732 (444) 1722 (313)	10	1729 (291) 1715 (437)	14
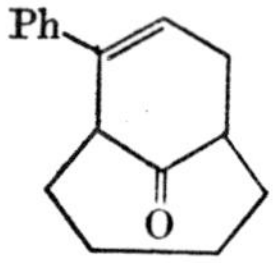 [structure: Ph-substituted tricyclic ketone]	1725 (252) 1710 (95)	15	1718 (474) 1710 (248)	8	1715 (222) 1705 (350)	10

TABLE 3

Comparison of ϵ and Δ values in solvents of different polarities[54, 89]

Compound	Hexane		CCl$_4$		CHCl$_3$		CH$_3$CN	
	ν (ϵ)	Δ	ν (ϵ)	Δ	ν (ϵ)	Δ	ν (ϵ)	Δ
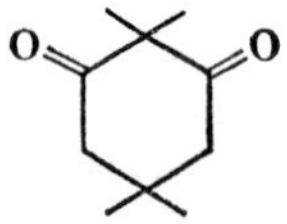 [structure: tetramethylcyclohexane-1,3-dione]	1736 (368) 1705 (817)	31	1734 (394) 1700 (980)	34	1730 (364) 1694 (930)	36	1730 (357) 1695 (945)	35
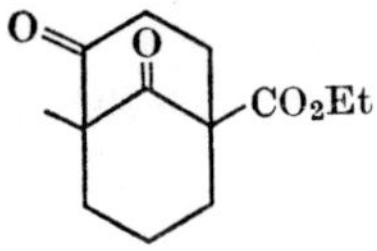 [structure: bicyclic dione, CO$_2$Et]	1748 (553) 1739[a] 1717 (1060)	31	1743 (497) 1734[a] 1711 (1008)	32	1739 (474) 1729[a] 1707 (895)	32	1741 (436) 1732[a] 1709 (734)	32
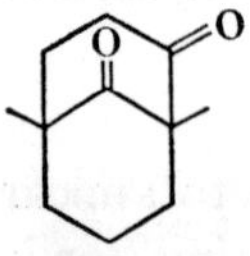 [structure: bicyclic dione]	1741 (245) 1712 (735)	29	1737 (230) 1707 (730)	30	1729 (255) 1701 (725)	28	1728 (250) 1702 (640)	26

[a] Ester absorption.

TABLE 4

I.r. carbonyl absorption of non-enolic 1,3-diketones

1,3-Dione	$\nu_{CO}^{CCl_4}$	Δ	Ref.	1,3-Dione	$\nu_{CO}^{CCl_4}$	Δ	Ref.
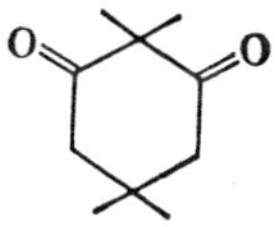	1734 1700	34	89	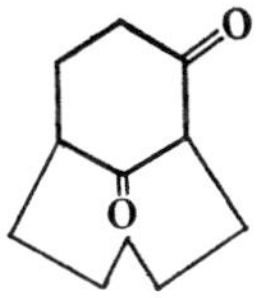	1724 1695 [a]	29	97
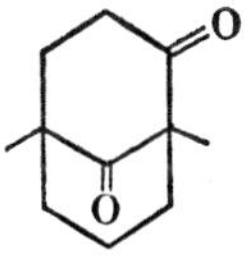	1737 1707	30	54	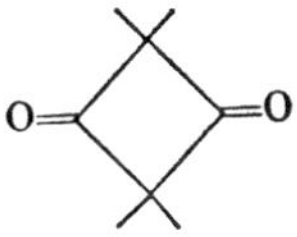	1754 1724	30	68
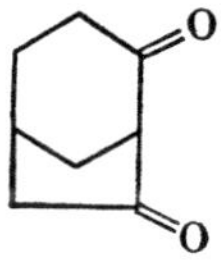	1745 1710 [a]	35	13		1802 1751	51	106

[a] Liquid film spectrum.

dipoles *do not intersect*. This circumstance, which is probably responsible for the unsplit ketone band, arises in a boat cyclohexanedione provided one carbonyl is located at a 'bow' or 'stern' position. In the

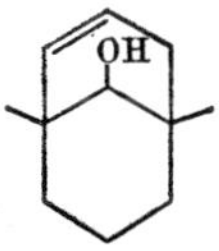

XVI 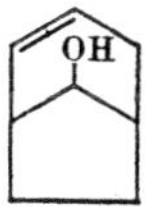 XVII 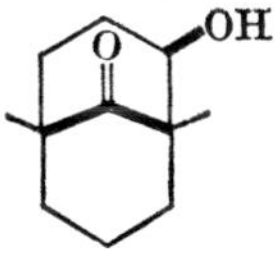 XVIII 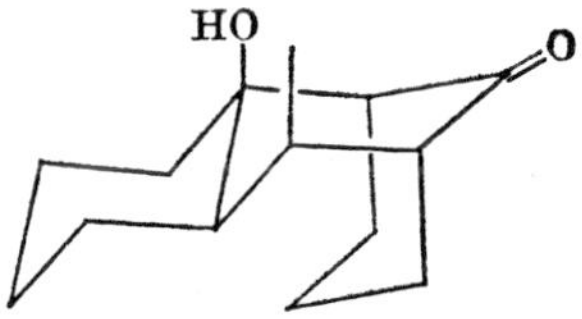 XIX

syn-alcohol XVI, interaction of the hydroxyl and olefin groups by OH-π bonding results in a broadening of the olefinic C—H deformation band and displacement of the O—H stretching absorption to 3584 cm^{-1}. The *anti*-isomer shows [54] normal characteristics (3640 and

3625 cm^{-1}), and a longer gas-liquid chromatography (g.l.c.) retention time. These effects are duplicated in the analogous bicyclo[3,2,1]-octenol XVII [91]. Analogous OH-carbonyl bonding was not observed in XVIII [54] but a parallel study on XIX found absorption due to both bonded and free C═O and OH [109].

There has been no systematic nuclear magnetic resonance (n.m.r.) study of the bicyclo[3,3,1]nonane skeleton, and few useful data are

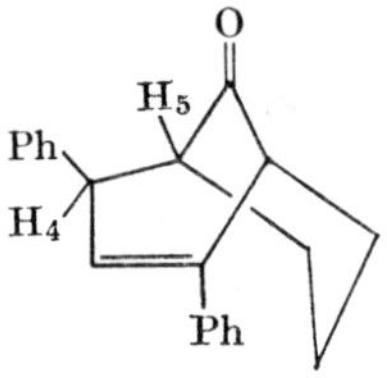

XX 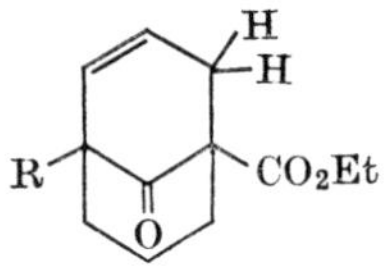 XXI 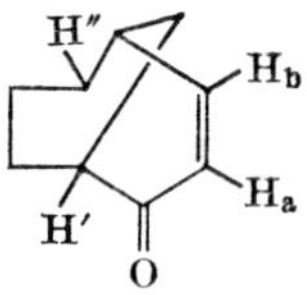 XXII

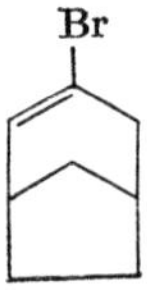

XXIII 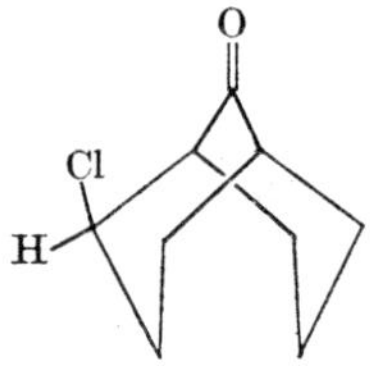 VI 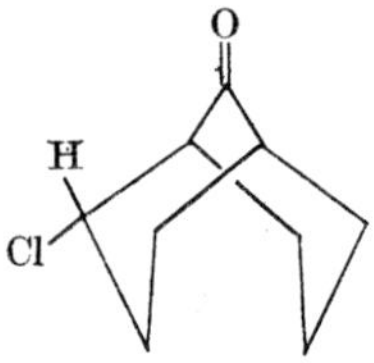 XXIV

available. In the bicyclo[3,3,1]non-2-en-9-one XX, the stereochemistry of a substituent at C(4) can be allocated by n.m.r., since an *endo*-proton (H(4) in XX) does not couple with H(5), with which it subtends an angle of c. 90°, whilst the epimer shows a 6 Hz coupling [77]. In the related keto-ester XXI (R = H or Me) one of the allylic protons is selectively deshielded by the ester group and appears at c. 6·6 τ, i.e. 1 ppm lower than the other. In the absence of the ester function both allylic protons absorb at 7·58 τ [33]. Long range couplings of 1–2 Hz have been noted in XXII and identified as H_aH' and H_bH'' [17]. Similar long range effects have been observed in XXIII [81] and in an analogous bicyclo[3,3,1]nonenone [4].

The rigidity of the bicyclic molecule makes it possible to assign the configuration of suitable substituents from the n.m.r. spectrum, by observing the J_{sum} (total width) or $W_{1/2}$ (width at half-height) of the adjacent proton signal. The epimers VI and XXIV show [58] J_{sum} 10·5 and 24 Hz respectively, and in the bicyclo[3,2,1]octane series, the use of $W_{1/2}$ is illustrated in Table 5[25]. In addition, the axial proton signal occurs at higher field.

TABLE 5

The use of $W_{1/2}$ in assigning stereochemistry[25]

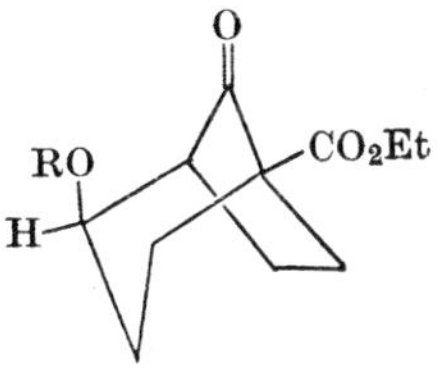

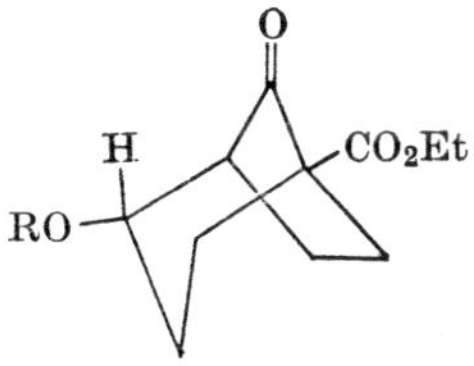

R	*CH*OR signal (τ)	$W_{1/2}$ (Hz)	R	*CH*OH signal (τ)	$W_{1/2}$ (Hz)
H	c. 5·75[a]	c. 12	H	c. 5·75[a]	c. 30
Ac	4·90	9·0	Ac	5·05	20·4
Tos	4·95	9·0	Tos	5·35	18·0

[a] Signal partly obscured.

D. CHEMICAL PROPERTIES

1. *General*

The unique V-shape of the bicyclo[3,3,1]nonane molecule, in the double chair conformation (see section C), leaves it exposed to attack from the *exo*-face but shielded from attack from the *endo*-face. It also brings C(3) and C(7) close together, and these geometric effects are reflected in the general chemistry of bicyclononanes.

The bridge C(9) position is exposed to attack from either side, the approach of the reagent being impeded only by the axial hydrogens at positions 2, 4, 6, and 8. The barrier presented by these hydrogen atoms is small, and is too small to render the carbonyl group inert[41]. Its effect becomes apparent only in the related ketones I and III. Reduction of the former yields only II[66] under conditions of kinetic control, and similar reduction of the enone III affords mainly (i.e. 75 per cent) the *anti*-alcohol IV[101]. Thus, lowering the barrier on one side of the molecule by ring-contraction or by ring-flattening facilitates attack from that side very significantly. Of these two factors, ring contraction appears to be the more powerful, for the bicyclo[3,2,1] octenone V yields mainly (c. 70 per cent) VI on reduction[66], although equilibration favours the *exo*-epimer[91]. An axial substituent, predictably, has a decisive effect; thus VII yields the *syn*-alcohol VIII exclusively whilst its equatorial isomer affords a mixture of IX and X in the ratio 2:1[5, 49]

It is clear that changing the hybridisation of the bridge carbon from sp^3 to sp^2 introduces some strain, and this effect becomes more apparent as the size of the rings is reduced. In bicyclo[3,2,1]octan-8-one and norbornan-7-one, not only is ν_{CO} raised (see section C), but semicarbazone formation, i.e. dehydration of the intermediate carbinolamine derivative, becomes progressively more difficult; norbornanone itself forms a stable hydrate and there are reports that in the bicyclo[3,2,1]octane series, difficulty has been experienced in hydrolysing a ketal and in oxidising an alcohol located at the bridge position[66, 91].

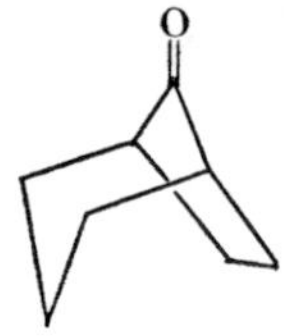

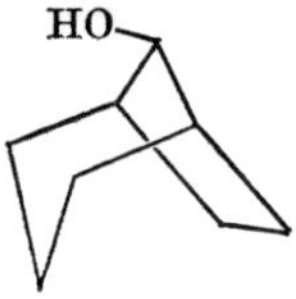

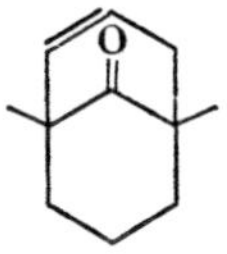

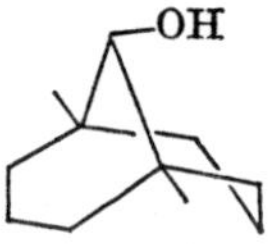

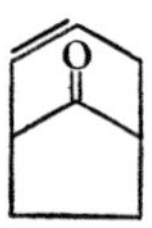

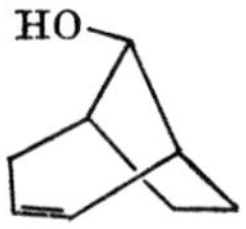

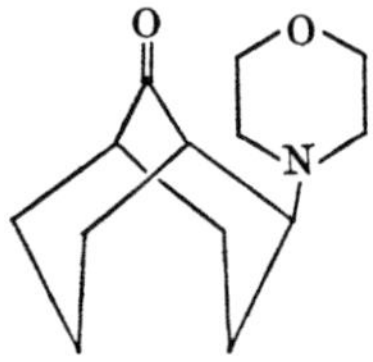

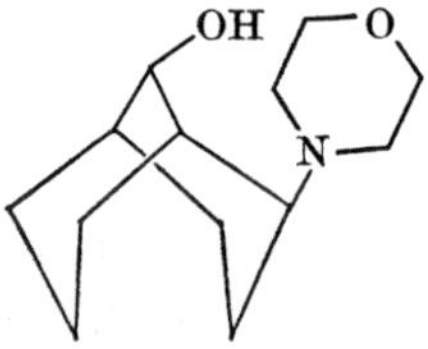

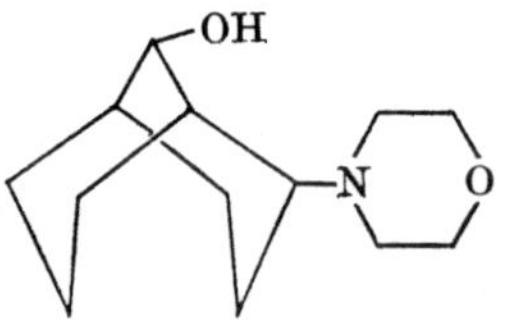

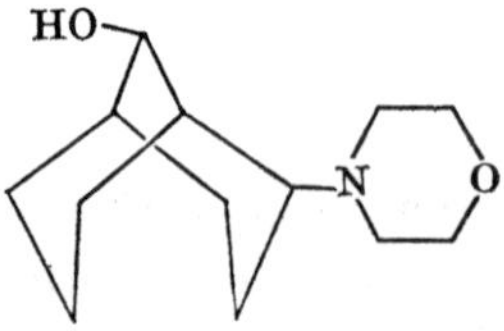

Of course, these generalisations apply to molecules existing in the double chair conformation. When one ring is in the boat form, and heavily substituted as in the bicyclononanolides (see section D.3, structures XXII and XXIV) the bridge carbonyl is inert even to hydride[1, 34, 35].

An interesting consequence of ring conformation is the formation of only a *mono*-bromoketone XI (R = Br) from the 7α-methylketone XI (R = H) whilst the 7β-methyl isomer (XII) is smoothly converted into XIII[4]. A bulky 3-α or 7-α substituent seems to force the cyclohexanone ring into the boat conformation, and gem-dibromination of XI (R = H) which would lead to a double boat conformation, becomes intolerable.

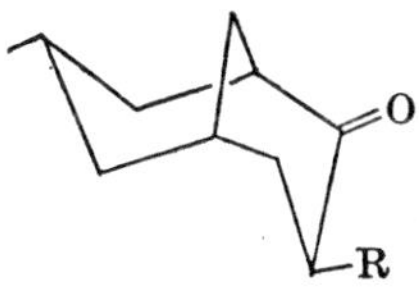

XI 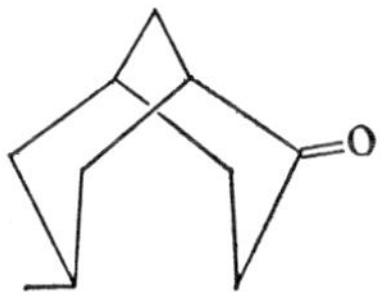 XII 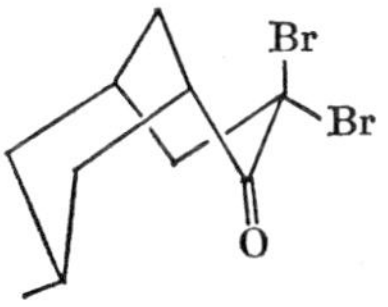 XIII

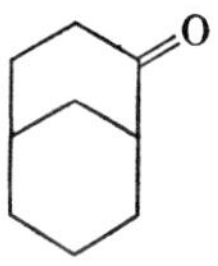

XIV 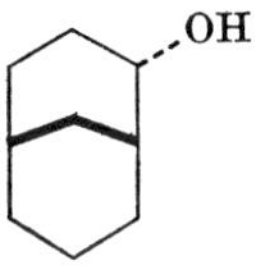 XV 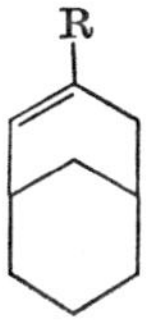 XVI 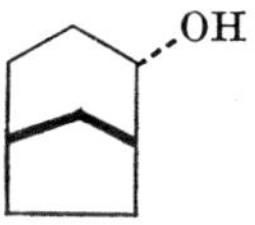 XVII

The approach of reagents to C(2) and C(3), only from the exposed β-face, is exemplified by the exclusive reduction of XIV to XV[6], β-epoxidation of XVI (R = H)[6] and β-reduction of its homologue (XVI; R = Me)[4]. It is of interest to note that under equilibrium conditions, the equatorial alcohol XV yields a surprising 30 per cent of its axial epimer. The relative destabilisation of the equatorial isomer is attributed[6] to a non-bonded interaction with the equatorial hydrogen on C(8), for the analogous bicyclo-octanol XVII, in which such an interaction is much reduced, affords but little of the epimer when similarly treated.

Reduction of the 2β,3β-epoxide of XVI (R = H) occurs exclusively[6] at the (more exposed) C(3) position leading to XVIII; selective reduction[102] of the dione XIX affords a 2:1 mixture of the ketols XX and

XXI, which reflects the accessibility of C(2) over C(9). Although the axial ketol (XXI) may arise from XX by equilibration (see above), the

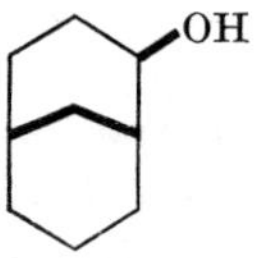

OH O O OH O O

XVIII 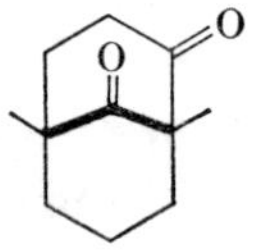 XIX 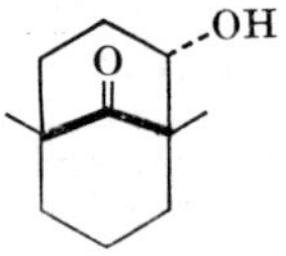 XX

OH O

O O

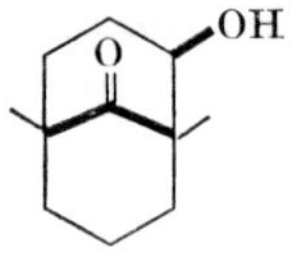

XXI 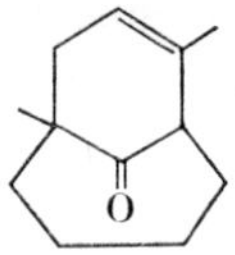 XXII 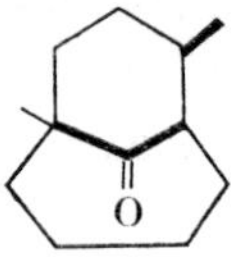 XXIII

formation of XVIII indicates that attack from the *endo*- (or α-) face is possible. *Endo*-attack is also implicit in the reduction[99] of XXII to XXIII, and (less surprisingly) in the reduction of XXIV to a mixture of XXV and XXVI[77]. The reduction of 3-methylene- to 3-methylbicyclo[3,3,1]nonane affords mainly the 3α-epimer, but 12 per cent of the 3β-isomer is also formed[5].

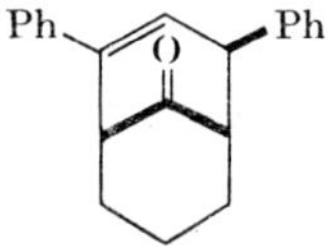

Ph Ph Ph Ph Ph Ph

O O O

XXIV 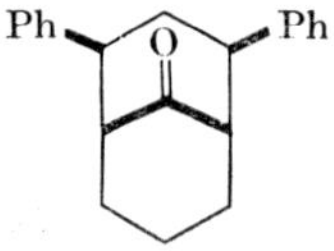 XXV 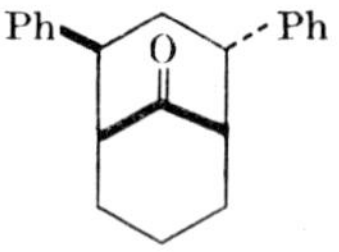 XXVI

Me Me

H

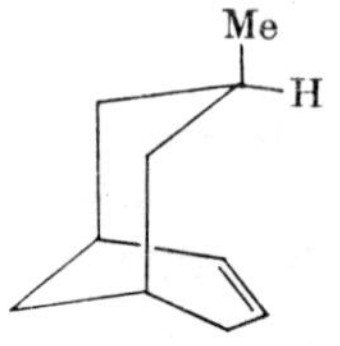

XXVII 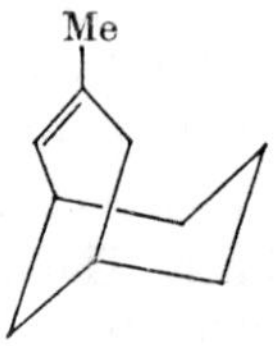 XXVIII

The proximity of C(3) and C(7) makes it easy to bridge these positions, and this subject is discussed in section D.4. It also facilitates C(3)–C(7) hydride transfer reactions. Thus boiling formic acid converts XXVII into XXVIII to the extent of 93 per cent[2] and formolysis of the

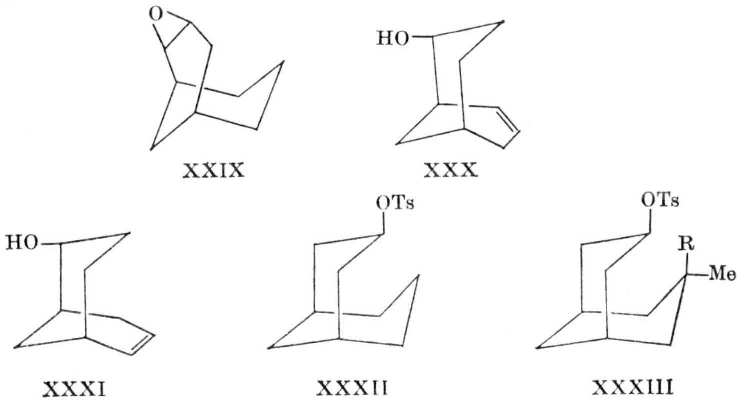

2β; 3β-epoxide XXIX gives a mixture of the isomeric olefins XXX and XXXI[3]. The same authors[3] have observed an interesting inhibition of C(3)–C(7) hydride transfer by an oxygenated function at C(9). Under kinetic conditions, i.e. solvolysis of the tosylate XXXII, labelled with ^{14}C at C(3), little hydride transfer is observed, but in XXXIII (R = H), the transformation of a secondary carbonium ion into a tertiary carbonium ion provides a more efficient (c. 50 per cent) hydride shift. The solvolysis of XXXIII (R = H or D) shows no significant deuterium isotope effect, and consequently the hydride shift is not synchronous with the ionisation step[52, 53].

2. *Bridge-fission*

Bridged bicyclic compounds undergo a variety of bridge-fission reactions which provide novel approaches to medium and large ring compounds. Since most bicyclic compounds are constructed from simple, readily available intermediates, this process provides a useful route to particularly substituted rings. For example the Beckmann[38, 39] and Baeyer-Villiger[75] reactions convert a bridge ketone into a cyclic amino- or hydroxy-acid; however more interesting transformations have come to light. Although bicyclononenones of type I are formed under acid conditions, and are stable to acid, the related bicyclo-octenones II are transformed by acid into a mixture of cycloheptenecarboxylic acids III via the intermediate IV[22, 26]. The fragmentation is promoted by ring strain, for although larger-ring homologues of I are stable to, and are readily prepared under, acid

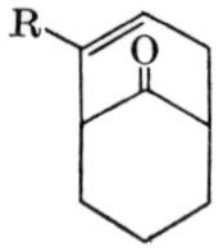

I

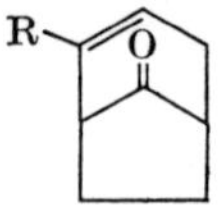

II

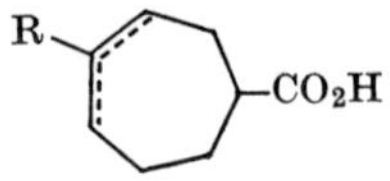

III

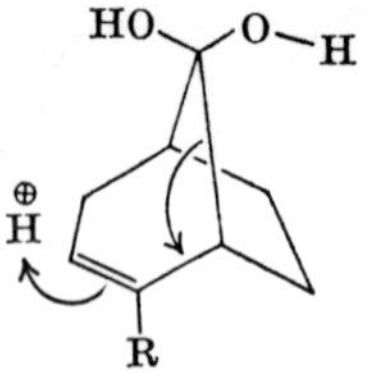

IV

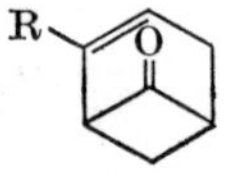

V

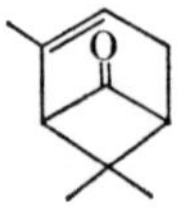

VI

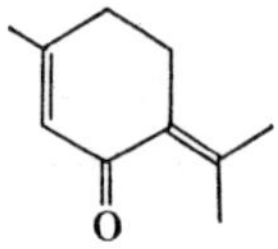

VII

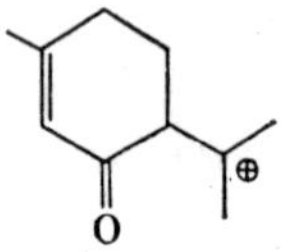

VIII

conditions, attempts to make smaller-ring analogues such as V (R = Ph) yield only the fragmentation product, phenylcyclohexene-4-carboxylic acid[26]. However, it should be noted that although V is

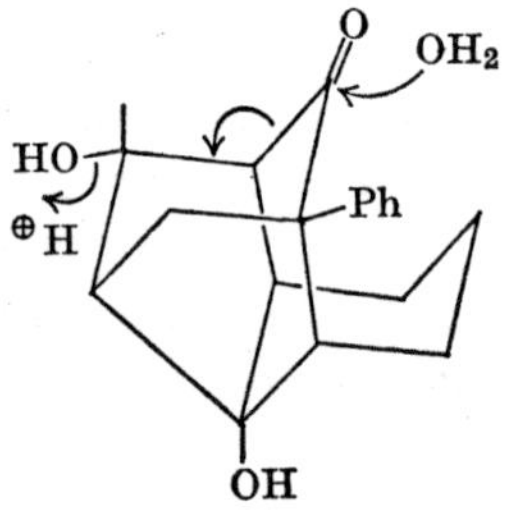

IX

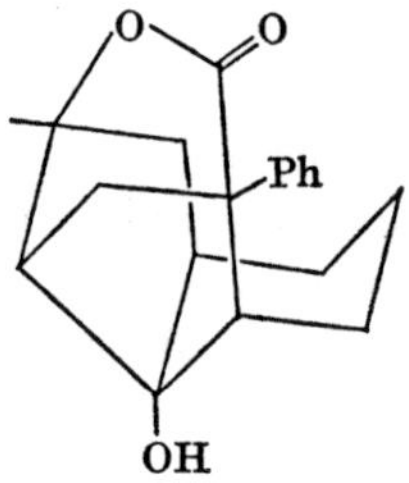

X

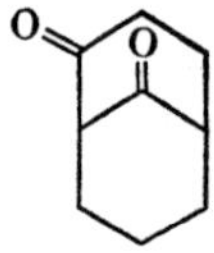

XI

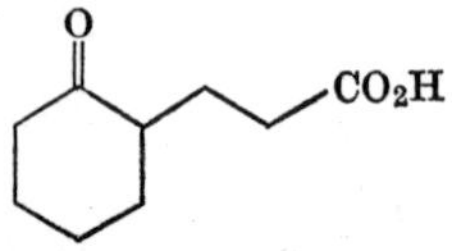

XII

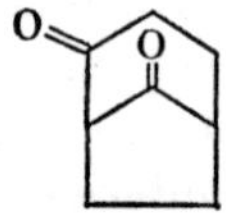

XIII

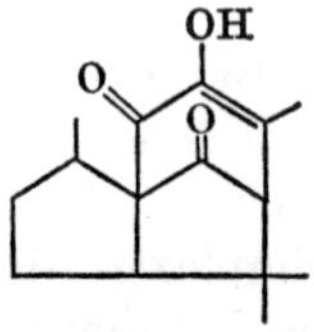

XIV

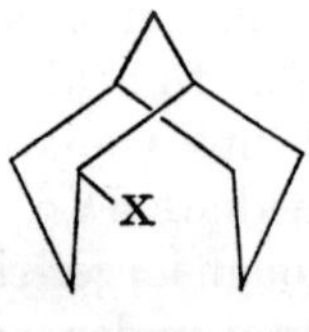

XV

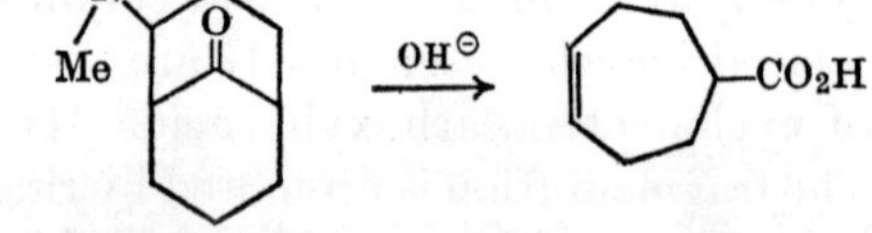

XVI XVII

highly unstable, chrysanthenone (VI) is a naturally occurring example of this type, which owes its stability to the gem-dimethyl grouping. Its conversion under acid conditions to piperitenone (VII), rather than a carboxylic acid, can be understood in terms of the stability of the carbonium ion VIII.

The role of ring stress in promoting bridge fission is further underlined by the case of IX. Although it is essentially a bicyclo[3,3,1]-nonanone, it is completely transformed into the lactone X by acid[24].

The same factors control the hydrolytic scission of bridged bicyclic β-diketones. The unstrained dione XI is cleaved to XII[71, 76], but the cyclopentanone analogue (XIII) affords cycloheptanone-4-carboxylic acid even under very mild conditions[74]. Once again it is necessary to point out that heavily substituted natural products behave anomalously. The pipitzols, stereoisomers of structure XIV, are alkali-stable[134], although they incorporate the same 1,3-dione system as XIII and it must be assumed that the stability of XIV is a feature of its special structure.

One of the most striking and most versatile fission reactions utilises the fact that a 2-(*endo*)-substituent in a bicyclo[3,3,1]nonane has the

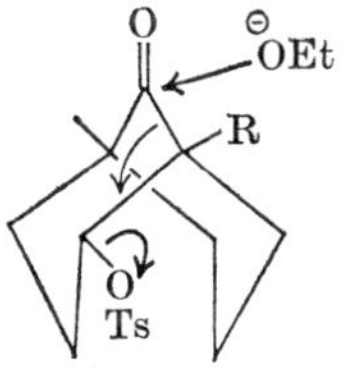

XVIII 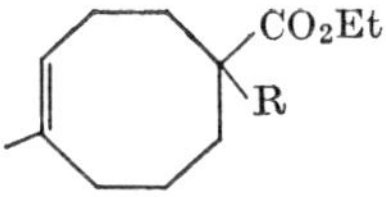 XIX 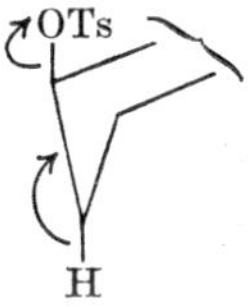 XX

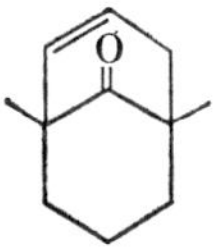

XXI 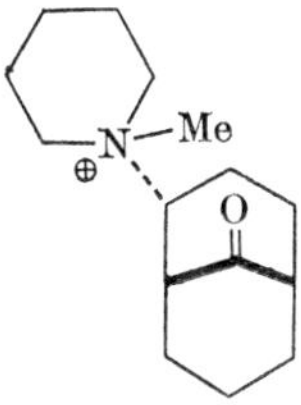 XXII 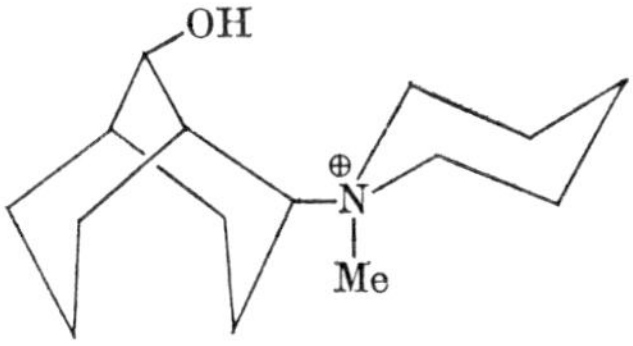 XXIII

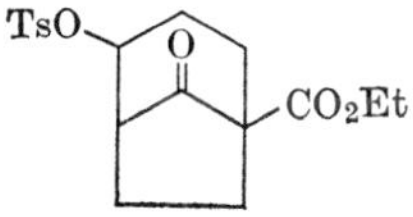

XXIV 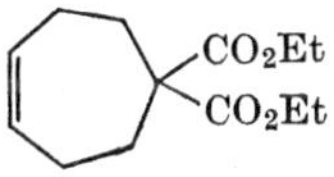 XXV

antiperiplanar geometry XV with respect to the C(1)–C(9) bond, which is essential to a smooth fragmentation reaction. The reaction was first demonstrated by the transformation XVI → XVII[49,131], but the process has been more extensively studied on the analogous tosylates. Thus it has been shown that the equatorial tosylate XVIII is specifically and very readily cleaved to the cyclooctene XIX (R = Me or CO_2Et) by mild base treatment, leaving the axial epimer unaffected[23,25,101]. Under more drastic conditions this epimer affords (cf. XX) *only* the enone XXI[101]. The reaction has been applied to appropriate bicyclo[3,2,1]octane[25] and bicyclo[4,3,1]decane[23] derivatives, giving a series of cycloalkene carboxylic acids. Similar conclusions have been reached in a very recent investigation[49] of the related quaternary salts XVI. The equatorial methohydroxide XXII yields the acid XVII; its axial epimer does not. The same authors went on to show that fragmentation of the alcohol XXIII affords the aldehyde corresponding to XVII, in low yield. A major side reaction appears to be fission of the heterocyclic ring.

In the fragmentation of the tosyloxy-β-ketoester XVIII, the role of the bridgehead methyl function was not appreciated until a parallel investigation of XXIV showed that *both* epimers give cycloheptene derivatives. The equatorial isomer affords XXV as expected, but the

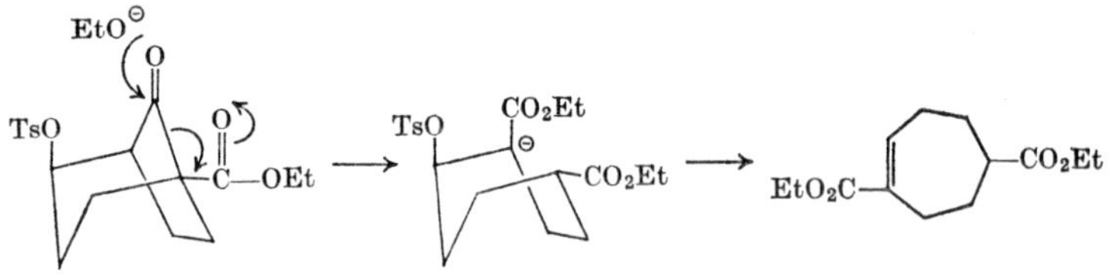

XXVI XXVII

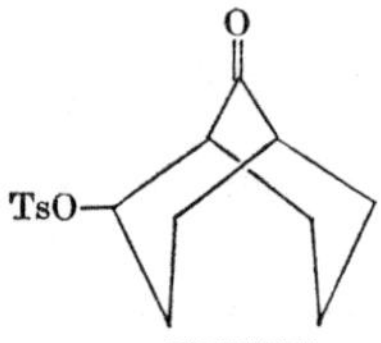

XXVIII 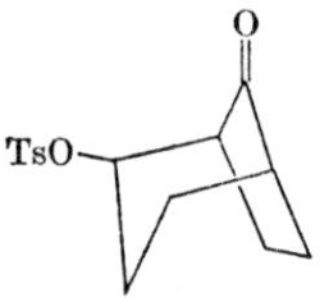 XXIX 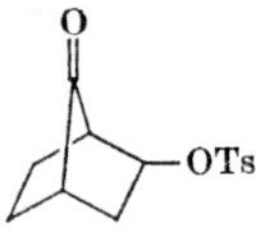 XXX 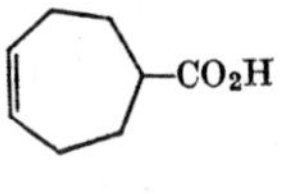 XXXI

axial tosylate XXVI gives rise to XXVII by the sequence shown[25]. A more precise picture of the mechanism of the tosylate fragmentation reaction has emerged from a recent study[85,86]. Acetolysis of XXVIII,

XXIX or XXX (both epimers), under anhydrous conditions, affords only the corresponding acetates; i.e. the absence of a nucleophile prevents the fragmentation. In aqueous acetic or formic acids the solvolysis of XXIX yields XXXI but its homologue XXVIII is merely hydrolysed. These experiments illustrate the complementary roles of ring strain and nucleophilicity in the fragmentation process, for the reaction XVIII → XIX can be brought about very smoothly by ethoxide. At the same time, the facile fission of the ethylene ketal XXXII to XXXIII by hydride, recently reported by the same authors[86], is difficult to reconcile with this simple picture of the mechanism.

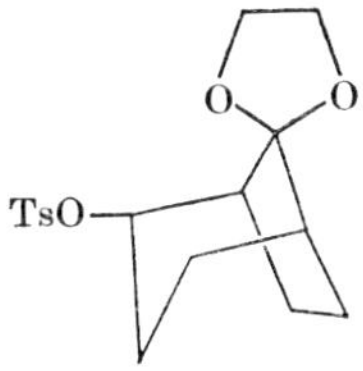

TsO, O, O — XXXII

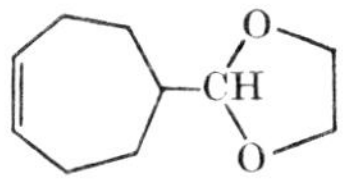

O, CH, O — XXXIII

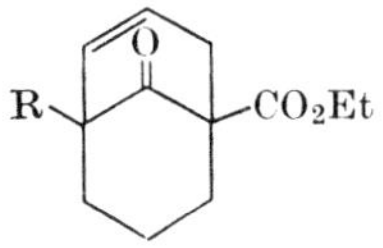

R, O, CO_2Et — XXXIV

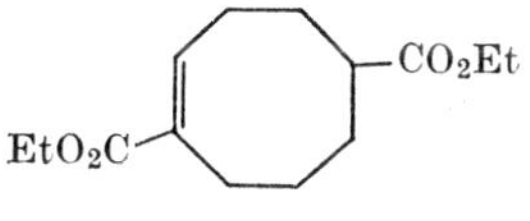

CO_2Et, EtO_2C — XXXV

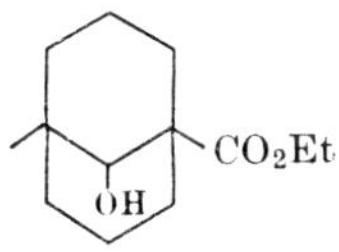

CO_2Et, OH — XXXVI

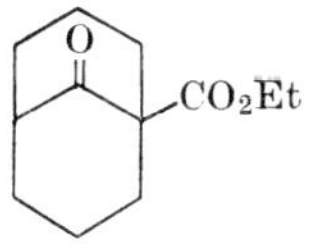

O, CO_2Et — XXXVII

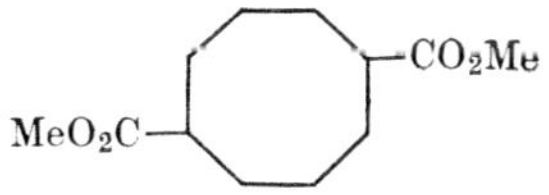

CO_2Me, MeO_2C — XXXVIII

Bridge scission has also been effected by a retro-Claisen reaction[42], e.g. XXXIV (R = H) → XXXV in which bridge opening is accompanied by migration of the double bond. This migration may well be vital to the success of the operation for the homologue XXXIV (R = Me) is inert under mild conditions, and more vigorous treatment leads only to the reduction product XXXVI[23]. Recently, this explanation has been challenged in the light of an observation that the saturated β-ketoester XXXVII is converted by alcoholysis to the diesters XXXVIII (mixture of isomers), in good yield[7].

3. *Skeletal rearrangements*

In describing synthetic approaches to the bicyclo[3,3,1]nonane skeleton (see section B.1) mention has already been made of rearrangements which lead to an aromatic acid and an enone (III and IV), during the conversion of I to II. One of these processes is clearly

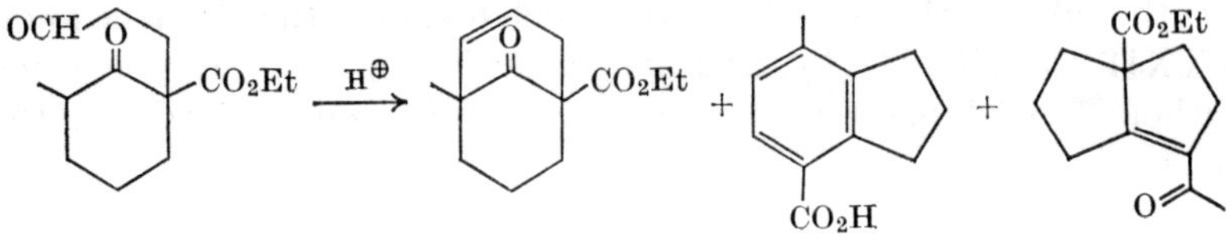

I II III IV

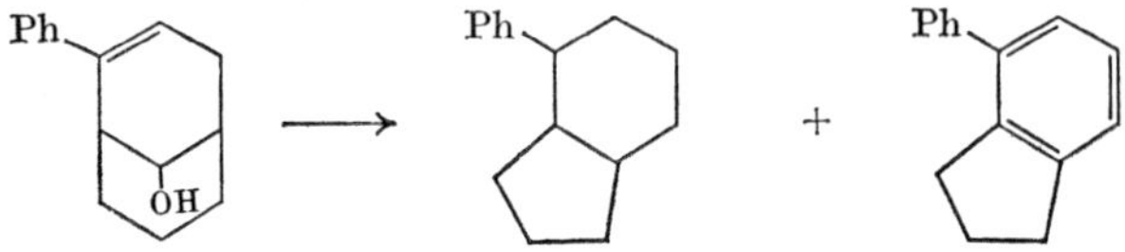

V VI VII

involved in the rearrangement of V to a mixture of VI and VII[11], under the influence of zinc chloride, and the same carbonium ion mechanism can be invoked for each. In similar fashion selagine VIII, an alkaloid of the lycopodium group is deaminated by nitrous acid, and the product, selaginol (IX) is rearranged by HCl to a ketone

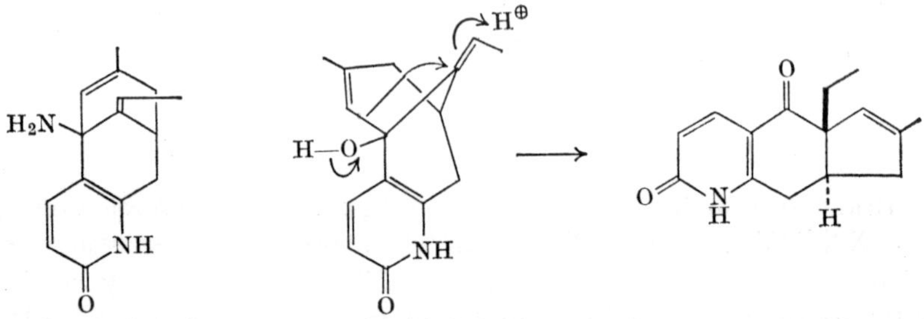

VIII IX X

which is probably represented by X[133, 141]. A similar transformation must take place during the dehydration of patchouli alcohol. The structure of this sesquiterpene was seemingly well established as XI by the time honoured processes of degradation and synthesis[28, 29] when its casual use in an X-ray crystallographic analysis of a chromate ester[50] revealed the error and showed it to be XII. It follows therefore that the degradative pyrolysis of patchouli acetate to a mixture of

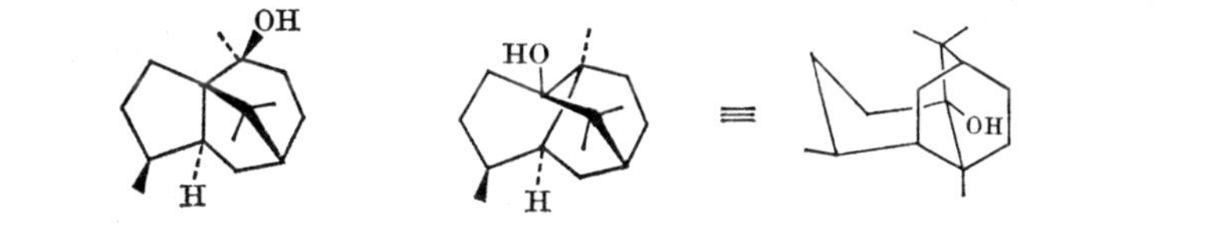

XI XII

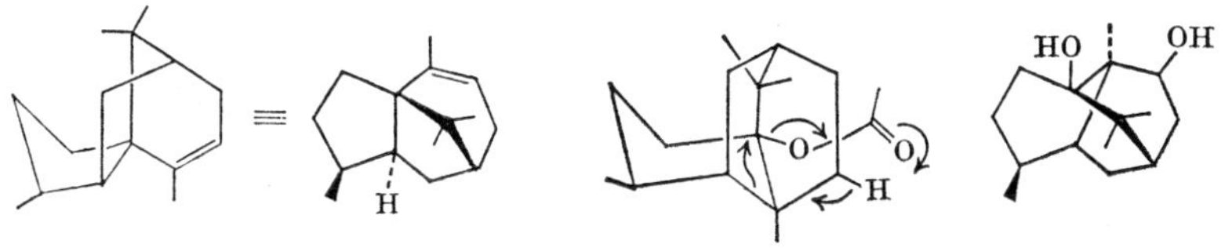

XIII XIV XV

(mainly) α-patchoulene XIII and its *exo*-methylene isomer (γ-patchoulene) involves a novel rearrangement. This could be interpreted in ionic terms or as an unusual concerted process XIV. Similarly, the synthesis of patchouli alcohol via α-patchoulene, must include an unplanned reverse rearrangement. It has been suggested[50] that the first step of this conversion, i.e. peracetic acid hydroxylation of XIII, must yield the 1,3-diol XV rather than the orthodox 1,2-diol which would be expected.

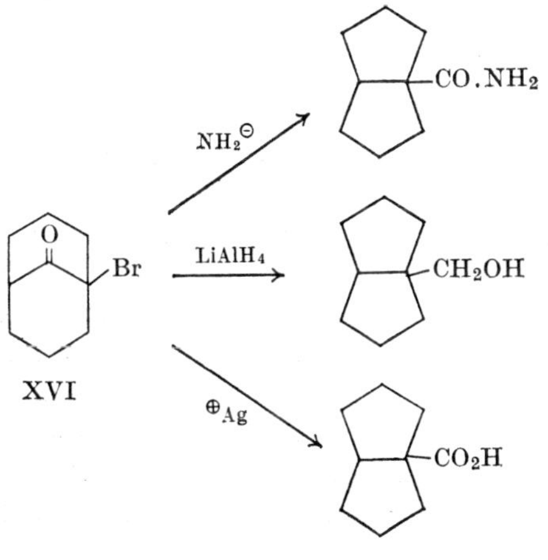

XVI 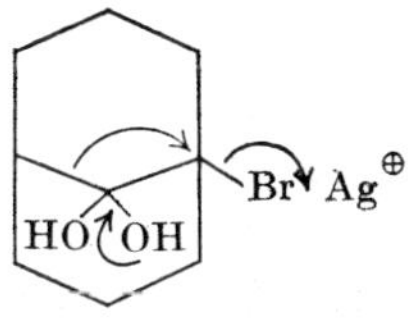 XVII

A rearrangement which is reminiscent of the Favorskii rearrangement takes place when the bromo-ketone XVI is treated as shown below. In the last case, the participation of water has been demonstrated and the mechanistic interpretation (XVII) has been proposed[41, 42].

Bridge migration can be effected under photolytic conditions, when the enones XVIII and XIX yield XX and XXI respectively[57] in

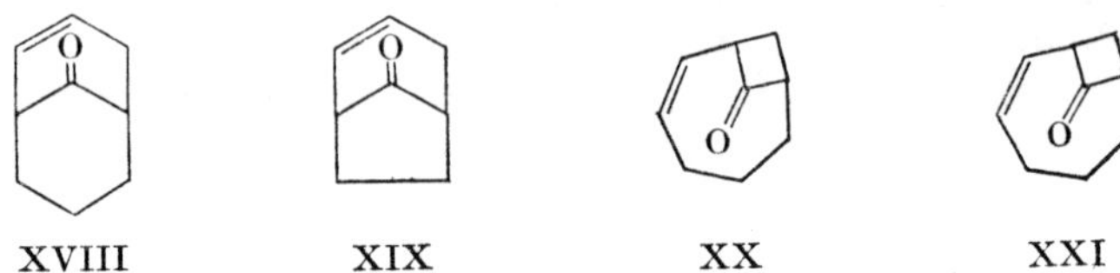

useful amounts. Neither structure would be readily accessible by classical routes.

The discovery and investigation of a group of tetranortriterpenoids which incorporate the bicyclo[3,3,1]nonane skeleton and are appropriately named bicyclononanolides[34, 35, 36, 37], has brought to light some fascinating skeletal rearrangements. These substances, which occur in woods of the Meliaceae family, are exemplified by mexicanolide XXII[35] and swietenine XXIV[34].

Despite their similarity, the two series show unexpected differences in reactivity. The action of dilute alkali on the β-diketone XXII (mexicanolide) brings about an extensive rearrangement to XXV (R = OH), initiated no doubt by a carbanion at C(15). The C(3) β-alcohol (XXIII) derived from mexicanolide by reduction is unaffected by alkali, but its acetate is cleaved by base to XXV (R = H) in a remarkably smooth manner. Strangely, the action of dilute alkali on swietenine (XXIV), which might be expected to resemble the above acetate in behaviour, merely hydrolyses the ester functions in the molecule and *epimerises* the resulting C(3) β-alcohol. The epimerisation is effected by a retro-aldol reaction followed by realdolisation, and transforms a C—OH bond which is axially oriented in a boat cyclohexanone ring into its equatorial isomer. It might be concluded that the ring-fragmentation reactions of mexicanolide and its reduction product are both initiated by a vinylogous retro-Michael reaction in which the final carbanion is specially stabilised in a β-diketone system or by ejection of acetate. On this basis, the location of the C═C would be vital, and the absence of a similar fragmentation in the swietenine series could be understood. However at least one other factor must

XXII; R = O (Mexicanolide)
XXIII; R = H, βOH

XXIV; T = Tigloyl
(Swietenine)

XXV

XXVI

XXVII

XXVIII

XXIX

be involved, for the hydrogenolysis product XXVI, derived from XXII, is unaffected by similar treatment, and the trigger which sets off the rearrangement may well be molecular strain originating from the distant lactone ring.

A more remarkable rearrangement occurs in the swietenine series. Its alkaline hydrolysis product, demethyl-detigloyl-*iso*swietenine (XXVII) is converted by lead tetra-acetate into the aldehyde XXVIII, and when treated with 0·25 *N* alkali, the latter undergoes the series of transformations XXVIII → XXIX. The overall effect is to convert a bicyclo[3,3,1]nonane into a bicyclo[3,2,1]octane derivative via retro-aldol, alternative aldol and Cannizarro reactions.

4. *Ring closure*

Trigonal substituents in positions 3 and 7 are in unusual proximity to each other, and bonding readily occurs since it leads to a strainless system, e.g. adamantane, which is free from non-bonded interactions.

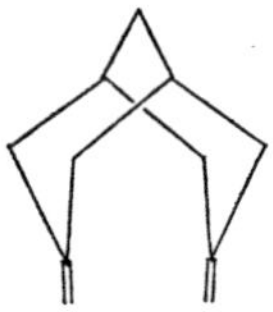

I 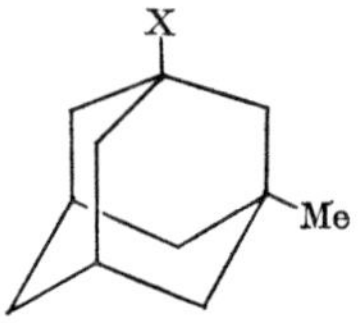 II 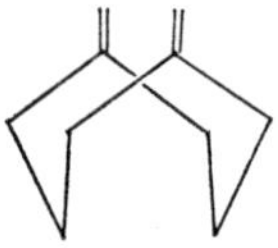 III 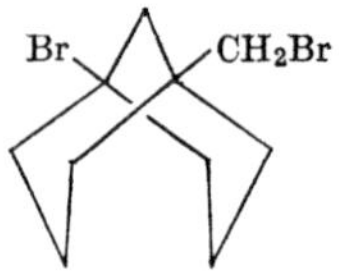 IV

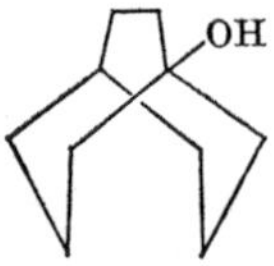

V 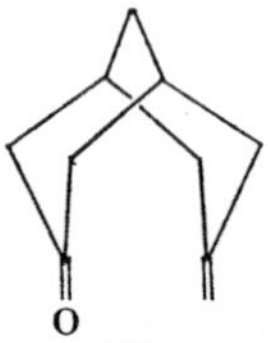 VI 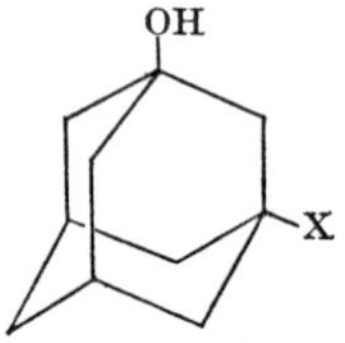 VII

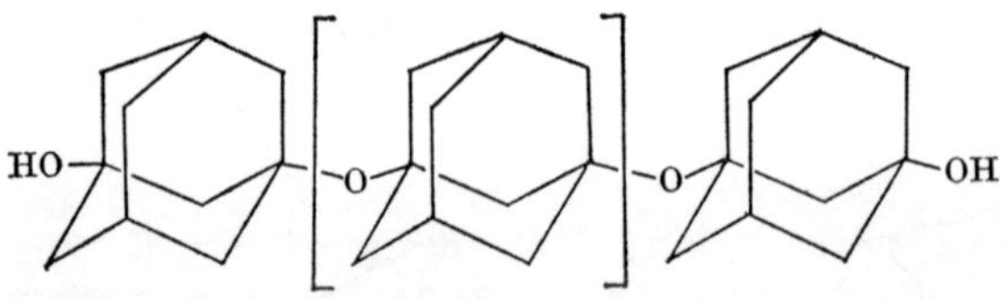

VIII 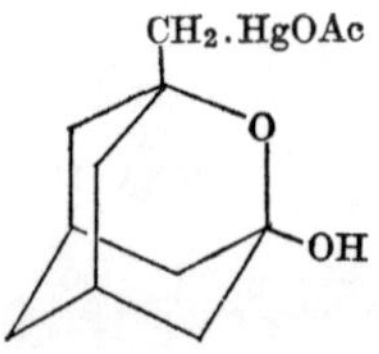 IX

The bis-methylene compound I is ring-closed to II (X = OH) by sulphuric acid, or to II (X = OMe) by acidified methanol[125], and similar cyclisations take place on bromination or mercuration[126]. The analogous bis-methylenecyclo-octane (III), which shows C—H 'scissoring' bands at 2976 and 1446 cm^{-1} reminiscent of bicyclo[3,3,1]-nonane (see section C), reacts similarly with bromine, affording IV, but abnormally with HCl giving the anti-Markovnikov product V[7]. Like I, the enone VI undergoes cyclisation to an adamantane VII (X = OH, OEt or Cl) according to the reagent employed[125], and in an inert solvent it is polymerised by acid to VIII[125]. These reactions are all initiated by protonation of the carbonyl group, but mercuration preferentially attacks the C=C and so leads to the hemi-acetal IX[128].

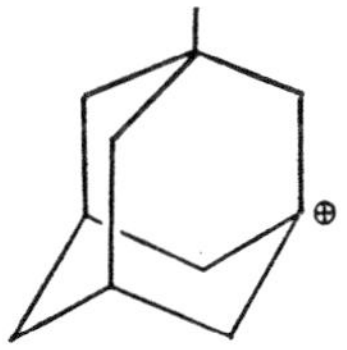

⊕

X 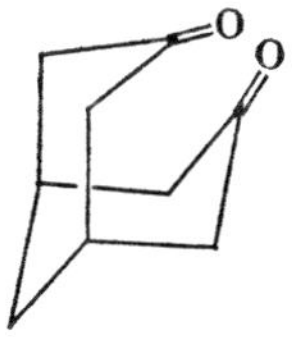 XI 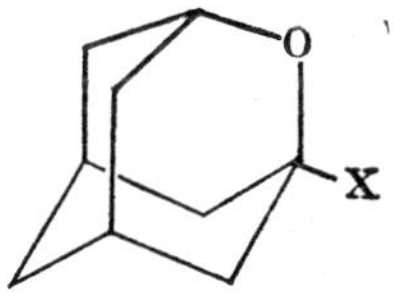 XII

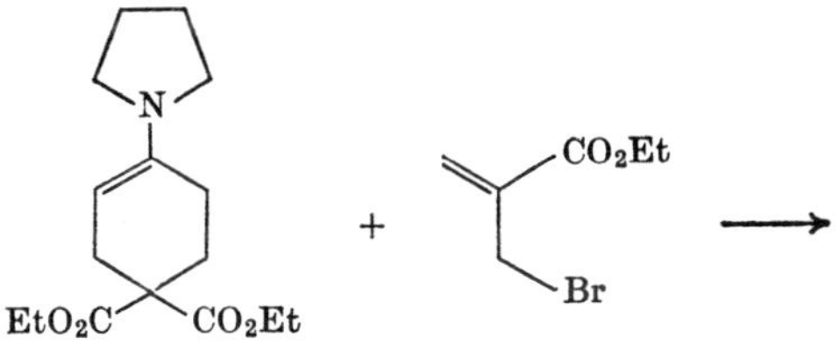

N

EtO_2C CO_2Et

XIII

CO_2Et

Br

XIV

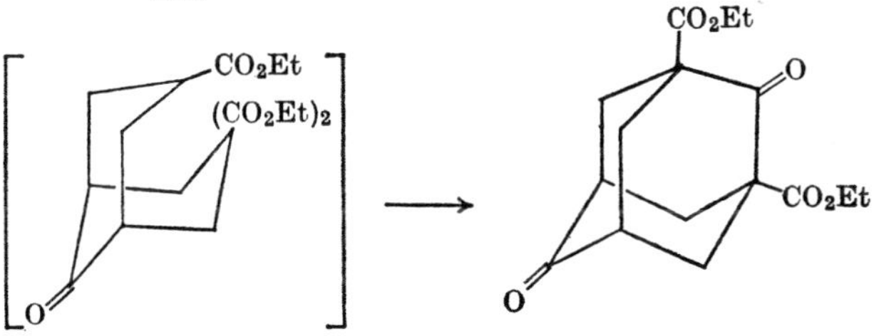

CO_2Et

$(CO_2Et)_2$

O

XV

CO_2Et

O

CO_2Et

O

XVI

These reactions also involve an intermediate adamantyl carbonium ion X, whose unique stability despite its obvious non-planarity has been acknowledged[67]. Such an intermediate must also be implicated in the reaction of I with anisole in the presence of acid to give II ($X = p\text{-}C_6H_4.OMe$)[125].

Even catalytic reduction of the dione XI brings about ring-closure. The reaction stops after uptake of one mole of hydrogen and yields XII ($X = OH$). The 3,7-diol expected from XI can only be obtained, as a mixture of isomers, by hydride reduction[124], and cyclises to XII ($X = H$) on treatment with acid[123]. Spontaneous cyclisation (XV → XVI) likewise occurs following the reaction of XIII with XIV[129].

If suitably activated, the 3 and 7 positions can be linked by alkylation. The bis-enamine XVII reacts with methyl dibromoacetate affording XVIII[122, 127] and Meerwein's ester XIX yields the adamantanedione XX[19, 110], processes which provide useful routes to

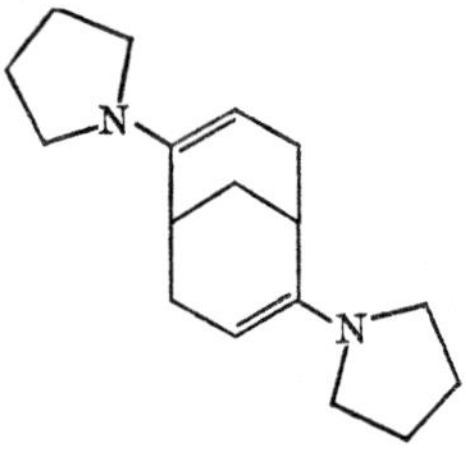

XVII 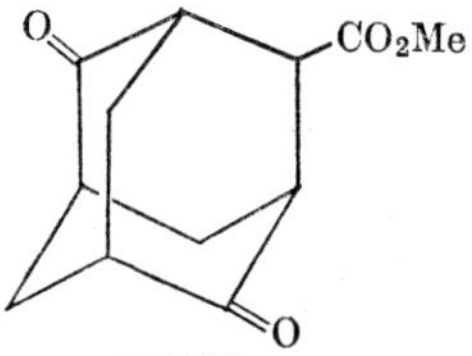 XVIII 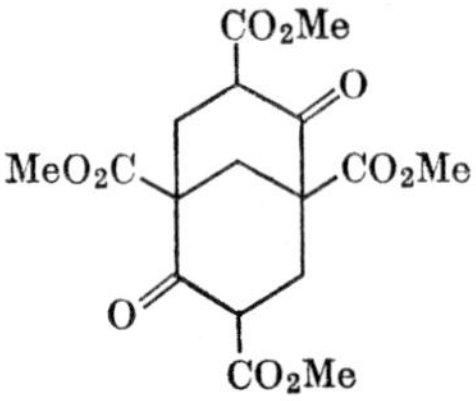 XIX

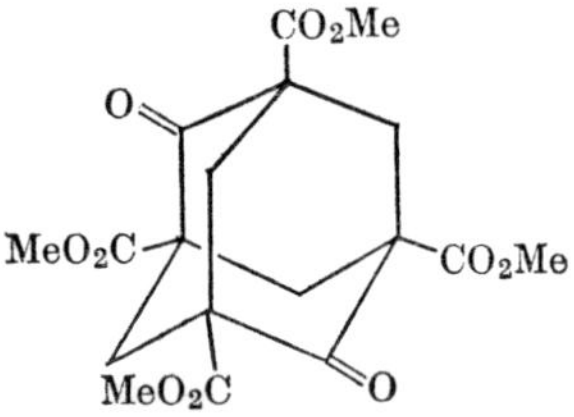

XX 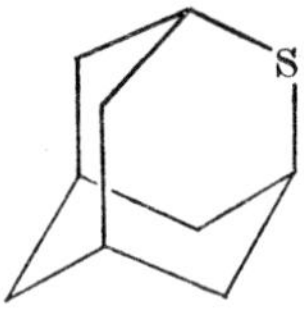 XXI 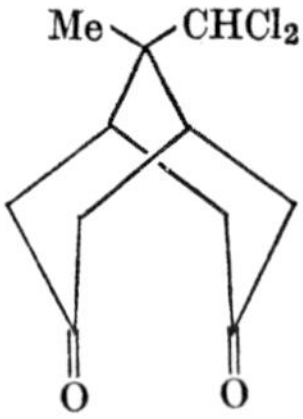 XXII

substituted adamantanes. Ring closure of XVII by means of sulphur dichloride, followed by removal of the carbonyl groups, yielded thia-adamantane XXI[121], identical with the product previously isolated[18] from South Iranian oil.

Cyclisation by a double aldol reaction between nitromethane and the dione XXII has been shown to give XXIII[122].

The bridging process, which has been the subject of this section, has been reversed by the fragmentation reaction XXIV. The resulting imine is hydrolysed under the reaction conditions and the final product is the enone VI[123].

Substituents in the 2 and 6 positions are less proximate than those in the 3 and 7 positions, but Meerwein[105] was able to prepare XXV from bicyclo[3,3,1]nonan-2,6-dione by a pinacol reduction.

XXIII XXIV XXV

5. *Bredt's Rule*

Of all the stereochemical effects that operate in bridged bicyclic molecules, that which limits the location of double-bonds is undoubtedly the best known. Enunciated by Bredt[20] over 40 years ago, it stated:

> 'Auf Grund unserer Vorstellungen über die Lage der Atome in Raum kann, in dem Systemen der Camphan- und Pinanreihe, sowie in ähnlich konstituierten Verbindungen von...den Brückenkopfen eine Kohlenstoff doppelbindung nicht ausgehen.'

The rule applies specifically to *bridged* systems in which the component rings are relatively *small*, and in forbidding I and II, it accounts for a number of chemical anomalies. Thus, the alcohol III cannot be dehydrated[114]; the β-keto acid IV resists decarboxylation or

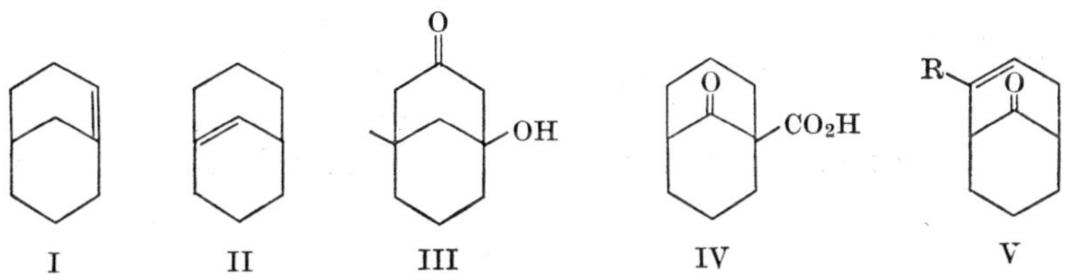

deuterium exchange[40, 63], processes which both involve enol or enolate intermediates; and the double bond in V cannot be brought into conjugation with the carbonyl[39, 113]. A host of similar examples can be cited, and are to be found in an excellent review[59].

The keto-dicarboxylic acid VI is an apparent exception to the rule, for unlike III it is smoothly decarboxylated in boiling 20% HCl, yielding the ketone VII which incorporates 2 atoms of deuterium

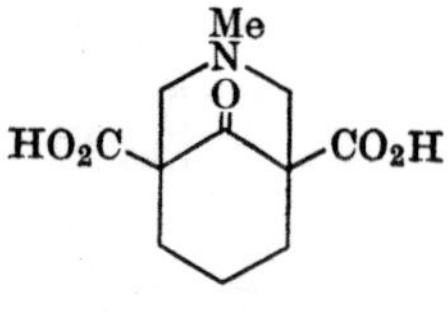

VI 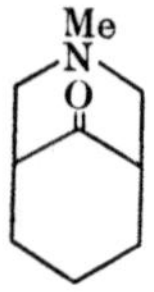 VII 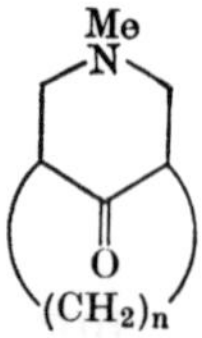 VIII

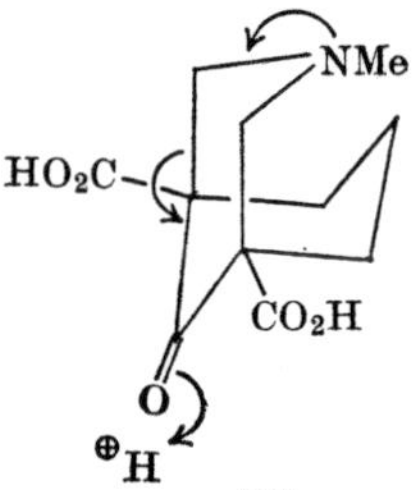

IX 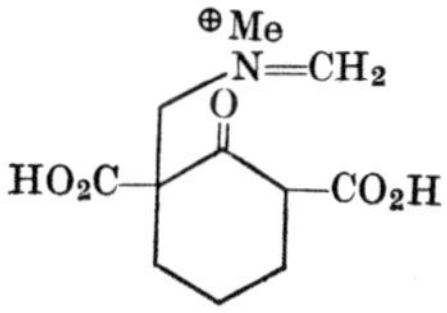 X

in DCl/D_2O. Even more remarkably, the rate of deuterium exchange in the series VIII ($n = 2$, 3 and 4) *falls* with increasing ring size. All of this can be rationalised in terms of a retro-Mannich process IX $\leftrightarrows$ X which allows decarboxylation to take place via the mono-cyclic β-ketoacid X. Deuterium exchange can be similarly explained[79].

Although small bicyclic molecules are unable to tolerate a bridgehead double bond, this is no longer true of larger-ring analogues. The classic investigations of Prelog[113] showed that ring-closure of the series XI ($n = 4$, 5 and 6) yielded only XII, if $n = 4$, a mixture of XII and XIII (36 per cent and 14 per cent respectively) if $n = 5$ and only XIII if $n = 6$. These experiments encouraged the view that the limits of Bredt's rule had been defined—at least for a bicyclo[n,3,1] system. Nor was it at variance with the report[144] that Dauben had been able

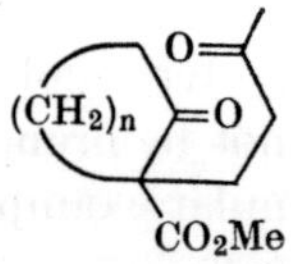

XI 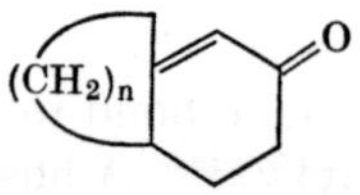 XII 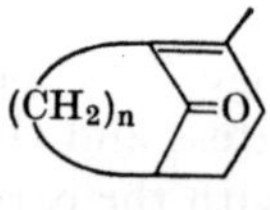 XIII 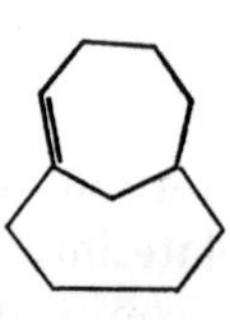 XIV

to prepare a bridgehead bicyclo[4,4,1]undecene derivative (cf. XIV), for Fawcett had introduced[59] the idea of a strain number S, for a bicyclo(x,y,z) system, defined as $S = x + y + z$, provided x, y or $z \neq 0$. It was then concluded that $X = 9$ represented the limit beyond which the rule became invalid, although evidence derived from decarboxylation studies—e.g. by Meerwein[105]—contradicts this conclusion.

In fact Prelog's cyclisations, which were effected under equilibrating conditions, only reflect the relative thermodynamic stabilities of the two series, XII and XIII. The formation of a π-bond depends on efficient overlap of two p-orbitals. This will occur ideally when the axes of the p-orbitals are parallel, but it is reasonable to expect that partial overlap will still be possible when they are not parallel; so it is not surprising that evidence for anti-Bredt double bonds has been found in the bicyclo[3,3,1]nonane molecule, i.e. where $S = 7$.

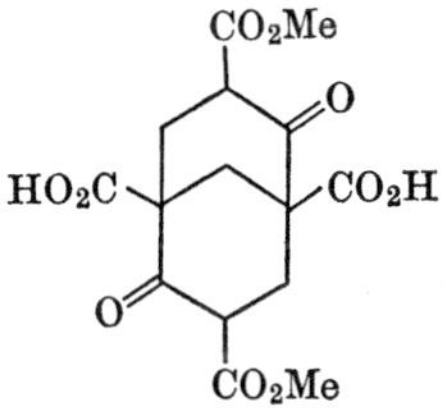

XV 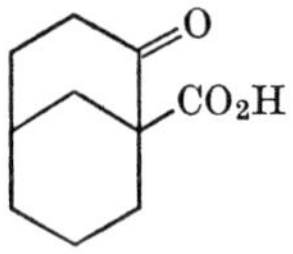 XVI

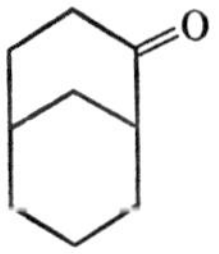

XVII 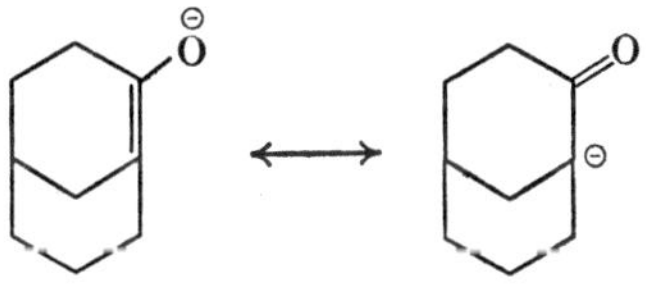 XVIII

The decarboxylation of XV[105] and XVI[63, 64] and the *bridgehead* deuteration of XVII[116] are all examples which require the intervention of the enolate XVIII. Molecular models indicate that the cyclohexene ring must be in the boat conformation XIX; in the chair conformation, the developing p-orbital at C(1) is orthogonal to the p-orbital on the carbonyl C—a fact which may well explain the stability[19] of XX.

If there remained any doubt that a bicyclo[3,3,1]non-1-ene was capable of existence it has been dispelled by two simultaneous syntheses of XXIII. In the presence of a strong base, the mesylate XXI affords[100] a mixture of XXII and XXIII. The reaction is

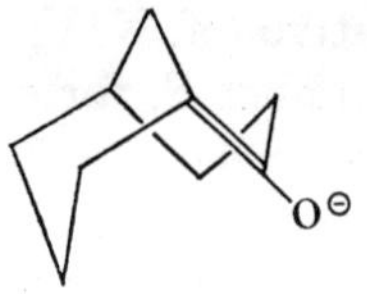

XIX 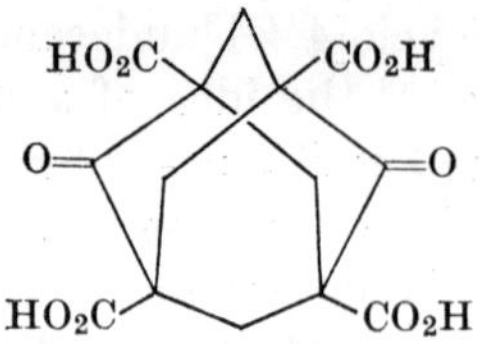 XX 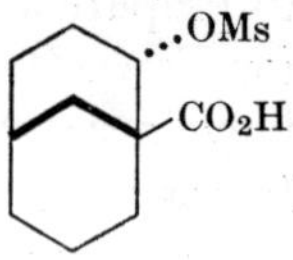 XXI

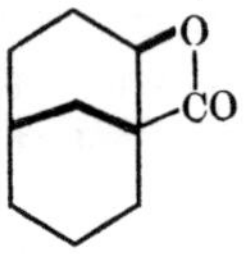

XXII 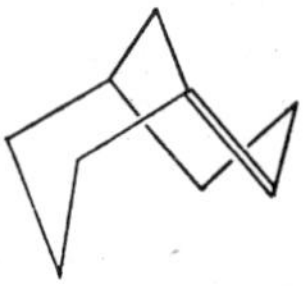 XXIII 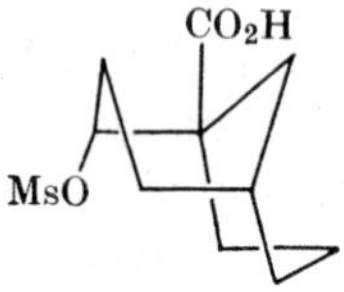 XXIV

visualised as proceeding via the twist boat conformer XXIV by two competing processes—intramolecular displacement and decarboxyla-

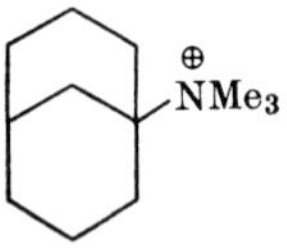

XXV 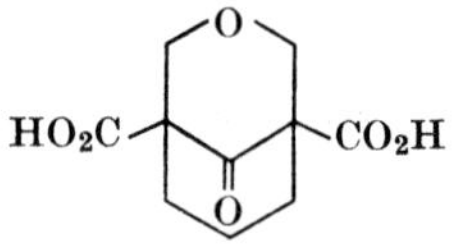 XXVI

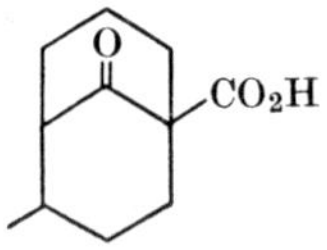

XXVII 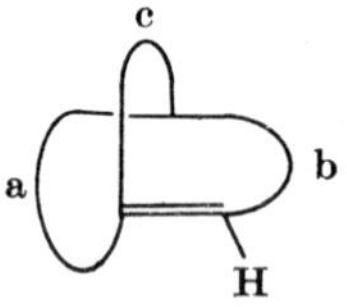 XXVIII

tive β-elimination; but an alternative view, that XXIII arises via the β-lactone by thermal degradation, has not been eliminated. A second and more efficient synthesis employs a Hofmann elimination of the quaternary salt XXV[139]. The 'anti-Bredt' olefin (XXIII) obtained by these reactions polymerises in air and is readily hydrated to the bridgehead alcohol.

Bredt made no distinction between alternative bridgehead positions when he formulated his rule, but there is now clear evidence that although a C(1)–C(2) double bond in a bicyclo[3,3,1]nonene is strained, a C(1)–C(9) double bond is much *more* strained. Indeed there is no indication that the latter can be formed even transiently. The acids IV, XXVI and XXVII all resist decarboxylation, and IV shows no tendency to exchange the bridgehead proton for deuterium [79]. Thus the rule needs qualification, and the simple intuitive correlation of strain with an *S* number (see above) is inadequate. A simple but satisfying solution has been provided by Wiseman[139], who relates strain in bridgehead double bonds to strain in *trans*-cycloalkenes, since in any bridged bicyclic alk-1-ene XXVIII, the double bond must be *trans* within one or other of the rings, *ab* or *bc*. On this basis, it is possible to predict that since *trans*-cyclooctene is known[143] and *trans*-cycloheptene is formed transiently [44], anti-Bredt double bonds should be feasible in molecules of structure XXVIII, where ring *ab* is 8-, or possibly 7-membered. The synthesis of XXIII provides the first indication that this interpretation of Bredt's rule is a valid one.

References

1. Adesogan, E. K., Bevan, C. W. L., Powell, J. W. and Taylor, D. A. W. (1966) *J. Chem. Soc.* (*C*) 2127.
2. Appleton, R. A. and Graham, S. H. (1965) *Chem. Communications* 297.
3. Appleton, R. A., Dixon, J. R., Evans, J. M. and Graham, S. H. (1967) *Tetrahedron* **23**, 805.
4. Appleton, R. A., Egan, C., Evans, J. M., Graham, S. H. and Dixon, J. R. (1968) *J. Chem. Soc.* (*C*) 1110.
5. Appleton, R. A., Baggaley, K. H., Egan, C., Davies, J. M., Graham, S. H. and Lewis, D. O. (1968) *J. Chem. Soc.* (*C*) 2032.
6. Baggaley, K. H., Dixon, J. R., Evans, J. M. and Graham, S. H. (1967) *Tetrahedron* **23**, 299.
7. Baggaley, K. H., Evans, W. H. and Graham, S. H. (1968) *Tetrahedron* **24**, 3445.
8. Bailey, D. M., Bowers, J. E. and Gutsche, C. D. (1963) *J. Org. Chem.* **28**, 607, 610.
9. Baisted, D. J. and Whitehurst, J. S. (1961) *J. Chem. Soc.* 4089.
10. Baker, W. and Leeds, W. G. (1948) *J. Chem. Soc.* 974.
11. Barbulescu, N. and Govela, M. (1963) *Chem. Abstr.* **59**, 1506b [(1961) *Analele Univ. C.I. Parhon, Ser. Stiint, Nat.* **10**, 151].
12. Bardhan, J. C. and Banerjee, R. C. (1956) *J. Chem. Soc.* 1809.
13. Becker, D. and Loewenthal, H. J. E. (1965) *Chem. Communications* 149.
14. Becker, D., and Loewenthal, H. J. E. (1965) *J. Chem. Soc.* 1338.
15. Beckmann, S. and Ling, O. S. (1961) *Chem. Ber.* **94**, 1899.
16. Bellamy, L. J. (1968) *Advances in Infrared Group Frequencies*, Methuen, London, p. 128.
17. Birch, A. J. and Hill, J. S. (1966) *J. Chem. Soc.* (*C*) 419.
18. Birch, S. F., Cullum, T. V., Dean, R. A. and Denyer, R. L. (1952) *Nature, Lond.* **170**, 629.

19. Bottger, O. (1937) *Chem. Ber.* **70**, 314.
20. Bredt, J. (1924) *Liebigs Ann.* **437**, 1.
21. Brown, W. A. C., Martin, J. and Sim, G. A. (1965) *J. Chem. Soc.* 1844.
22. Buchanan, G. L., Maxwell, C. and Henderson, W. (1965) *Tetrahedron* **21**, 3273.
23. Buchanan, G. L., McKillop, A. and Raphael, R. A. (1965) *J. Chem. Soc.* 833.
24. Buchanan, G. L., Ferguson, G., Lawson, A. M. and Pollard, D. R. (1966) *Tetrahedron Letters* 5303.
25. Buchanan, G. L. and McLay, G. W. (1966) *Tetrahedron* **22**, 1521.
26. Buchanan, G. L., Curran, A. C. W., McCrae, J. M. and McLay, G. W. (1967) *Tetrahedron* **23**, 4729.
27. Buchi, G., Biemann, K., Vittimberger, B. and Stoll, M. (1956) *J. Amer. Chem. Soc.* **78**, 2622.
28. Buchi, G., Erickson, R. E. and Wakabayashi, N. (1961) *J. Amer. Chem. Soc.* **83**, 927.
29. Buchi, G. and McLeod, W. D. (1962) *J. Amer. Chem. Soc.* **84**, 3205.
30. Buchta, E. and Billenstein, S. (1964) *Naturwissenschaften* **51** (16), 383.
31. Clunie, J. S., and Robertson, J. M. (1960) *Proc. Chem. Soc.* 82.
32. Clunie, J. S. and Robertson, J. M. (1961) *J. Chem. Soc.* 4382.
33. Colvin, E. W. and Parker, W. (1965) *J. Chem. Soc.* 5764.
34. Connolly, J. D., Henderson, R., McCrindle, R., Overton, K. H. and Bhacca, N. S. (1965) *J. Chem. Soc.* 6935.
35. Connolly, J. D., McCrindle, R. and Overton, K. H. (1968) *Tetrahedron* **24**, 1489.
36. Connolly, J. D., McCrindle, R. and Overton, K. H. (1968) *Tetrahedron* **24**, 1497.
37. Connolly, J. D., McCrindle, R. and Warnock, W. D. C. (1968) *Tetrahedron* **24**, 1507.
38. Cope, A. C., Fawcett, E. S. and Munn, G. (1950), *J. Amer. Chem. Soc.* **72**, 3399.
39. Cope, A. C. and Hermann, E. C. (1950) *J. Amer. Chem. Soc.* **72**, 3405.
40. Cope, A. C. and Synerholm, M. E. (1950) *J. Amer. Chem. Soc.* **72**, 5228.
41. Cope, A. C. and Graham, E. S. (1951) *J. Amer. Chem. Soc.* **73**, 4702.
42. Cope, A. C., Graham, E. S. and Marshall, D. J. (1954), *J. Amer. Chem. Soc.* **76**, 6159.
43. Cope, A. C., Nealy, D. L., Scheiner, P. and Wood, G. (1965) *J. Amer. Chem. Soc.* **87**, 3130.
44. Corey, E. J., Carey, F. A. and Winter, R. A. E. (1965) *J. Amer. Chem. Soc.* **87**, 934.
45. Corey, E. J. and Nozoe, S. (1965) *J. Amer. Chem. Soc.* **87**, 5728.
46. Danishefsky, S., Koppel, G. and Levine, R. (1968) *Tetrahedron Letters* 2257.
47. Dauben, W. G. and McFarland, J. W. (1960) *J. Amer. Chem. Soc.* **82**, 4245.
48. Dauben, W. G., Boswell, G. H. and Templeton, W. H. (1961) *J. Amer. Chem. Soc.* **83**, 5006.
49. Dean, C. S., Dixon, J. R., Graham, S. H. and Lewis, D. O. (1968) *J. Chem. Soc.* (*C*) 1491.

50. Dobler, M., Dunitz, J. D., Gubler, B., Weber, H. P., Buchi, G. and Padilla, O. J. (1963) *Proc. Chem. Soc.* 383.
51. Doyle, P., Maclean, I. R., Murray, R. D. H., Parker, W. and Raphael, R. S. (1965) *J. Chem. Soc.* 1344.
52. Eakin, M. A., Martin, J., Parker, W., Egan, C. and Graham, S. H. (1968) *Chem. Communications* 337.
53. Eakin, M. A., Martin, J. and Parker, W. (1968) *Chem. Communications* 298.
54. Eglinton, G., Martin, J. and Parker, W. (1965) *J. Chem. Soc.* 1243.
55. Eistert, B., Haupter, F. and Schank, K. (1963) *Liebigs Ann.* **665**, 55.
56. Eliel, E. L. (1962) *Stereochemistry of Carbon Compounds*, McGraw-Hill, New York, p. 296.
57. Erman, W. F. and Kretschmar, H. C. (1967) *J. Amer. Chem. Soc.* **89**, 3842.
58. Erman, W. F. and Kretschmar, H. C. (1968) *J. Org. Chem.* **33**, 1545.
59. Fawcett, F. S. (1950) *Chem. Rev.* **47**, 219.
60. Fell, B., Seidl, W. and Asinger, F. (1968) *Tetrahedron Letters* 1003.
61. Ferguson, G., Hawley, D. M., McKillop, T. F. W., Martin, J., Parker, W. and Doyle, P. (1967) *Chem. Communications* 1123.
62. Ferguson, G., Macrossan, W. D. K., Martin, J. and Parker, W. (1967) *Chem. Communications* 102.
63. Ferris, J. P. and Miller, N. C. (1963) *J. Amer. Chem. Soc.* **85**, 1325.
64. Ferris, J. P. and Miller, N. C. (1966) *J. Amer. Chem. Soc.* **88**, 3522.
65. Foote, C. S. (1964) *J. Amer. Chem. Soc.* **86**, 1853.
66. Foote, C. S. and Woodward, R. B. (1964) *Tetrahedron* **20**, 687.
67. Fort, R. C. and Schleyer, P. von R. (1966) in *Advances in Alicyclic Chemistry*, edited by Hart, H. and Karabatsos, G. J. Vol. I, Academic Press, New York, p. 283.
68. Freeman, S. K. (1965) *Interpretive Spectroscopy*, Reinhold, New York, p. 111.
69. Fujimoto, G. (1951) *J. Amer. Chem. Soc.* **73**, 1856.
70. Fujimoto, G. and Zwahlen, K. D. (1960) *J. Org. Chem.* **25**, 445.
71. Fujimoto, G. and Parlos, J. (1965) *Tetrahedron Letters* 4477.
72. Ghatak, U. R. and Chakravarty, J. (1966) *Tetrahedron Letters* 2449.
73. Gould, E. S. (1959) *Mechanism and Structure in Organic Chemistry*, Rinehart & Winston, New York, p. 325.
74. Grob, C. A. and Hostynek, J. (1963) *Helv. Chim. Acta* **46**, 2212.
75. Gutsche, D., Smith, T. D., Sloan, M. F., van Ufford, J. J. Q. and Jorgan, D. E. (1958) *J. Amer. Chem. Soc.* **80**, 4117.
76. Hargreaves, J. R. and Hickmott, P. W. (1966) *Tetrahedron Letters* 4173.
77. Hellmann, H. M., Jerussi, R. A. and Lancaster, J. (1967) *J. Org. Chem.* **32**, 2148.
78. Hickmott, P. W. and Hargreaves, J. R. (1967) *Tetrahedron* **23**, 3151.
79. House, H. O. and Muller, H. C. (1962) *J. Org. Chem.* **27**, 4436.
80. House, H. O. (1965) *Modern Synthetic Reactions*, W. A. Benjamin, New York, p. 210.
81. Jefford, C. W., Waegell, B. and Ramay, K. (1965) *J. Amer. Chem. Soc.* **87**, 2191.
82. Johnston, W. S., Korst, J. A., Clement, R. A. and Dutta, J. (1960) *J. Amer. Chem. Soc.* **82**, 614.

83. Julia, S. A. (1954) *Bull. Soc. Chim. Fr.* 780.
84. Knoevenagel, E. (1894) *Liebigs Ann.* **281**, 39; (1903) *Chem. Ber.* **36**, 2144.
85. Kraus, W. and Rothenwohrer, W. (1968) *Tetrahedron Letters* 1007.
86. Kraus, W. and Rothenwohrer, W. (1968) *Tetrahedron Letters* 1013.
87. Landa, S. and Kamycek, Z. (1959) *Coll. Czech. Chem. Comm.* **24**, 1320.
88. Laszlo, I. (1965) *Rec. Trav. Chim.* **84**, 251.
89. Lawson, A. M. (1966) Ph.D. Thesis, Glasgow University.
90. LeBel, N. A. and Spurlock, L. A. (1964) *J. Org. Chem.* **29**, 1337.
91. LeBel, N. A. and Spurlock, L. A. (1964) *Tetrahedron* **20**, 215.
92. Loewenthal, H. J. E. and Neuwirth, Z. (1967) *J. Org. Chem.* **32**, 517.
93. Maclay, G. W. (1965) Ph.D. Thesis, Glasgow University.
94. McPhail, A. T. and Sim, G. A. (1964) *Tetrahedron Letters* 2599.
95. Macrossan, W. D. K., Martin, J. and Parker, W. (1965) *Tetrahedron Letters* 2589.
96. Macrossan, W. D. K. and Ferguson, G. (1968) *J. Chem. Soc.* (*B*) 242.
97. Marshall, J. A. and Scanio, C. J. V. (1965) *J. Org. Chem.* **30**, 3019.
98. Marshall, J. A. and Schaeffer, D. J. (1965) *J. Org. Chem.* **30**, 3642.
99. Marshall, J. A. and Partridge, J. J. (1966) *Tetrahedron Letters* 2545.
100. Marshall, J. A. and Faubl, H. (1967) *J. Amer. Chem. Soc.* **89**, 5965.
101. Martin, J., Parker, W. and Raphael, R. A. (1964) *J. Chem. Soc.* 289.
102. Martin, J., Parker, W., Shroot, B. and Stewart, T. (1967) *J. Chem. Soc.* 101.
103. Maxwell, C. (1965) Ph.D. Thesis, Glasgow University.
104. Mayer, R., Wenschuh, G. and Topelmann, W. (1958) *Chem. Ber.* **91**, 1616.
105. Meerwein, H. (1922) *J. Prakt. Chem.* (2) **104**, 161.
106. Meinwald, J., Shelton, J. C., Buchanan, G. L. and Courtin, A. (1968) *J. Org. Chem.* **33**, 99.
107. Murray, R. D. H., Parker, W., Raphael, R. A. and Jhaveri, D. B. (1962) *Tetrahedron* **18**, 55.
108. Opitz, G. and Holtmann, H. (1965) *Liebigs Ann.* **684**, 79.
109. Pitha, J., Plesek, J. and Horak, M. (1961) *Coll. Czech. Chem. Comm.* **26**, 1209.
110. Prelog, V. and Seiwerth, R. (1941) *Chem. Ber.* **74**, 1644.
111. Prelog, V., Barman, P. and Zimmermann, M. (1949) *Helv. Chim. Acta* **32**, 1284.
112. Prelog, V. and Zimmermann, M. (1949) *Helv. Chim. Acta* **32**, 2360.
113. Prelog, V. (1950) *J. Chem. Soc.* 420.
114. Rabe, P., Ehrenstein, R. E. and Jahr, M. (1908) *Liebigs Ann.* **360**, 265.
115. Sands, R. D. (1963) *J. Org. Chem.* **28**, 1710.
116. Schaeffer, J. P., Endres, L. S. and Moran, M. D. (1967) *J. Org. Chem.* **32**, 3963.
117. Schut, R. N. and Liu, T. M. H. (1965) *J. Org. Chem.* **30**, 2845.
118. Spencer, T. A., Newton, M. D. and Baldwin, S. W. (1964) *J. Org. Chem.* **29**, 787.
119. Stetter, H., Bander, O. E. and Neumann, W. (1956) *Chem. Ber.* **89**, 1922.
120. Stetter, H. and Mayer, J. (1959) *Chem. Ber.* **92**, 2664.
121. Stetter, H., Held, H. and Schulte-Oestrich, A. (1962) *Chem. Ber.* **95**, 1687.

122. Stetter, H., Held, H. and Mayer, J. (1962) *Liebigs Ann.* **658**, 151.
123. Stetter, H., and Tacke, P. (1963) *Chem. Ber.* **96**, 694.
124. Stetter, H., Tacke, P. and Gartner, J. (1964) *Chem. Ber.* **97**, 3480.
125. Stetter, H., Gartner, J. and Tacke, P. (1965) *Angew. Chemie* (*Internat. Ed.*) **4**, 153.
126. Stetter, H. and Gartner, J. (1966) *Chem. Ber.* **99**, 925.
127. Stetter, H. and Thomas, H. G. (1966) *Chem. Ber.* **99**, 920.
128. Stetter, H., Gartner, J. and Tacke, P. (1966) *Chem. Ber.* **99**, 1435.
129. Stetter, H. and Thomas, H. G. (1968) *Chem. Ber.* **101**, 1115.
130. Stoll, M., Willhalen, B. and Buchi, G. (1955) *Helv. Chim. Acta* **38**, 1573.
131. Stork, G. and Landesmann, H. K. (1956) *J. Amer. Chem. Soc.* **78**, 5129.
132. Stork, G., Kretchmer, R. A. and Schlessinger, R. H. (1968) *J. Amer. Chem. Soc.* **90**, 1647.
133. Valenta, Z., Yoshimura, H., Rogers, E. F., Ternbah, M. and Wiesner, K. (1960) *Tetrahedron* **10**, 26.
134. Walls, F., Padilla, J., Joseph-Nathan, P., Giral, F. and Romo, J. (1965) *Tetrahedron Letters* 1577.
135. Warnhoff, E. W., Wong, C. M. and Tai, W. T. (1967) *J. Org. Chem.* **32**, 2664.
136. Webb, N. C. and Becker, M. R. (1967) *J. Chem. Soc.* (*B*) 1317.
137. Wichterle, O. and Hudlicky, M. (1947) *Coll. Czech. Chem. Comm.* **12**, 101.
138. Wichterle, O. (1948) *Coll. Czech. Chem. Comm.* **13**, 300.
139. Wiseman, J. R. (1967) *J. Amer. Chem. Soc.* **89**, 5966.
140. Yogev, A. and Mazur, Y. (1965) *J. Amer. Chem. Soc.* **87**, 3520.
141. Yoshimura, H., Valenta, Z. and Wiesner, K. (1960) *Tetrahedron Letters* **12**, 14.
142. Zbinden, R. and Hall, H. K. (1960) *J. Amer. Chem. Soc.* **82**, 1215.
143. Ziegler, K. and Wilms, H. (1950) *Liebigs Ann.* **567**, 1.
144. Ref. 100, footnote 4.

4

FEIST'S ACID

DOUGLAS LLOYD

Department of Chemistry, United College of St. Salvator and St. Leonard, University of St. Andrews

A. INTRODUCTION

Although Feist's Acid is a small molecule with molecular formula $C_6H_6O_4$, almost sixty years passed from its first preparation before its correct structure was elucidated. The unfolding of the story makes an entertaining historical study while the authenticated properties of this compound are also of much interest. Perhaps most surprising is its considerable stability, in view of the large steric strain which must be anticipated for the molecule, but samples remain unchanged for many years.

Feist (1893) first prepared the acid by the action of hot concentrated aqueous potassium hydroxide on 3-bromo-5-ethoxycarbonyl-4,6-dimethyl-2-pyrone. It was initially assigned structure (I). This was

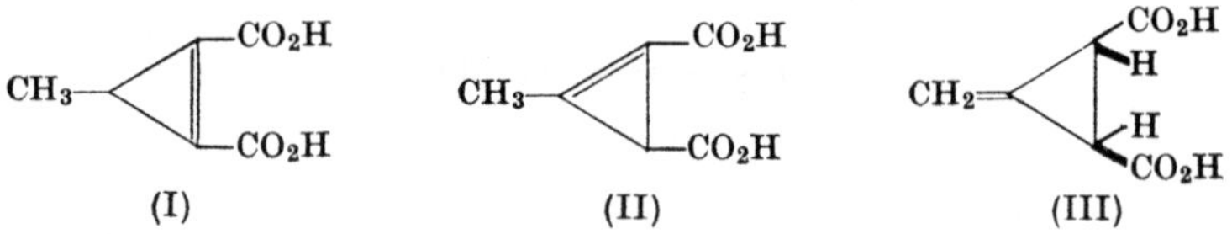

(I) (II) (III)

later modified to (II) (Ingold, 1922; Goss, Ingold and Thorpe, 1923; Feist, 1924). Finally Ettlinger (1952) showed that its structure should be reformulated as (III) and all evidence, both physical and chemical, which has accrued since then is in accord with this formulation.

B. Earlier Work (Pre-1952)

As mentioned above, Feist first prepared this acid in 1893 by the following sequence of reactions:

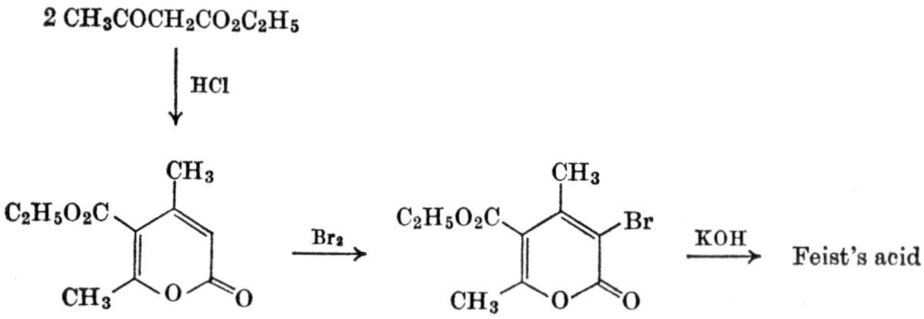

$2\ CH_3COCH_2CO_2C_2H_5 \xrightarrow{HCl}$ (pyrone) $\xrightarrow{Br_2}$ (bromopyrone) $\xrightarrow{KOH}$ Feist's acid

He showed that it was a dicarboxylic acid, m.p. 200°, and that it decolorised solutions of bromine and permanganate. He ascribed the symmetrical methylcyclopropenedicarboxylic acid structure (I) to the molecule. Feist also isolated a dibromo addition product, and claimed that when this was treated with sodium amalgam an isomer of (I), m.p. 189°, was produced.

Jones (1905) prepared the methyl and ethyl esters of Feist's acid. He showed that both esters added one molar equivalent of bromine and that, in contrast to Feist's report in the case of the acid, the original esters could be regenerated from the bromine adducts by means of zinc dust in acetic acid.

Some years later evidence was obtained that formula (I) was incorrect and it was replaced by formula (II). Formula (I) was untenable since the acid could be resolved by means of its quinine salt (Feist, 1924); the alternative formula was also suggested on the basis of studies of the chemical properties of the acid (Ingold, 1922; Goss, Ingold and Thorpe, 1923).

Goss, Ingold and Thorpe also insisted that Feist's acid of m.p. 189° was identical with that of m.p. 200°. This was disputed by Feist (1924) but is obviously correct. However Goss *et al.* caused even more confusion by suggesting that two forms of Feist's acid existed, the 'normal' form which they described as having a mobile hydrogen

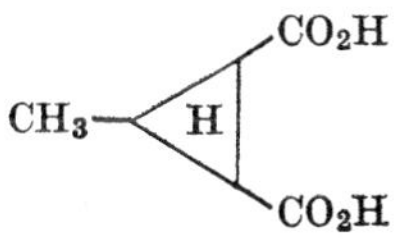

(IV) 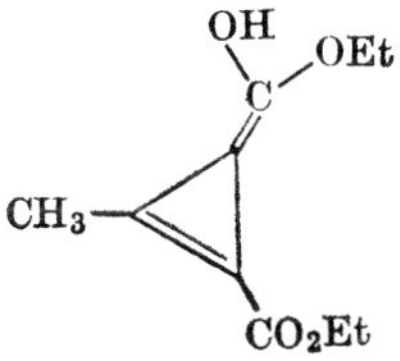 (V)

atom and represented as (IV), and a 'labile' form, which they identified with structure (II) and considered to be present in solution. Furthermore they considered that there were three forms of the esters of Feist's acid, two corresponding to the 'normal' and 'labile' esters and a third enolic form, which they formulated as in (V).

In a later reinvestigation of the problem Kon and Nanji (1932) suggested that the so-called 'normal' acid and esters had structure (II); they were unable to obtain any evidence for the existence of an enol ester, and showed that the 'labile' ester was not an ester of Feist's acid at all. They could only obtain it by heating the normal ester and demonstrated that this caused deep-seated changes in the molecular structure. It is fascinating to read their comment that 'there was usually a flash of blue flame above the liquid just before it began to boil ; Feist's acid and its derivatives were indeed strange

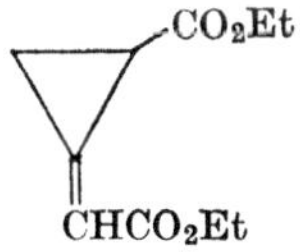

(VI)

compounds! Kon and Nanji suggested that the ‘labile’ ester was diethyl but-1-yne-1,4-dicarboxylate but more recent work has established that its formula is (VI) (Ettlinger, 1952; Ullman, 1959); this is discussed in more detail below (Section D.11).

From 1932 until 1952 formula (II) for Feist’s acid was never challenged publicly. Before passing to more recent work it is worth mentioning two other earlier comments on its structure.

Lowry and Burgess (1923) reinterpreted the ‘normal’ formula of Goss *et al.* in terms of a ‘polar’ formula (VII) with the hydrogen atom co-ordinated between the methylene and ester groups:

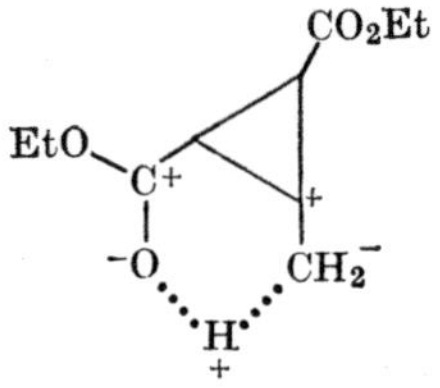

(VII)

‘A much more plausible view is that the hydrogen in this form of the ester is held with a crab-like firmness as part of a co-ordinated complex, in which it is gripped, not by one other atom, but by at least two’.

More strikingly, when Goss, Ingold and Thorpe’s paper (1923a) was read before a meeting of the Chemical Society, a questioner suggested that there was no proof that the ester did not have the methylenecyclopropane structure (III). This led Goss *et al.* to undertake ozonisation experiments which, as confirmed by a number of later workers, gave as product ethyl acetyloxaloacetate but no formaldehyde, thus misleading the investigators into accepting the incorrect formula (see also below, section D.9).

C. The Enlightenment (1952 and After)

In 1952 Ettlinger suggested that Feist’s acid must be reformulated as methylenecyclopropane-*trans*-2,3-dicarboxylic acid (III), and reinterpreted its chemistry in these terms. He showed that it contained no *C*-methyl group, and that the structures of the products from its reactions with bromine and with methanol/sodium methoxide (see below, sections D.10 and 12) were consistent with the reformulation.

His suggestion has been amply confirmed by X-ray crystallographic methods (Lloyd, Downie and Speakman, 1954; Downie and Speakman, 1954; Petersen, 1956), by nuclear magnetic resonance (n.m.r.) spectroscopy (Ettlinger and Kennedy, 1956; Kende, 1956; Bottini and Roberts, 1956) (it was one of the earliest examples of a compound having its structure confirmed by this technique), and by chemical degradation (Lloyd and McOmie, 1956).

There is evidence, however, (Ettlinger and Kennedy, 1957) that the cyclopropene form (II) can exist, since not only do the α-hydrogen atoms in the acid undergo rapid alkali-catalysed H/D exchange, but in addition the γ-hydrogen atoms (i.e. those of the methylene group) also exchange, but much more slowly. This suggests that the cyclopropene form may be present to some extent in equilibrium with the methylenecyclopropane form, but the equilibration of the anions is only slowly established and must be unfavourable to the cyclopropene form. It must be mentioned that some other workers (Bottini and Roberts, 1956; Bottini and Davidson, 1965) record that H/D exchange at the methylene group could not be observed.

The chemistry of Feist's acid, as presently interpreted, will now be outlined systematically.

D. Properties of Feist's Acid

1. *Preparation*

Virtually the only method of preparation of Feist's acid to date is that of Feist himself (1893) as described in the previous section, although more recent workers (Blomquist and Longone, 1959) have increased the yield by modification of the work-up procedure. Ingold (1922) stated that he obtained it in small yield among other products by the action of alkali on ethyl α,α'-dibromo-β-methylglutarate. Attempted syntheses by the addition of diazoacetic ester to unsaturated esters have failed to give the desired product (Feist, 1906; Gilchrist, 1964; Lloyd, unpublished work).

Buchner and Schröder (1902) attempted to extend the reaction and investigated the action of alkali, under a variety of conditions, on 3-bromo-5-ethoxycarbonyl-6-methyl-4-phenyl-2-pyrone. No cyclopropane derivative was obtained; this is not surprising in view of the now accepted structure of Feist's acid.

The reaction has recently been studied in much detail (Gilchrist,

1964; Gilchrist and Rees, 1968a). Feist (1893) had noted that under milder alkaline conditions the product was not a cyclopropane derivative but instead 2,4-bis(ethoxycarbonyl)-3,5-dimethylfuran. Gilchrist and Rees confirmed this observation; their findings may be summarised by the following reaction scheme:

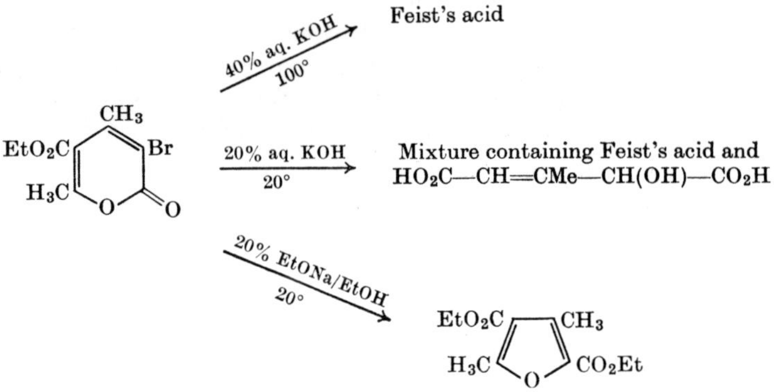

CH_3 / EtO_2C / Br / H_3C / O / O

40% aq. KOH, 100° → Feist's acid

20% aq. KOH, 20° → Mixture containing Feist's acid and $HO_2C—CH{=}CMe—CH(OH)—CO_2H$

20% EtONa/EtOH, 20° → EtO_2C / CH_3 / H_3C / O / CO_2Et

The furan ester was readily hydrolysed by alkali to the corresponding acid but could not be converted into Feist's acid when heated to reflux in 40 per cent aqueous potassium hydroxide solution. It is thus not an intermediate in the formation of Feist's acid.

The later workers were also unable to obtain any cyclopropane derivative from the corresponding 4-phenylbromopyrone, but by reaction of 3-bromo-5-ethoxycarbonyl-4-ethyl-6-methyl-2-pyrone with 40 per cent aqueous alkali they isolated a product to which, on the basis of its n.m.r. spectrum, they assigned structure (VIII).

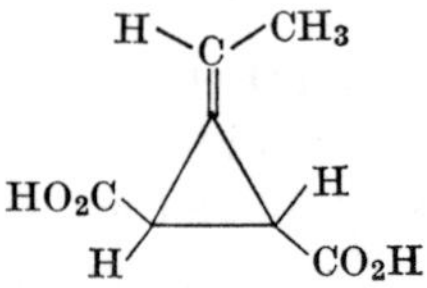

H / C / CH_3 / HO_2C / H / H / CO_2H

(VIII)

This compound (like Feist's acid itself) was surprisingly stable to hot concentrated alkali, and did not isomerise to a vinylcyclopropane.

A mechanism which has been proposed (Ullman, 1959) for the formation of Feist's acid from the bromopyrone is as follows:

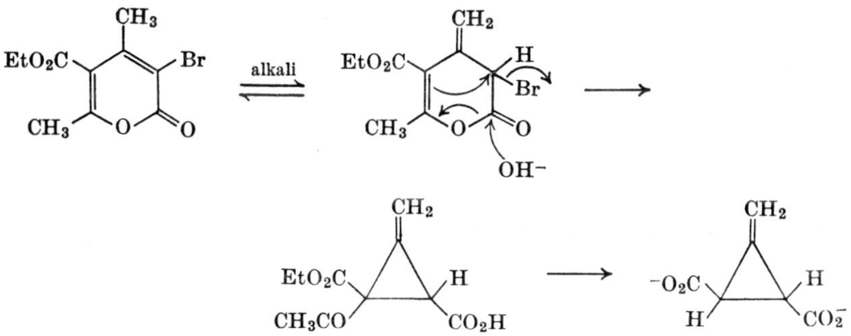

2. *General Properties*

Feist's acid is a strong acid ($pK_a^1 = 3{\cdot}2$). It is interesting to note that its ΔpK value (i.e. the difference between the pK values for the two carboxyl groups) is 1·23 pK units, which implies that these groups are separated by a distance of between 5·0 and 6·2 Å (Downie, 1952); this is in agreement with the *trans* structure for the acid. It is only very slightly soluble in cold water, but readily soluble in hot water, in ether and in warm ethanol, chloroform or acetone. It forms crystalline barium and calcium salts (Feist, 1893) and crystalline potassium, strontium and thallium salts (Lloyd, unpublished work). The potassium salt is highly hygroscopic; the thallium salt was utilised for X-ray crystallographic examination (see below).

3. *Derivatives*

Feist's acid forms esters in the usual way (Jones, 1905; Goss, Ingold and Thorpe, 1923a; Feist, 1924). The methyl ester (m.p. 32°; b.p. 122°/20 mm) and ethyl ester (m.p. 38°; b.p. 135°/15 mm) are both low-melting solids. The acid is converted into an acid chloride (b.p. 85–86°/9·9 mm) by thionyl chloride (Blomquist and Longone, 1959). This acyl chloride reacts with aniline to give a dianilide (m.p. 244·5–245°) and with dimethylamine in benzene to give a bis(dimethylamide) (m.p. 70–71°), which may be reduced by lithium aluminium hydride to 1,2-bis(dimethylaminomethyl)-3-methylenecyclopropane (b.p. 77–79°/9·5 mm) (Blomquist and Longone, 1959). Reaction of the ethyl ester with hydrazine hydrate gives a hydrazide (m.p. 233°) (Liebermann *et al.*, 1954).

4. *X-ray Crystallography*

X-ray crystallographic examinations have been made of the acid and of its thallium salt (Lloyd, Downie and Speakman, 1954; Downie and Speakman, 1954; Petersen, 1956). They confirm the methylene-cyclopropane structure and the *trans* arrangement of the carboxyl groups. In the case of the free acid, intermolecular hydrogen bonding between carboxyl groups holds the molecules in the form shown in

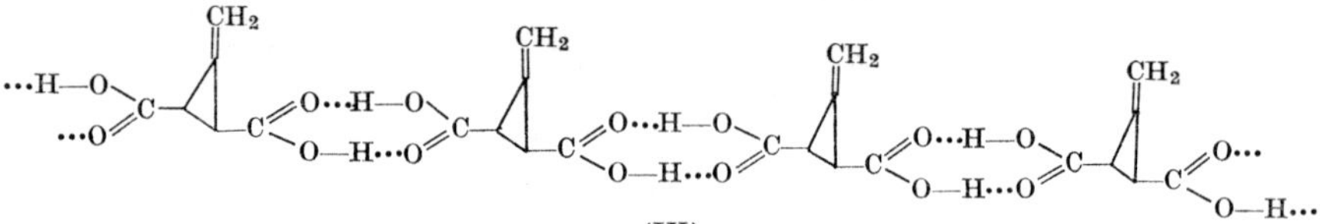

(IX)

(IX). The length of the exocyclic double bond, 1·32 Å, is that of a normal unconjugated double-bond, while the length of the ring bonds (~1·50 Å for $CH_2{=}C{-}C{-}CO_2H$, and 1·55 Å for $HO_2C{-}C{-}C{-}CO_2H$) are also normal.

5. *Spectra*

The ultra-violet spectrum of Feist's acid has maxima at 195 mμ ($\epsilon = 4800 \pm 300$) and 205 mμ ($\epsilon = 1800 \pm 200$), showing that there is no conjugation between the olefinic and carboxyl groups (Ettlinger and Kennedy, 1956).

The infra-red spectrum (Ettlinger and Kennedy, 1956; Blomquist and Longone, 1959) entirely accords with the suggested structure. The spectra of the acid and a number of its derivatives have been analysed in detail by Blomquist and Longone, as follows:

Infra-red Spectra of Derivatives of Feist's Acid (μ)

Compound	$>C{=}O$	$>C{=}CH_2$	*Attributed to cyclopropane ring, but unreliable as diagnostic*
Feist's acid (KBr)	5·88 s	11·0 s	9·75 w
Diacid Chloride (neat)	5·60 s	10·83 m	9·68, 9·96 m
Dianilide (KBr)	6·04 s	11·09 m	9·73, 9·97 m
Bis(dimethylamide) (nujol)	6·07 s	10·96 s	—

The absorption at $\sim 11 \cdot 0$ represents out-of-plane deformations in the terminal methylene group. It is shifted to abnormally high frequencies for this type of group; for normal unconjugated examples such absorption lies in the range $11 \cdot 22$–$11 \cdot 27\ \mu$. It is suggested that this must be due to interaction with the carbonyl groups since methylene-cyclopropane itself, and derivatives having no such —COX groups, absorb at normal frequency. Thus, when the bis(dimethylamide) is reduced to the corresponding bis(dimethylaminomethyl) derivative the absorption is then found to be at $11 \cdot 28\ \mu$. This interaction cannot however be ascribed to simple conjugation since the carbonyl group absorbs at a normal frequency in each case. Furthermore other evidence, such as ultra-violet spectra, gives no indication of any simple conjugation. Blomquist and Longone suggest that it may be associated with an inductive effect due to the carbonyl group which is transmitted through the ring. In accord with this suggestion, the most electronegative group, namely the acid chloride group, is associated with the largest shift. There is also an absorption at $3 \cdot 23\ \mu$ due to the methylene group but no absorption that could be associated with a methyl group.

The nuclear magnetic resonance spectrum has been analysed by a number of workers (Ettlinger and Kennedy, 1956; Kende, 1956; Bottini and Roberts, 1956; Ullman, 1959; Graham and Rogers, 1962). It shows quite unequivocally that the acid has the methylenecyclopropane structure. Both the signals for the methylene protons and for the hydrogen atoms attached to the ring appear as triplets, the coupling constant being 2·5 cps. This relatively large coupling constant between protons separated from one another by four bonds is of interest. It may be compared with the corresponding values for substituted propenes ($J = 1 \cdot 4$–$1 \cdot 8$ cps) and allenes ($J = 6 \cdot 1$–$7 \cdot 0$ cps). Graham and Rogers suggest that a possible explanation is that the nature of the electronic structure of the cyclopropane ring imparts some measure of allene-like character to the molecule. However the role which the geometry of the molecule plays is difficult to ascertain and this might also be a controlling factor.

6. *Hydrogen–Deuterium Exchange*

The protons attached to the cyclopropane ring undergo deuterium exchange with DO^-/D_2O which is catalysed by excess alkali (Bottini and Roberts, 1956; Ettlinger and Kennedy, 1957; Bottini and Davidson, 1965). The latter workers studied the rates for both Feist's acid and its *cis*-isomer (for preparation of the *cis*-isomer see section D.7). The *cis*-isomer exchanges about a hundred times faster than the *trans*, and both exchange thousands of times faster than *cis*- and *trans*-cyclopropane-1,2-dicarboxylic acids. Bottini *et al.* (1956, 1965) could observe no H/D exchange of the methylene protons but Ettlinger and Kennedy (1957) state that both the i.r. and n.m.r. spectra indicate that these protons undergo exchange, but much more slowly than the ring protons. They suggest that this provides evidence for the existence of the methylcyclopropene form (II) of the acid in solution but that the equilibration is only slowly established and is unfavourable to form (II).

7. *Stereomutation*

Reaction with acetic anhydride and potassium acetate readily converts Feist's acid into the anhydride of its *cis*-isomer, m.p. 21·5–23° (Ettlinger and Kennedy, 1957). This anhydride may in turn be converted into an anilic acid (m.p. 139–139·5°), and a dianilide (m.p. 123·5–124°), and into the corresponding *cis*-dicarboxylic acid (m.p. 120·5–121·5°). Spectra determinations confirm that the latter is an isomer of Feist's acid; its stereochemistry is proved by its hydrogenation (H_2/Pt/HOAc) to pure all-*cis*-methylcyclopropane-2,3-dicarboxylic acid.

This *cis*-isomer is stable to concentrated hydrochloric acid at room temperature but reverts readily to the *trans*-isomer in the presence of base. The intermediate anion is apparently converted into the *cis*- and *trans*-forms at equal rates, but the *cis*-acid is converted into this anion much faster than is the *trans*-acid (Bottini and Davidson, 1965). The equilibria may be represented as follows:

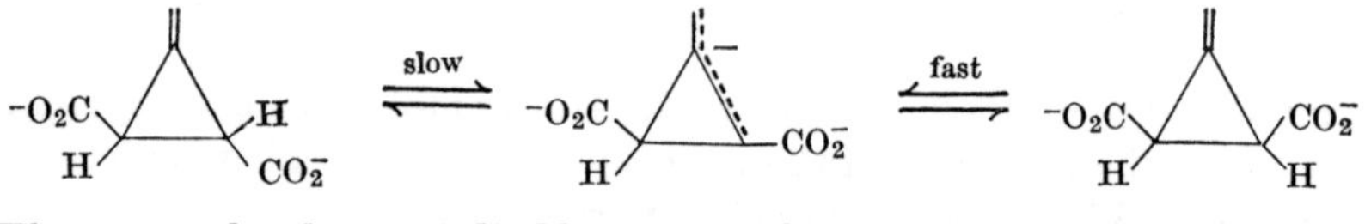

^-O_2C, H, H, CO_2^- $\underset{}{\overset{\text{slow}}{\rightleftharpoons}}$ ^-O_2C, H, CO_2^- $\underset{}{\overset{\text{fast}}{\rightleftharpoons}}$ ^-O_2C, H, CO_2^-, H

The system has been studied by means of rate studies on H/D exchange of the *cis*- and *trans*-acids and on racemisation of the *trans*-acid (Bottini and Davidson, 1965).

8. *Reduction*

Feist's acid and its esters are readily hydrogenated, for example over colloidal palladium or palladium-charcoal (Feist, 1924; Ettlinger,

Harper and Kennedy, 1957; Ullman, 1959; Lloyd, unpublished work). Somewhat over one mole of hydrogen is taken up and the product is the corresponding methylcyclopropanedicarboxylic acid. Chromatographic examination of the reduction product from the acid, and vapour phase chromatography of the product from the dimethyl ester, have shown, respectively, the presence of ethylsuccinic acid and dimethyl ethylsuccinate in addition to the methylcyclopropane derivatives (Ullman, 1959).

Lithium aluminium hydride reduces Feist's acid to the corresponding dialcohol (Dorko, 1965).

Goss, Thorpe and Ingold (1923) claimed that they obtained β-methylglutaconic acids by the action of hydriodic acid on Feist's acid.

9. *Oxidation*

In a direct attempt to distinguish between the methylcyclopropene structure (II) and the methylenecyclopropane structure (III) for Feist's acid, Goss, Ingold and Thorpe (1923) investigated the ozonisation of its diethyl ester. No formaldehyde could be detected and the product was identified as ethyl acetyloxaloacetate (X, R = Et).

$$RO_2C{-}CO{-}\underset{\displaystyle COCH_3}{\underset{|}{C}}HCO_2R$$

(X)

This result, which has been confirmed by other workers (Feist, 1924; Lloyd, Downie and Speakman, 1954a), bedevilled the elucidation of the correct structure for Feist's acid. More recently it has been pointed out (Gragson, Greenlee, Derfer and Boord, 1953) that methylenecyclopropane itself gives only 2 per cent formaldehyde on ozonisation, although it has been reported that methylenecyclopropanes which have no electron-withdrawing groups do give normal ozonisation products (Schwan, 1958; Erickson, 1966).

A rationalisation of the ozonisation reaction has been made in terms of the mechanism shown on page 260 (Bottini and Roberts, 1956).

Some recent workers (Schwan, 1958; Erickson, 1966) did not find (X) as a product in their ozonisation experiments but instead obtained (XI) by a method involving neither oxidative nor reductive

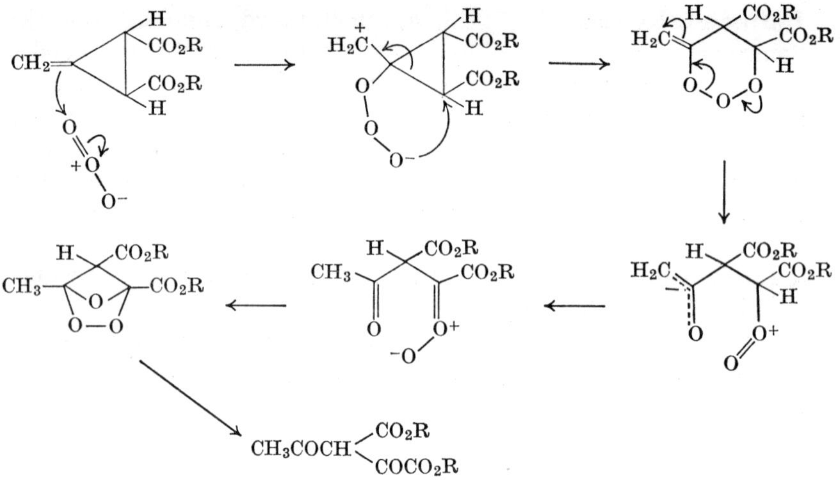

$CH_3COCH(CO_2R)COCO_2R$

(X)

work-up. It was suggested (Erickson, 1966) that it was formed as follows:

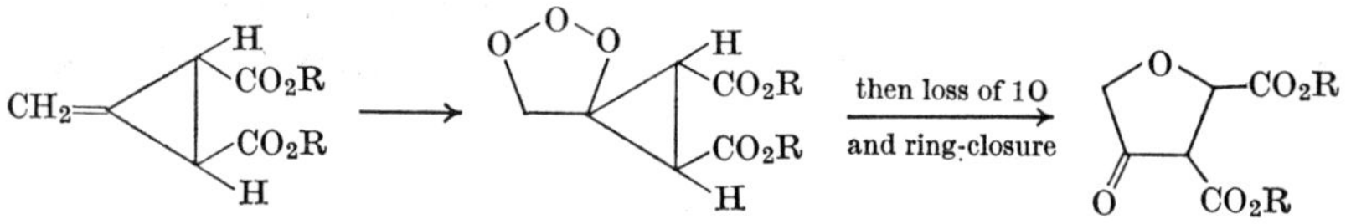

then loss of 1O and ring-closure

(XI)

The fate of the lost oxygen atom was not ascertained, and Erickson comments that it is difficult to reconcile this reaction with normal ozonisation mechanisms.

Boreham, Goss and Minkoff (1955) tried to obtain proof for the methylcyclopropene structure (II) by oxidising Feist's acid with chromic acid but could not detect any acetic acid among the products.

An earlier attempt to investigate the alternative and now accepted structure (III) by chemical means (Lloyd, unpublished work) was unsuccessful in that no formaldehyde could be detected in the oxidation of either the acid or its diethyl ester with dilute neutral permanganate. However by means of the elegant method of Lemieux and Rudloff (1955), which utilises as oxidising agent sodium periodate in the presence of a catalytic amount of potassium permanganate, Lloyd and McOmie (1956) were able to obtain formaldehyde in high yield from both acid and ester.

Very recently it has been shown (Gilchrist and Rees, 1968b) that the diethyl ester will not give an epoxide on treatment with perbenzoic acid.

10. *Bromination*

The addition of bromine to Feist's acid or its esters to give dibromo-derivatives was noted by the earliest workers (Feist, 1893; Jones, 1905; Goss, Ingold and Thorpe, 1923) but the correct structure of the adducts as 1-bromo-1-(bromomethyl)cyclopropane-2,3-dicarboxylic acid or ester inevitably awaited the final reappraisal of the formula of Feist's acid itself (Ettlinger, 1952).

Feist (1893, 1924) claimed that reaction of the dibromo-compounds with sodium amalgam produced isomers of the original acid and esters, but subsequent work (Goss, Ingold and Thorpe, 1923a) showed that the products were in fact Feist's original acid or ester. The dibromo-esters are also reconverted into esters of Feist's acid by zinc, but are stable to diethylaniline (Jones, 1905).

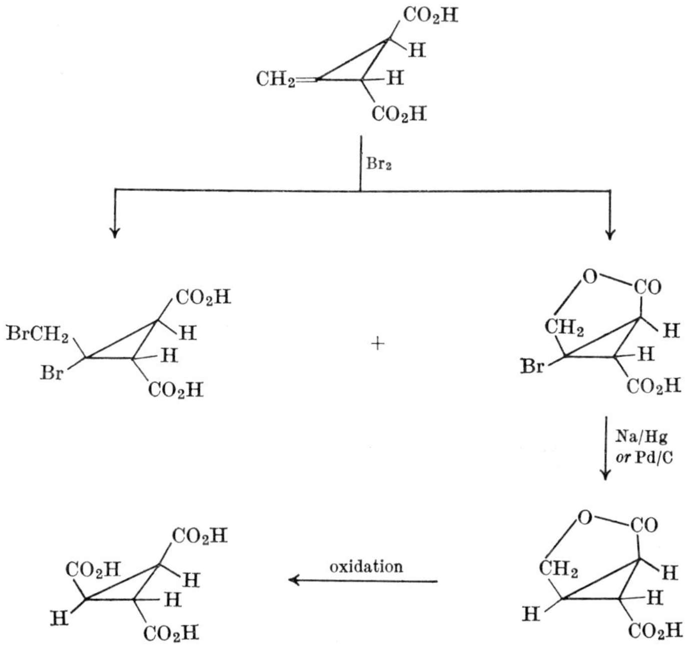

Feist (1893, 1924) found that treatment of the dibromo-acid with water gave a monolactone of a monobromo-dicarboxylic acid which was, in turn, hydrolysed to a monobromo-monohydroxy-dicarboxylic acid. Goss, Ingold and Thorpe (1923) obtained the latter compound by the action of bromine-water on Feist's acid and converted it into the lactone. The lactone could be converted into the dibromo-adduct of Feist's acid by treatment with bromine, phosphorus pentachloride and iron.

The bromination was reinvestigated in detail by Ettlinger (1952), and his findings may be summarised by the diagram at the foot of page 261.

Lactone (Y) was inert to both permanganate and hot silver nitrate in accord with the general behaviour of halocyclopropanes. Lactone (Z) could be reversibly hydrolysed to a hydroxydicarboxylic acid; it did not revert to an unsaturated acid with alkali. Both lactones showed carbonyl absorption in their infra-red spectra at 5·55 μ.

Gilchrist and Rees (1968b) have shown that reaction of the dimethyl dibromo-ester with lithium chloride in formdimethylamide gives a mixture of a bromochloro-ester and the already known bromolactone:

They suggest that the chloro-compound is formed first and then undergoes intramolecular displacement of chlorine:

The corresponding diethyl dibromo-ester gave only the chloro-compound and no lactone under these conditions (Gilchrist, 1964).

11. *Pyrolysis*

Kon and Nanji (1932) showed that pyrolysis of the diethyl ester of Feist's acid caused an irreversible isomerisation. They formulated the product as diethyl but-1-yne-1,4-dicarboxylate, but Ettlinger pointed out (1952) that it showed no triple-bond absorption in its infra-red spectrum and suggested that its most probable structure was (XII). Ullman (1959) obtained an analogous product by vapour phase chromatography of the dimethyl ester of Feist's acid. Spectral

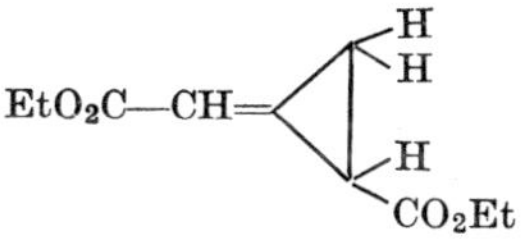

(XII)

evidence (n.m.r., u.v., i.r.) strongly supported a structure as in (XII), as did its catalytic reduction to the corresponding saturated acid. Ullman suggested that its formation involved an intermediate such as (XIII) which could collapse to give either the starting material or

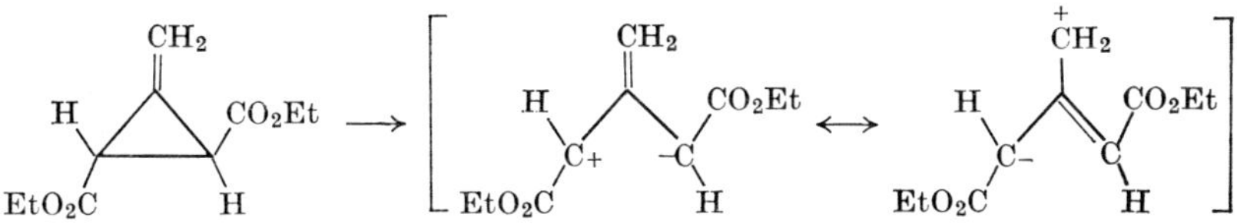

(XIII)

(XII). On the basis of stereochemical evidence Ullman later proposed (1960) that the formation of the pyrolysis product proceeded by two competitive mechanisms, viz.,

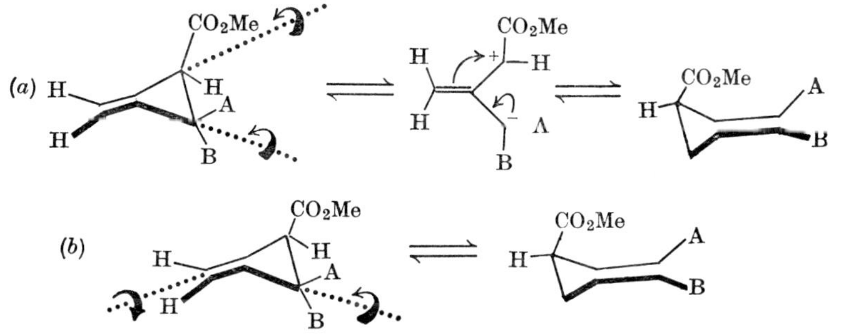

(a)

(b)

either A = H, B = CO_2Me

or A = CO_2Me, B = H

Mechanism (a) involves rotation of both ester-bearing carbon atoms with scission of the bond linking them to give a zwitterion which then recombines in an alternative way. Mechanism (b) is a direct valence-bond tautomerism involving simultaneous cleavage and reformation of the ring-bonds with concomitant rotation of one ester-bearing carbon atom and of the terminal methylene group, to give the pyrolysis product directly. Alternatively this reaction might involve a diradical intermediate which would be a substituted trimethylenemethane species.

Similar rearrangement reactions have been observed with other methylenecyclopropane derivatives, e.g. methyl (2′-methylenecyclopropyl) acetate is converted into methyl β-cyclopropylidenepropionate (Ullman and Fanshawe, 1961), and 2-methyl-methylenecyclopropane into ethylidenecyclopropane (Cheswick, 1963).

12. *Addition of Alcohols*

The reaction of methanol or ethanol in the presence of methoxide or ethoxide ions with esters of Feist's acid (Goss, Ingold and Thorpe, 1923a) appeared to provide evidence for the conjugated methylcyclopropene type of structure for Feist's acid. The structure of the adduct (XIV) was substantiated (Goss, Ingold and Thorpe, 1925) by its hydrolysis with concentrated hydrochloric acid to laevulic acid

$$(CH_3)(CH_3O)C\langle CH(CO_2R)\text{—}CH(CO_2R)\rangle \text{ (XIV)} \xrightarrow{HCl} \left[CH_3CO\text{—}CH(CO_2H)\text{—}CH_2CO_2H \right] \longrightarrow CH_3COCH_2CH_2CO_2H \text{ (XV)}$$

(XV). This reaction has been confirmed by later workers (Kon and Nanji, 1932; Ettlinger, 1952).

Goss and Ingold also reported (1925) the breakdown of Feist's acid to succinic acid and acetic acid, which they attributed to a similar type of reaction:

$$\text{Feist's acid} \xrightarrow{OH^-} (CH_3)(HO)C\langle CH(CO_2^-)\text{—}CH(CO_2^-)\rangle \longrightarrow CH_3CO\text{—}CH(CO_2^-)\text{—}CH_2CO_2^- \longrightarrow CH_3CO_2^- + CH_2CO_2^-\text{—}CH_2CO_2^-$$

Both reactions involve the equivalent of an enol-keto change followed by hydrolysis of the resultant β-keto-acid.

By using methanolic hydrogen chloride to cleave the methoxide adduct, an ester of the assumed breakdown intermediate, dimethyl acetosuccinate, has been isolated (Ettlinger, 1952):

$$\text{(cyclopropane: } CH_3, CH_3O; H, CO_2CH_3; H, CO_2CH_3) \xrightarrow[\text{in } CH_3OH]{\text{hot dil. HCl}} CH_3COCH(CO_2CH_3)CH_2CO_2CH_3$$

This addition reaction is easier to formulate in terms of the outdated structure (II) and Ettlinger and Kennedy (1956) have suggested that it could be rationalised either by the intermediacy of the methylcyclopropene form (II) (cf. section D.6, above), or by direct attack of the alkoxide ion at the unsaturated ring carbon atom of the methylenecyclopropane form (III) with stabilisation of the resultant negative charge on the terminal methylene group by the influence of the ester groups through the three-membered ring.

Goss and Ingold (1928) also claimed that the cyanoacetate carbanion [$^-CH(CN)CO_2Me$] would add to an ester of Feist's acid, although in an earlier paper (Goss, Ingold and Thorpe, 1923a) it had been stated that no such addition occurred.

13. *Attempted Diels-Alder Reaction*

Neither Feist's acid nor its ethyl ester would undergo a Diels-Alder reaction with cyclopentadiene either at room temperatures or at elevated temperatures in a sealed tube (Lloyd, unpublished results). In each case the acid or ester was recovered in essentially quantitative yield. This is in accord with its presently accepted structure, but reaction might be expected had the earlier proposed formula been correct.

14. *Radical Reactions*

The esters of Feist's acid appear to undergo straightforward radical reactions with great ease (Gilchrist and Rees, 1968b), e.g.

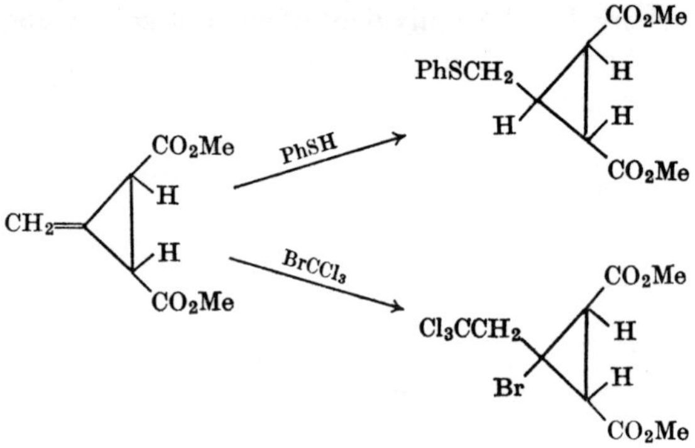

In each case the reaction was carried out in the presence of di-t-butyl peroxide.

Irradiation of a mixture of the ester with benzaldehyde in the presence of oxygen also apparently gave a 1:1 adduct by free radical addition, although in low yield.

This ready participation in radical addition reactions may be connected with the relief in steric strain which results from the formation of the adducts. No evidence was found for the participation of a non-classical cyclopropylmethyl radical (Gilchrist and Rees, 1968b).

15. *Carbenoid Addition Reactions*

The diethyl ester of Feist's acid reacted readily with diazoacetic ester at 90–100° in the presence of copper bronze to give an oil which appeared to be a mixture of two stereoisomeric spirans (Gilchrist and Rees, 1968b):

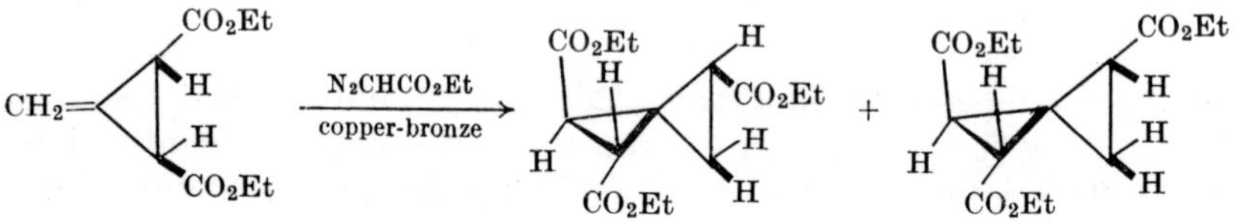

Gas-liquid chromatography showed that one isomer was present in large excess (19:1) but their stereochemistry was not determined.

E. Compounds Related to Feist's Acid

Only very few compounds closely related to Feist's acid have been prepared and attempts to extend the range have been few (Buchner and Schröder, 1902; Gilchrist and Rees, 1968a). The latter workers prepared the homologous compound (XVI) but have not reported any study of its chemistry. Ettlinger and Kennedy (1957) obtained the *cis*-isomer (XVII) (see above, section D.7), and Ullman (1959) prepared the isomer (XVIII) by the action of potassium t-butoxide on the bromo-ester (XIX). The ester (XVIII) was reduced catalytically to the corresponding methylcyclopropane ester. The mono-

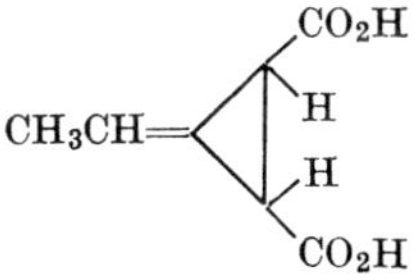

(XVI) 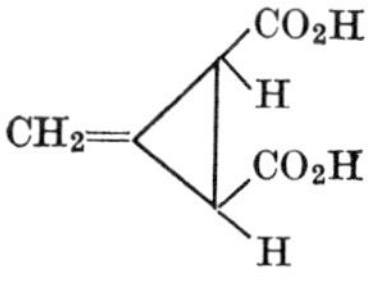 (XVII) 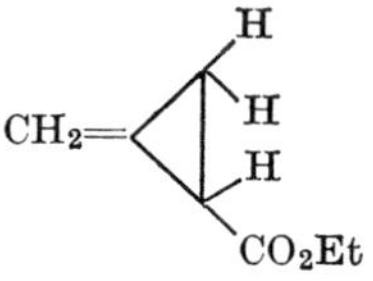 (XX)

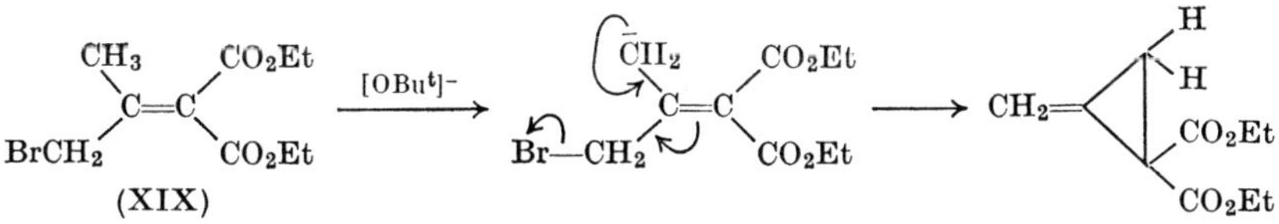

(XIX) $\xrightarrow{[OBu^t]^-}$ $\longrightarrow$ (XVIII)

carboxylic acid derivative (XX) has also been described, having been obtained both by treatment of ethyl 2-bromo-2-methylcyclopropane-carboxylate with sodium hydride, and by reaction of ethyl diazo-acetate with allene (Carbon, Martin and Swett, 1958), but, with these few exceptions, Feist's acid remains as an isolated but most interesting example of this type of methylenecyclopropane derivative.

Acknowledgments

I am most grateful to Professor C. W. Rees and to Dr. T. L. Gilchrist for making their results available to me prior to their publication. I am also grateful to my colleague, Dr. J. M. F. Gagan, for reading the original manuscript and offering valuable critical comments and suggestions.

References

Blomquist, A. T. and Longone, D. T. (1959) *J. Amer. Chem. Soc.* **81**, 2012.
Boreham, G. R., Goss, F. R. and Minkoff, G. J. (1955) *Chemy. Ind.* 1354.
Bottini, A. T. and Davidson, A. J. (1965) *J. Org. Chem.* **30**, 3302.
Bottini, A. T. and Roberts, J. D. (1956) *J. Org. Chem.* **21**, 1169.

Buchner, E. and Schröder, H. (1902) *Ber.* **35**, 782.
Carbon, J. A., Martin, W. B. and Swett, L. R. (1958) *J. Amer. Chem. Soc.* **80**, 1002.
Cheswick, J. P. (1963) *J. Amer. Chem. Soc.* **85**, 2720.
Dorko, E. A. (1965) *J. Amer. Chem. Soc.* **87**, 5518.
Downie, T. C. (1952) Ph.D. Thesis, University of Glasgow.
Downie, T. C. and Speakman, J. C. (1954) *Acta Cryst.* **7**, 647.
Erickson, R. E. (1966) *Tetrahedron Letters* 1753.
Ettlinger, M. G. (1952) *J. Amer. Chem. Soc.* **74**, 5808.
Ettlinger, M. G., Harper, S. H. and Kennedy, F. (1957) *J. Chem. Soc.* 922.
Ettlinger, M. G. and Kennedy, F. (1956) *Chemy. Ind.* 166.
Ettlinger, M. G. and Kennedy, F. (1957) *Chemy. Ind.* 891.
Feist, F. (1893) *Ber.* **26**. 747.
Feist, F. (1906) *Liebigs Ann.* **345**, 103.
Feist, F. (1924) *Liebigs Ann.* **436**, 125.
Gilchrist, T. L. (1964) Ph.D. Thesis, University of London.
Gilchrist, T. L. and Rees, C. W. (1968a) *J. Chem. Soc.* (*C*) 769.
Gilchrist, T. L. and Rees, C. W. (1968b) *J. Chem. Soc.* (*C*) 776.
Goss, F. R. and Ingold, C. K. (1925) *J. Chem. Soc.* **127**, 2776.
Goss, F. R. and Ingold, C. K. (1928) *J. Chem. Soc.* 1268.
Goss, F. R., Ingold, C. K. and Thorpe, J. F. (1923a) *J. Chem. Soc.* **123**, 327.
Goss, F. R., Ingold, C. K. and Thorpe, J. F. (1923b) *J. Chem. Soc.* **123**, 3342.
Goss, F. R., Ingold, C. K. and Thorpe, J. F. (1925) *J. Chem. Soc.* **127**, 460.
Gragson, J. T., Greenlee, K. W., Derfer, J. M. and Boord, C. E. (1953) *J. Amer. Chem. Soc.* **75**, 3344.
Graham, J. D. and Rogers, M. T. (1962) *J. Amer. Chem. Soc.* **84**, 2249.
Ingold, C. K. (1922) *J. Chem. Soc.* **121**, 2676.
Jones, D. T. (1905) *J. Chem. Soc.* **87**, 1062.
Kende, A. S. (1956) *Chemy. Ind.* 437, 544.
Kon, G. A. R. and Nanji, H. R. (1932) *J. Chem. Soc.* 2557.
Lemieux, R. U. and Rudloff, E. von (1955) *Canad. J. Chem.* **33**, 1701, 1710.
Liebermann, D., Rist, N., Grumbach, F., Moyeux, M., Gauthier, B., Rouaix, A., Maillard, J., Himbert, J. and Cals, S. (1954) *Bull. soc. chim. France* 1430.
Lloyd, D. unpublished work.
Lloyd, D., Downie, T. C. and Speakman, J. C. (1954a) *Chemy. Ind.* 222.
Lloyd, D., Downie, T. C. and Speakman, J. C. (1954b) *Chemy. Ind.* 492.
Lloyd, D. and McOmie, J. F. W. (1956) *Chemy. Ind.* 874.
Lowry, T. M. and Burgess, H. (1923) *J. Chem. Soc.* **123**, 2111.
Petersen, D. R. (1956) *Chemy. Ind.* 904.
Schwan, U. (1958) Dissertation, Tech. Hoch., Karlsrühe.
Ullman, E. F. (1959) *J. Amer. Chem. Soc.* **81**, 5386.
Ullman, E. F. (1960) *J. Amer. Chem. Soc.* **82**, 505.
Ullman, E. F. and Fanshawe, W. J. (1961) *J. Amer. Chem. Soc.* **83**, 2379.

5

ELECTRONIC STRUCTURE AND SPECTRAL PROPERTIES OF ANNULENES AND RELATED COMPOUNDS

HUBERT P. FIGEYS

Université Libre de Bruxelles, Service de Chimie Organique, Faculté des Sciences, 50 Avenue F. D. Roosevelt, Brussels 5, Belgium

A. INTRODUCTION

The formal name of the monocyclic unsaturated hydrocarbons made up of an even number of carbon atoms and of general formula $C_N H_N$ is '[N]annulenes'.

This article is concerned with the electronic structure of annulenes and their benzo-derivatives in their ground-state, but benzene and higher polycyclic benzenoid hydrocarbons, such as anthracene, are excluded. Biphenylene will be treated as a dibenzo-[4]annulene. The spectral properties of these compounds will be discussed only as far as they throw some light on the problems under consideration. This article is thus not intended as a review but rather as an up-to-date synthesis of knowledge in this field.

Although annulenes and some of their benzo-derivatives aroused the interest of several theoretical chemists some time ago, their synthesis and the subsequent study of their spectral properties has, for a great part, only been achieved in very recent years. Both theoretical and experimental efforts were focused on the notion of 'aromaticity'. The problem is a fundamental one whose experimental study must be approached very carefully. In general, aromaticity has been related to some type of chemical reactivity or to some characteristic features of electronic absorption spectra. However, this concerns not only the properties of the ground state of the molecule but also the properties of transition states or electronically excited states. Modern definitions of aromaticity are based on purely electronic ground state properties such as heats of formation and magnetic behaviour; these are discussed in the following sections.

B. Aromaticity and Anti-Aromaticity

1. *The Thermodynamic Criterion*

Several decades ago, Hückel formulated his famous $(4n + 2)$ rule[144], which states that 'Those monocyclic coplanar systems of trigonally hybridised atoms which contain $(4n + 2)$ π-electrons will

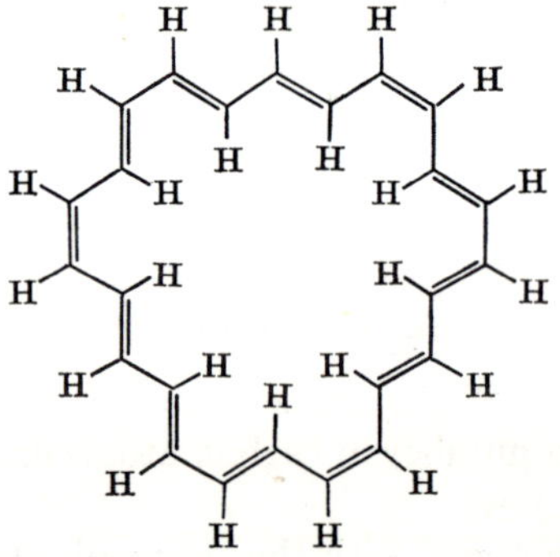

Fig. 1. Probable geometry of [24]annulene.

possess relative electronic stability'[263]. His argument is based on the assumption that, if a molecule has occupied anti-bonding or non-bonding π-orbitals, its stability should be considerably reduced; on the other hand, if the electrons can be put into bonding orbitals, stability should be gained. As has been pointed out by several authors[78, 85, 168, 226, 263], experimental verification of this important rule includes, for example, the great stability shown by benzene, the cyclopentadienide anion and the cycloheptatrienium ('tropylium') cation.

However, in the framework of simple Hückel theory, no sound theoretical reason can be given why the higher members of the [4n]annulene series should not be aromatic! A compound such as [24]annulene, for example (Fig. 1), should be reasonably planar without angular distortions and can be shown to have appreciable conventionally calculated delocalisation energy. Coulson[75] and Polansky[207] have indeed shown that the energy of the π-molecular orbitals for an [N]annulene with equal bond lengths (such as in benzene) are given in the linear combination of atomic orbitals (LCAO) approximation by:

$$\epsilon_j = 2 . \beta . \cos \frac{2j\pi}{N} \qquad (j = 0, 1, 2, \ldots N - 1). \tag{1}$$

Application of this relation to cyclobutadiene ($N = 4$) gives us the energy diagram shown in Fig. 2. There are four electrons which can

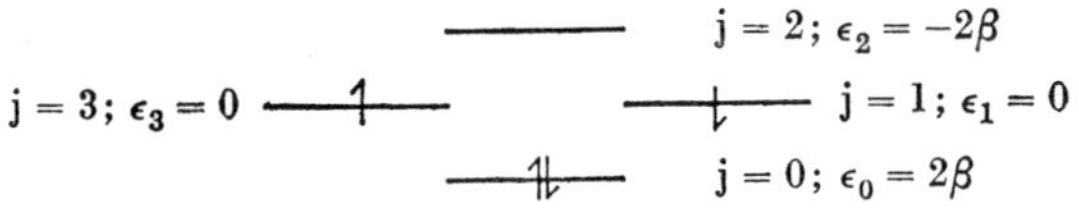

$j = 2;\ \epsilon_2 = -2\beta$

$j = 3;\ \epsilon_3 = 0$ $\quad$ $j = 1;\ \epsilon_1 = 0$

$j = 0;\ \epsilon_0 = 2\beta$

Fig. 2. Energy diagram of square cyclobutadiene.

occupy one bonding and two non-bonding orbitals. One pair of electrons fills the bonding orbitals; the two others enter, according to Hund's rule, one into each non-bonding orbital with parallel spins. Thus square cyclobutadiene should, undoubtedly, have a triplet ground state. The same conclusion can be obtained for the higher members of the [4n]annulene series, all of them having a pair of degenerate non-bonding orbitals. Their total π-bonding energy can easily be obtained by the relation:

$$E_{\pi b} = \sum_{j}^{\text{occ.}} n_j \epsilon_j \tag{2}$$

where n_j is the number of π-electrons in orbital j. This quantity, and the corresponding delocalisation energy (DE) calculated with respect to the hypothetical structure made up of alternating pure single and pure double bonds, are given in Table 1. It appears that the higher

TABLE 1

Total π-bonding energy and delocalisation energy of [4n]-annulenes with equal bond lengths (1·397 Å)

[N]annulene	$E_{\pi b}$	DE ($\|\beta\|$)
4	4·000	0
8	9·657	1·657
12	14·928	2·928
16	20·109	4·109
20	25·255	5·255
24	30·383	6·383
28	35·378	7·378

members of the series should have a much greater delocalisation energy than, for example, benzene ($DE = 2|\beta|$) and thus should show extraordinary stability! It seems therefore that Hückel's argument was based rather on the existence of closed shells of electrons, with some kind of comparison with the especially stable electron shells of the inert gases, $(4n + 2)$ being in this case the magic number of electrons.

Recently, Breslow[45] and Dewar[88] proposed a thermodynamic criterion by which to define *aromaticity*, namely 'cyclic conjugated systems are considered aromatic if cyclic delocalisation of electrons makes a negative contribution to their heat of formation'. By this definition, an opposite phenomenon can be imagined in which cyclic delocalisation of electrons is destabilising; such a phenomenon should then be called *anti-aromaticity*. As a result, *non-aromaticity* should be a simple lack of stabilisation energy in the cyclic compound as compared with the linear one.

Much of the controversy over bond fixation, delocalisation and resonance has been based on misunderstandings in the fundamental concepts involved[91]. There is, in fact, no such thing as a localised bond; even in saturated molecules such as paraffins, the valence electrons must, according to current quantum theory, be completely delocalised. However, such a molecule may behave in certain respects *almost as if* the electrons were localised in definite bonds. This implies that the criterion of bond fixation must be based on an artificial concept: 'a molecule can be regarded as having localised bonds if its collective properties can be regarded as additive functions of bond properties'[91] (bond energies, bond dipole moment, etc.). On this basis, molecules may be represented in terms of localised bonds even if interactions between them are not negligible, for it may be possible to include a correction for these interactions into 'empirical' bond properties.

Recently, it has been shown, both by Pople's self-consistent field (SCF) molecular orbital (MO) theory[88] and by Hückel theory[110], that, provided the variation of the core resonance integrals β_{ij}^{c} or the exchange integrals β_{ij} with bond length are taken into account by an iterative procedure[88, 115], the classical linear even polyenes might be represented in terms of localised 'single' and 'double' bonds. In other words, if we assume that σ-bond energy is an additive property of the number of formal C_{sp2}—C_{sp2} single and double bonds and C_{sp2}—H bonds, this means that the total π-energy E_{π} of these molecules can be written:

$$E_{\pi} = n \, . \, E_{\pi}^{s} + n' \, . \, E_{\pi}^{d} \tag{3}$$

where n and n' are the number of CC 'single' and 'double' bonds respectively, E_{π}^{s} and E_{π}^{d} being their π-energies.

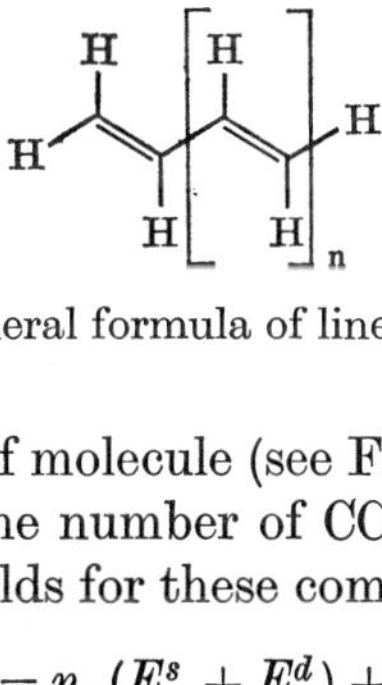

FIG. 3. General formula of linear polyenes.

However, in this type of molecule (see Fig. 3), if n is the number of CC 'single' bonds, then the number of CC 'double' bonds should be $(n + 1)$. If equation (3) holds for these compounds, then

$$E_{\pi} = n \, . \, (E_{\pi}^{s} + E_{\pi}^{d}) + E_{\pi}^{d}, \tag{4}$$

and a plot of E_{π} against n should be a straight line of slope $(E_{\pi}^{s} + E_{\pi}^{d})$ and intercept E_{π}^{d}. Figure 4 shows the result obtained in the improved Hückel approximation just mentioned: the calculated values for E_{π}^{s} and E_{π}^{d} are $0{\cdot}5209\beta$ and $2{\cdot}0127\beta$ respectively. A similar plot is obtained with the improved Pople method[91].

These results were applied to the calculation of the resonance energy in the different members of the annulene series up to $C_{30}H_{30}$ [89, 110]. In this series, we have the same number (n) of formal 'single' and 'double' bonds; accordingly, the π-energy of the 'localised' structure may be written:

$$E_{\pi}(\text{loc}) = n.(E_{\pi}^{s} + E_{\pi}^{d}). \tag{5}$$

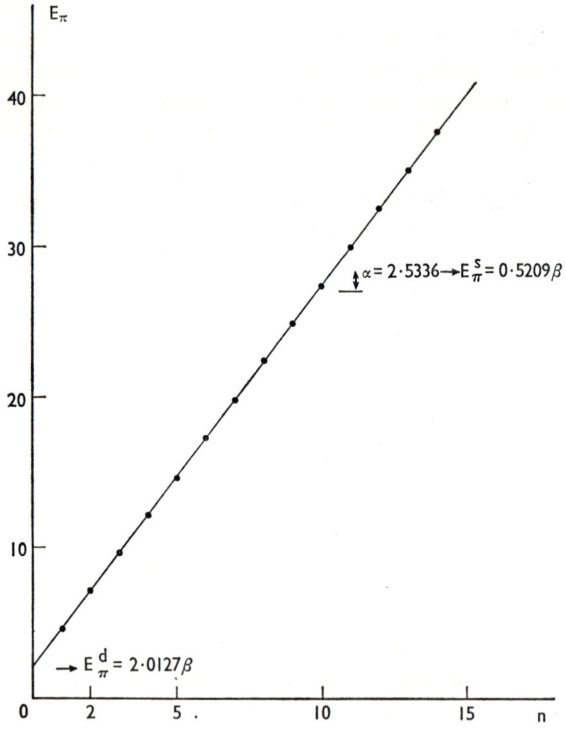

FIG. 4. Total π-binding energy in linear polyenes.

Thus, the π-contribution to the resonance energy $E_R(\pi)$ of the [2n]annulene is:

$$E_R(\pi) = E_{\pi} - E_{\pi}(\text{loc}), \tag{6}$$

where E_{π} is the total molecular π-energy calculated by the improved Hückel theory for this annulene. Table 2 lists resonance energies calculated in this way for the series of compounds investigated by the improved Hückel method. Figure 5 shows the plot obtained for $E_R(\pi)$ as a function of the dimension of the ring [N].

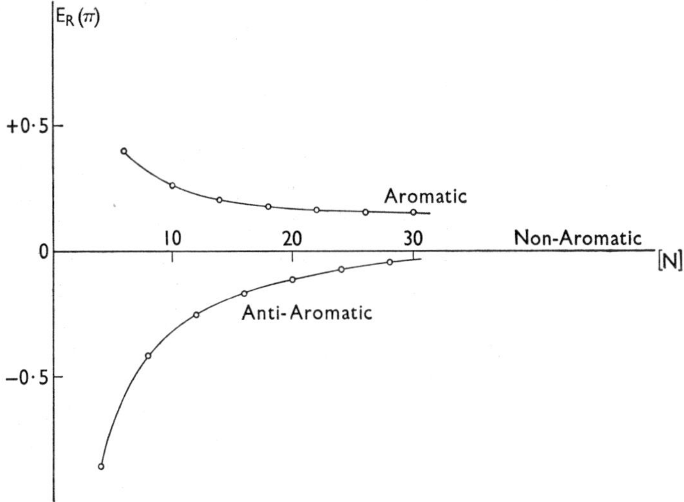

FIG. 5. π-Electron contribution to the resonance energy in the annulene series.

Obviously, the thermodynamic criterion divides the annulenes into two well separated series: the $(4n + 2)$ π-electron ring-systems with stabilising resonance energy (positive values of $E_R(\pi)$ expressed in $|\beta|$-units) and the $(4n)$ π-electron ring-systems showing decreased

TABLE 2

π-Contribution to the resonance energy of annulenes according to Breslow's and Dewar's definition

[N]annulene	$E_R(\pi)$ in $\|\beta\|$ units	[N]annulene	$E_R(\pi)$ in $\|\beta\|$ units
4	−0·8588	18	0·1752
6	0·3992	20	−0·1160
8	−0·4148	22	0·1616
10	0·2608	24	−0·0760
12	−0·2550	26	0·1556
14	0·2024	28	−0·0452
16	−0·1704	30	0·1538

stabilisation as compared with the 'localised' model compounds. Thus, according to Breslow's and Dewar's definition, the first should be called *aromatic*, the latter *anti-aromatic*. The higher members in each case converge to *non-aromaticity* ($E_R(\pi) = 0$).

The same results can be obtained in a more qualitative way by simple MO-perturbation theory[87, 88, 121]. It should be pointed out, however, that these conclusions are reached assuming planar geometries without any distortion energy or steric effects and do not include the difference in σ-bond compression energy between the 'localised' and delocalised structures. This last point has been taken into account by Dewar[88] in the calculation of the resonance energy for a number of unstrained planar annulenes by the improved Pople SCF method; his conclusions are essentially identical to those obtained by the improved Hückel method. It should also be noted that both methods[88, 110] take into account the Jahn–Teller distortions which may occur in this series of molecules (see section B.2).

2. *The Jahn–Teller Effect*

In 1937, Jahn and Teller[149] stated that, except in some special cases, symmetrical molecular systems having an orbitally degenerate ground state are geometrically unstable with respect to a displace-

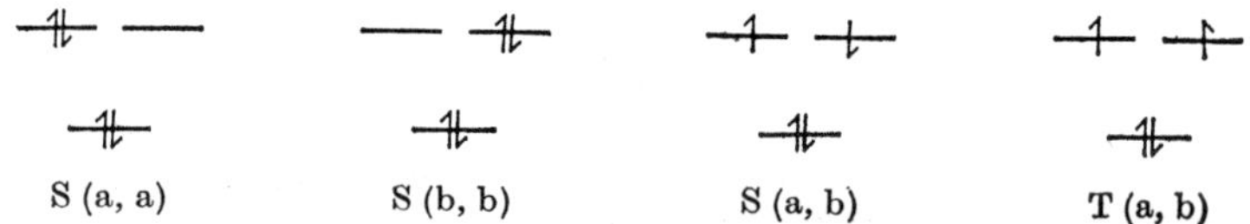

FIG. 6. Lowest singlets and triplet states of square cyclobutadiene.

ment of atoms which removes that degeneracy. The reason is that displacement of the nuclei from their symmetrical positions produces a splitting of the degenerate level with the consequence that one of the resulting new levels is below the original one; the system can thus lower its energy by becoming less symmetrical.

According to Fig. 6, square cyclobutadiene would not have a closed shell of electrons. When two electrons are placed in one or the other of the degenerate non-bonding orbitals it gives two singlet states which we shall call S(aa) and S(bb); if one electron is placed in each orbital it gives a singlet and triplet ground state denoted S(ab) and T(ab). Simple MO-theory makes all four degenerate.

However, when electron repulsion is included, these states are split, but they remain near-degenerate so that the possibility of mixing of the singlet states by a finite nuclear displacement may result in a lowering of the total molecular energy [228]. The closer the states, the more effective the mixing. These distortions are called *pseudo-Jahn–Teller effects*. Such effects are thus to be expected in the whole series of [4n]annulenes.

A characteristic feature of Jahn–Teller distortions is the small energy gain which they generally produce, and hence experimental

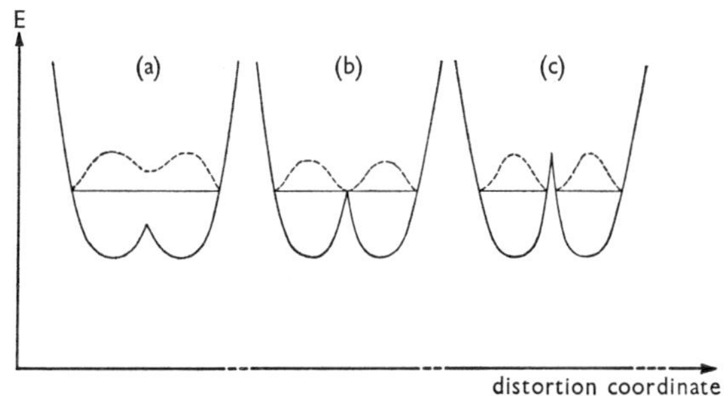

FIG. 7. The different cases of the Jahn–Teller effect.

evidence of their existence is scarce. Broadly, there are three different cases to consider [138, 194, 227, 273]:

(a) The energy the molecule can gain by distortion (ΔE) is much smaller than the zero point energy ($\frac{1}{2}h\nu$) of the lowest vibrational mode which can remove the degeneracy; there is a continual *dynamic interconversion* of the distorted forms as the molecule vibrates back and forth (Fig. 7(a));

(b) The stabilising distortion energy is just about the magnitude or is slightly larger than this lowest zero-point vibrational energy; any external perturbation will cause a tunnelling through the potential barrier by mixing the two degenerate vibrational components and the result is again a *dynamic interconversion* (Fig. 7(b));

(c) Only when the energy that can be gained by distortion is much larger than the involved zero point vibrational energy level, will the molecule be permanently in a distorted state; we then obtain a *static Jahn–Teller effect*. In this case, the system has nevertheless a kind of degeneracy arising from the fact that there are two or more configurations which are geometrically equivalent and have equal energies.

As early as 1937, Lennard-Jones and Turkevich[165] showed how compression energy could be introduced into the simple LCAO theory in order to determine the most stable geometry of a conjugated hydrocarbon, provided the variation of the exchange integrals β_{ij} with bond length was taken into account; they were able to show

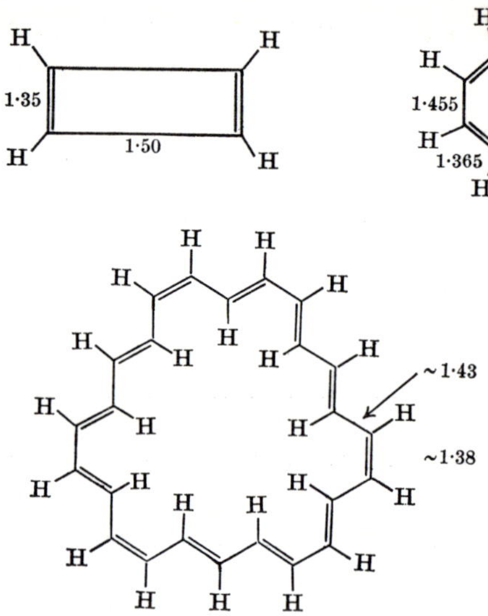

FIG. 8. Stable equilibrium structures of some [4*n*]annulenes[126, 241] (bond lengths in Ångströms).

that cyclic planar polyenes of formula $C_{2n}H_{2n}$ with n even (i.e. the [4*n*]annulenes) should consist of links of alternating length. Their conclusions were that C_4H_4 would contain two 'pure' double and two 'pure' single bonds in a rectangular shape with zero conventional resonance energy; cyclo-octatetraene on the other hand should exist at room temperature as a mixture of two rapidly interchanging equilibrium configurations separated by an energy barrier of about 8 kcal. mole^{-1}. Several years later, Liehr[167] re-examined the cyclobutadiene problem with the aid of MO-perturbation theory, using again Lennard-Jones' parameters; the rectangular form was found to be more stable than the square by 20·9 kcal. mole^{-1}. In more recent calculations using Longuet-Higgins and Salem's parameters[172], a stabilising distortion energy of about 11·44 kcal. mole^{-1} is obtained [139, 241] for this molecule. Similar calculations on cyclo-octatetraene [241] and [24]annulene[126] gave a distortion energy gain of 7·213 and about 4·75 kcal. mole^{-1} respectively. The distorted shapes are shown in Fig. 8 and reveal the interesting result that the geometry of lowest energy of these molecules is a structure with alternating 'single' and 'double' bonds. The antisymmetric carbon–carbon stretching vibration may be expected at about 1600 cm^{-1}; this gives a zero point vibrational energy level removing the degeneracy at $\sim$ 800 cm^{-1} or 2·28 kcal. mole^{-1}. These molecules should thus show the characteristic feature of *bond alternation* in a static Jahn–Teller structure.

Recently, the problem of bond-alternation in planar strainless $(4n)$ π-electron annulenes up to [28]annulene has been reinvestigated by Figeys using a self-consistent-β-iterative procedure[111, 115] minimising the total $(\sigma + \pi)$ energy. The results are gathered in Table 3

TABLE 3

Bond-alternation in [$4n$]*annulenes*

N	Energy barrier ΔE (kcal. mole^{-1})	Bond lengths (Å)	
		R_1	R_2
4	12·24	1·517	1·358
8	6·50	1·447	1·372
12	4·61	1·431	1·379
16	3·57	1·423	1·383
20	2·94	1·419	1·386
24	2·58	1·416	1·388
28	2·08	1·414	1·390
C_{30} polyene	—	1·4077	1·3945

as a function of ring-size N; they clearly indicate that with increase in N, there is a gradual alteration in the annulene molecules from a static to a dynamic Jahn–Teller bond-alternating structure, with bond-lengths R_1 and R_2 converging to the values calculated for the middle of a long open chain polyene. Similar conclusions with regard to this last point have also been reached by Dewar [89] using the iterative Pople SCF method. In the static case, this phenomenon should lead to the existence of *valence isomers produced by bond-shift*; e.g.[165]:

Ia $\overset{K}{\rightleftharpoons}$ Ib

Such equilibria have indeed been observed (see section C).

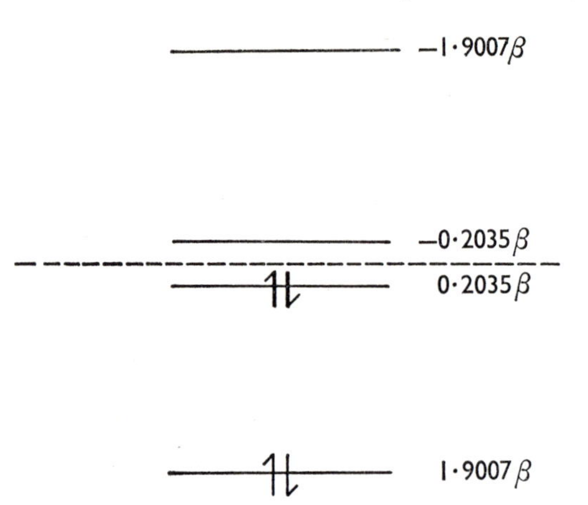

Fig. 9. Orbital energies of rectangular cyclobutadiene.

The orbital energy pattern of the rectangular cyclobutadiene molecule obtained by the improved Hückel method[111] after introduction of the Jahn–Teller distortion is shown in Fig. 9.

In this approximation, rectangular cyclobutadiene should thus have a *singlet ground state*, while square cyclobutadiene should be a triplet T(ab) (see Fig. 6) since this state has a lower electron repulsion energy than the corresponding singlets S(aa), S(bb) or S(ab). The problem of the spin multiplicity of cyclobutadiene in its ground state has been treated by several authors in various more elaborate ways [6, 51, 90, 92, 127, 180, 193, 237, 295]. Although several different conclusions were obtained, it now seems very likely that in its ground state, cyclobutadiene should be a rectangular singlet; the lowest triplet state, however, will most certainly be square.

In their early work, Lennard-Jones and Turkevitch[165] stated that in the series of cyclic (planar) polyenes of general formula $C_{4n+2}H_{4n+2}$, minimum total $(\sigma + \pi)$ energy would occur when the molecule adopts a regular polygonal geometry. Recently, however, Longuet-Higgins and Salem[172] showed that, provided n is very large, marked bond-alternation should occur even in this series of compounds. Using simple Hückel LCAO theory and parameters chosen to fit selected benzene force constants, they showed that instability of the totally symmetric configuration should set in when $n = 8$, that is, when the ring contains 34 atoms; in the case of the infinite polyene the 'short' and 'long' bonds would have lengths of 1·4227 Å and 1·3872 Å respectively. Similar results were obtained by Figeys[111], who found no bond alternation in the series of the $[4n + 2]$annulenes up to 30 carbon atoms.

The Jahn–Teller distortions in planar unstrained annulenes have also been treated by several authors, using either the SCF-LCAO theory including electron repulsion energy[71, 89] or an improved valence-bond approximation[94]. Some results are shown in Table 4. There seems to be considerable disagreement between the results

TABLE 4

Bond-alternation and distortion energies ΔE in [N]annulenes

	Valence-bond theory[94]			SCF-LCAO theory[71]		
N	r_1^*	r_2^*	ΔE†	$r_1^{(*)}$	$r_2^{(*)}$	ΔE†
4	1·393	1·393	0·0	—	—	—
6	1·400	1·400	0·0	1·397	1·397	0·0
8	1·378	1·432	0·013	—	—	—
10	1·365	1·455	0·16	1·400	1·400	0·0
12	1·358	1·465	0·294	—	—	—
14	—	—	—	1·363	1·447	4·93
18	—	—	—	1·358	1·453	14·49
22	—	—	—	1·357	1·455	24·53

* r_1 and r_2 are adjacent bond lengths, in Ångström units.
† ΔE in kcal. mole^{-1}.
(*) r_1 and r_2 obtained from the calculated bond-orders and the relation: $r_{ij}(\text{Å}) = 1{\cdot}520 - 0{\cdot}185 p_{ij}$[88].

obtained by the improved Hückel methods[111, 172] and those obtained by more elaborate calculations. Valence-bond theory apparently does not indicate a fundamental difference in molecular geometry between the $(4n)$- and $(4n+2)$-π electron series, bond alternation and distortion energy becoming more and more important with ring size. A similar evolution is predicted by the SCF-LCAO method for the $[4n+2]$annulenes, but in the series of the $[4n]$annulenes, an inverse evolution may be expected[71, 89]. If N is sufficiently large ($N \geqslant 18$), the annulene should in all cases show a bond alternation similar to that calculated for an infinite linear polyene[89]. The limit from which bond-alternation should appear in the $(4n+2)$annulenes seems to depend strongly on the parameters used; e.g., Cizek and Paldus[71] found that, if the parameters of Mataga and Nishimoto[183] were used instead of the completely 'theoretical' parameters[204], instability of the symmetric solution does not occur until $N = 26$.

A more precise knowledge of the geometry and distortion energies of these molecules should be provided by the proper analysis of their spectroscopic properties (see section C).

3. *Magnetic Behaviour as a Criterion of Aromaticity*

When a diamagnetic molecule is placed in a magnetic field $\mathscr{H}$, it acquires an induced magnetic moment

$$M = \chi \cdot \mathscr{H},$$

where χ is defined as the *molecular susceptibility*. The molar susceptibility of the compound is then given by:

$$\chi_M = N_0 \cdot \chi$$

where N_0 is Avogadro's number. For molecules containing no cyclic delocalised π-electrons, the molar susceptibility can easily be obtained by Pascal's rule as the sum of atomic susceptibilities and certain structural corrections λ[137, 215]:

$$\chi_M = \sum \chi_{at} + \sum \lambda.$$

For benzenoid aromatic hydrocarbons however *the molar susceptibilities are larger than one would expect on the basis of the additivity rules.* Recently *the susceptibility exaltation* of a compound has been defined as 'the difference between the susceptibility exhibited by that compound and that predicted for the identical but non-cyclically delocalised structural counterpart'[84]:

$$\Lambda = \chi_M - \chi'_M.$$

This exaltation seems to be almost entirely due to some special property of the π-electrons*. Indeed, for non-spherical molecules, it is possible to measure the molar susceptibilities χ_x, χ_y and χ_z along the three directions of a system of reference; if the z-axis is taken normal to the molecular plane, the measured susceptibility χ_z for benzenoid aromatic compounds is always much greater than those measured in the plane of the molecule, i.e.:

$$\chi_x \simeq \chi_y \ll \chi_z.$$

χ_x and χ_y give the contribution of the σ-electrons to the molar susceptibilities in these directions, and if this contribution is supposed isotropic, a *diamagnetic anisotropy*

$$\Delta\chi_M = \chi_z - \tfrac{1}{2}\cdot(\chi_x + \chi_y)$$

may be defined.

The experimental observation either of appreciable exaltations of susceptibility or of significant diamagnetic anisotropies has been used commonly as a criterion of aromatic character[79, 84, 284]. Several years ago, Pauling[203] suggested that these special properties of aromatic molecules may be due to interatomic *ring-currents*, wherein the π-electrons, under the influence of the applied external field, flow freely from carbon atom to carbon atom all around the rings. If the applied field $\mathscr{H}$ is directed along the z-axis through the molecular plane, a large induced moment should be produced, which will cause an additional negative susceptibility perpendicular to the plane, thus explaining the anomalous exaltations and anisotropies shown by these compounds. The semi-classical theory of Pauling was soon followed by a quantum-mechanical treatment due to London[169]. Correlations between experimental data and the calculated contributions χ_π of the π-electronic ring-currents to the diamagnetic susceptibility of the molecules studied are fairly successful; some results obtained for a series of polycyclic aromatic hydrocarbons are gathered in Table 5 and their relations shown on Fig. 10. Pauling's hypothesis thus seems to explain the observed phenomena very well; confirmation of the assumption that the existence of interatomic

*See however references[83,209,229].

TABLE 5

Diamagnetic anisotropy, susceptibility exaltation, and π-electronic susceptibility χ_π of aromatic hydrocarbons

Hydrocarbon	$\Delta\chi_M \cdot 10^6$ [Ref. 229]	$\Lambda \cdot 10^6$ [Ref. 84]	$\chi_\pi \cdot 10.^{17}\beta$ [Ref. 215]
Benzene	–59·7	–13·7	1·293
Naphthalene	–114·0	–30·5	2·826
Phenanthrene	–166·0	–46·2	4·198
Anthracene	–182·6	–48·6	4·460
Pyrene	–232·9	–57·3	5·923

π-electronic currents is related to *cyclic* delocalisation of π-electrons is provided by the theoretical demonstration that this current will be zero for open-chain conjugated polyenes[215, 229].

Another important magnetic property of aromatic molecules, which may be explained[178, 208] by the existence of ring-currents, is the large chemical shift of the high-resolution magnetic resonance

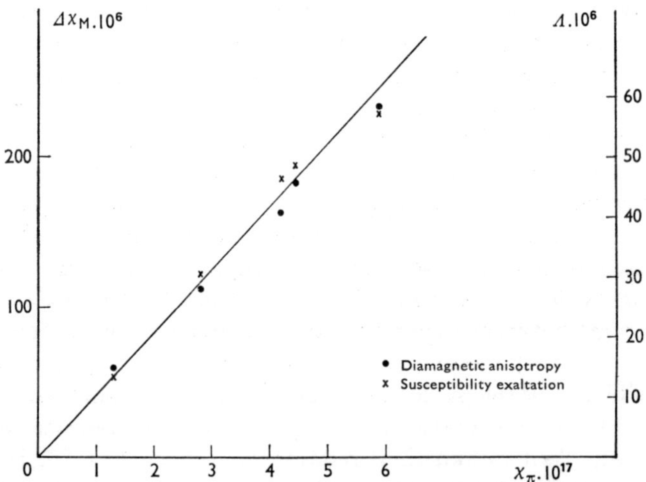

FIG. 10. Relation between experimental susceptibility exaltations Λ, diamagnetic anisotropies $\Delta\chi_M$, and calculated π-electronic susceptibilities χ_π for a series of polycyclic aromatic hydrocarbons.

TABLE 6

Ring-current intensities and proton chemical shifts in polycyclic aromatic hydrocarbons

Molecule	Ring	Ring-current intensity	Proton	Chemical shift Observed*	Chemical shift Calculated
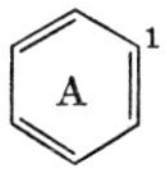	A	1·000	1	7·342†	7·276
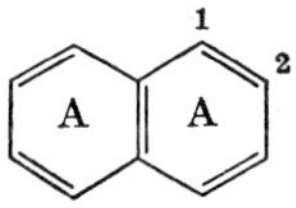	A	1·079	1	7·81	7·747
			2	7·46	7·489
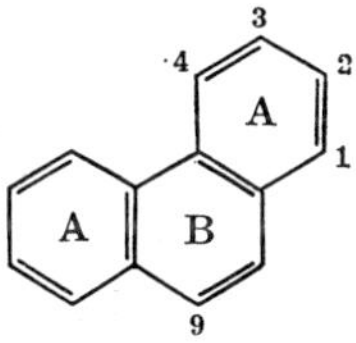	A	1·125	1	7·855‡	7·828
	B	0·937	2	7·570	7·577
			3	7·612	7·604
			4	8·648§	8·104
			9	7·702	7·764
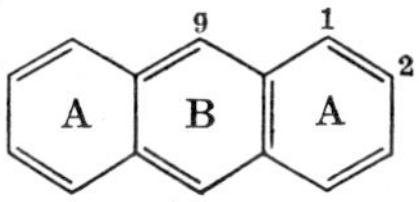	A	1·059	1	7·91	7·896
	B	1·274	2	7·39	7·539
			9	8·31	8·360
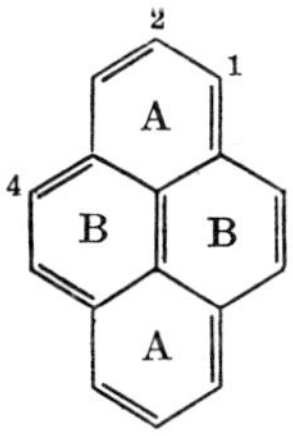	A	1·315	1	8·16	7·982
	B	0·927	2	7·99	7·928
			4	8·06	8·181

* In ppm with respect to TMS; experimental values obtained from ref. 153 unless otherwise stated.

† Obtained in 5% $CDCl_3$ solution[108].

‡ R. C. Fahey and G. C. Graham, *J. Phys. Chem.* **69**, 4417 (1965).

§ Discrepancy between experimental and calculated shift is due to a mutual Van der Waals effect between the angular hydrogen atoms.

signals of aromatic protons[1, 83, 102, 108, 152, 153, 184, 209, 211]. The chemical shift δ_H(ppm) of a proton in the vicinity of the rings is given by

$$\delta\ (\text{ppm}) = 2\beta.(2\pi e/hc)^2.(S^2/a^3).\sum_{\mu} J_{\mu}.K_{\mu},$$

where S is the area of the benzene ring, β the benzene exchange integral and a the length of the benzene CC bond; J_{μ} is the ring-current intensity of ring μ, and K_{μ} a distribution function of the effect of the anisotropy in space[108, 178, 229]. The result is a *large deshielding* of the outer protons of the benzenoid hydrocarbons as compared with the 'olefin-like' proton which resonates at 6·129 ppm[108]. As shown in Table 6, calculated values fit very well with the experimental ones, the correspondence being in most cases better than 0·08 ppm.

The theoretical investigation of the magnetic properties of annulenes (and benzo-annulenes) other than benzene and benzenoid hydrocarbons has received attention only in very recent years. First it was pointed out by Longuet-Higgins[171] and by Nakajima and Kohda[197] that, when a cyclic conjugated molecule is placed in an external magnetic field, the π-electronic ring-current is the difference between a *diamagnetic part* depending only on the electronic structure in the ground-state of the molecule, and a *paramagnetic part* which is a sum of contributions implying excited states. The paramagnetic part will be particularly large if the molecule possesses suitable very low-lying excited states. This is exactly the situation obtained by introducing the Jahn–Teller effect in the series of [4n]annulenes (see section B.2). As a result, *paramagnetic ring-currents* should occur in these molecules, which should give rise to a *positive* susceptibility exaltation or anisotropy.

Under these circumstances, the usual diamagnetic n.m.r. rules have to be reversed: protons outside the ring will show an increased shielding relative to the 'olefinic' reference compound, and those inside the ring should be deshielded. Similar conclusions were obtained by Pople and Untch[212] and by Baer, Kuhn and Regel[18].

The influence of bond alternation on the magnetic properties of the annulenes was investigated by several authors[18, 173, 197, 212, 225]. For all sizes of rings, the theory predicts that the magnitude of the calculated ring-current should be partially quenched if bond alternation occurs. In the [$4n + 2$] as well as in the [4n]annulene series, it appears that the larger the ring, the more effective is the quenching due to a given amount of alternation. This is very clearly shown in Fig. 11, where the variable part of the ring-current J_μ is plotted for certain annulenes as a function of the bond-alternation parameter $\beta_1/\beta_2 = \lambda$, β_1 and β_2 being the exchange integrals of the 'short' and 'long' bonds respectively. As a result, the calculated contribution χ_π of the π-electrons to the diamagnetic susceptibility of the annulenes becomes negligible if bond alternation is significant[225] (Table 7), and the

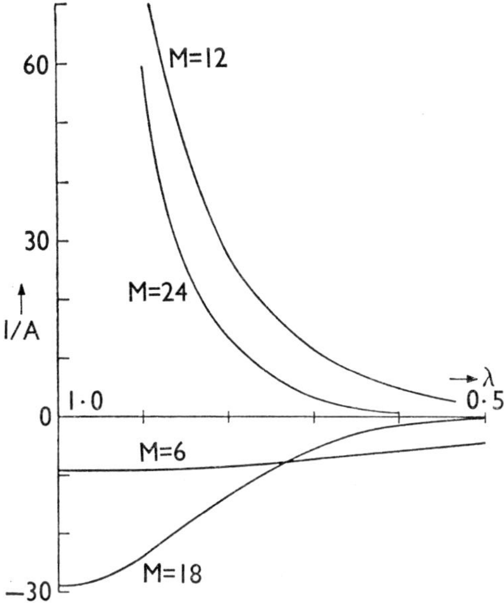

Fig. 11. Dependence of ring-current on alternation parameter for some annulenes [$J\mu = A.f(N,\lambda)$]. Reprinted from the *Journal of the American Chemical Society* (1966), **88**, 4811; reprinted by permission of the copyright owner.

chemical shifts for the inner and outer protons tend rapidly to the 'olefinic' frequency[173] (Table 8).

These tables show that the magnetic properties of the annulenes are exceedingly sensitive to the amount of distortion from the symmetrical structure. Nevertheless, several authors have used proton n.m.r. spectroscopic arguments in order to decide whether or not an uncharged cyclic conjugated molecule is aromatic[148, 242(b)]. The presence of an induced *diamagnetic ring-current* (as in benzene), detected by n.m.r. spectroscopy or susceptibility measurements, should be used as a criterion of *aromatic* character. *Anti-aromaticity*

TABLE 7

Evolution of the π-electron contribution to the diamagnetic susceptibility χ_π with the alternation parameter λ

(χ_π^0 is the contribution for the symmetric undistorted polyene)

λ	χ_π/χ_π^0 C_4H_4	$C_{18}H_{18}$	$C_{30}H_{30}$
1·0000	1·0000	1·0000	1·0000
0·9657	0·9978	0·9799	0·9448
0·8107	0·9244	0·5117	0·2000
0·6500	0·7291	0·1104	0·0107
0·4903	0·4540	0·0115	0·0002
0·2174	0·0723	0·0000	0·0000

Reprinted by kind permission from L. Salem, *Proc. Cambridge Phil. Soc.* (1961) **57**, 353.

TABLE 8

Calculated proton chemical shifts in $C_{18}H_{18}$ *and* $C_{30}H_{30}$ *as a function of bond alternation* λ

	δ_H(ppm)*				
	$C_{18}H_{18}$		$C_{30}H_{30}$		
				External protons	
λ	Internal protons	External protons	Internal protons	At corners	At edges
1·0000	−7·44	+10·42	−12·42	+13·26	+14·71
0·9657	−7·21	+10·27	−11·43	+12·88	+14·26
0·8107	−0·85	+8·28	+2·45	+7·58	+7·82
0·6500	+4·60	+6·59	+5·90	+6·21	+6·21
0·4903	+5·98	+6·21	+6·13	+6·13	+6·13

* δ (ppm) is the chemical shift expressed in ppm relative to tetramethylsilane (TMS); the 'olefinic' resonance frequency is evaluated at 6·129 ppm and the ring-current effect for the benzene protons is assumed to be 1·15 ppm[108].

can then be defined on the basis of n.m.r. or susceptibility arguments; it is shown by the presence of an induced *paramagnetic* ring-current [84, 243]. Thus [$4n$]annulenes should be considered theoretically as anti-aromatic molecules on the basis of their expected magnetic behaviour; [$4n + 2$]annulenes on the other hand should be aromatic and thus

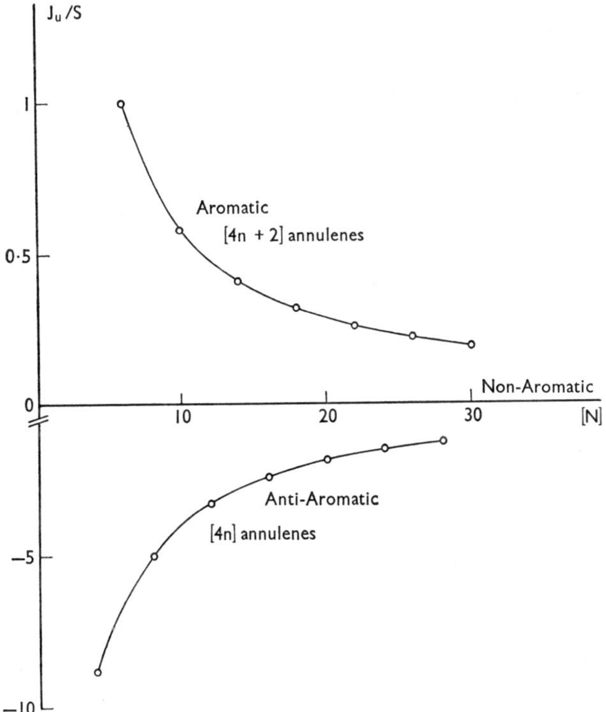

FIG. 12. Ring-current intensity per unit benzene area ($J\mu/S$) in the annulenes as a function of ring-size.

will show benzene-like magnetic properties. The results of a recent calculation of the induced ring-currents in the annulenes up to $C_{30}H_{30}$ are gathered in Table 9[112]. Figure 12 shows the evolution of the ring-currents per unit benzene area as a function of ring-size; qualitatively, the conclusions concerning the aromaticity of these compounds are in excellent agreement with those obtained from the thermodynamic criterion (see Fig. 5).

Comparison between experimental behaviour and these theoretical values proves to be very instructive with regard to the electronic structure of this series of molecules (see section 3).

TABLE 9

Ring-current intensity per unit benzene area (J_μ/S) *in the annulenes*

Annulene	Ring-current intensity J_μ	J_μ/S	Annulene	Ring-current intensity J_μ	J_μ/S
4	–3·682	–8·814	18	2·234	0·320
6	1·000	1·000	20	–14·860	–1·858
8	–9·394	–4·984	22	2·346	0·261
10	1·164	0·581	24	–18·124	–1·510
12	–9·795	–3·265	26	3·085	0·221
14	1·647	0·412	28	–19·000	–1·266
16	–11·916	–2·383	30	3·536	0·191

C. Properties and Spectra of the Annulenes

In the preceding section, a rather detailed theoretical analysis of the electronic structure of the annulenes was set up. Important conclusions with regard to their thermodynamic stability, geometry and magnetic behaviour were reached, and the problem which now arises is to see whether experiment significantly confirms or contradicts these results. As has been pointed out in the Introduction, the sole valuable criteria of aromaticity or anti-aromaticity should be pure electronic ground state properties and as a result, experimental data concerning heats of formation, molecular geometry and magnetic properties will largely be discussed in this section. However, for some of the annulenes which have been discussed theoretically, these molecular properties are, in the present state of experimentation, very difficult if not impossible to measure. On the other hand, some very useful conclusions can be reached by the proper analysis of their reactivities and/or electronic spectra, taking into consideration, however, that these properties may depend to a large extent on transition or electronically excited states.

In order to make a coherent series of comparisons between theory and experiment, the various experimental properties of the annulenes will now be discussed, starting with cyclobutadiene and treating the larger rings in the order of the increasing number of electrons [N].

1. *Cyclobutadiene*

Although cyclobutadiene has been postulated as a transient species in several reaction paths by a number of authors, a very large number of unsuccessful approaches to the synthesis of cyclobutadiene or its substituted derivatives has been recorded in the literature[62]. Very recently the preparation of 1,3-dicarboethoxy-2,4-bis(dimethyl-amino)-cyclobutadiene has been described[125].

The synthesis of cyclobutadiene iron tricarbonyl has also been described recently[100, 224] and free cyclobutadiene may be liberated from this complex by the action of ceric ions at low temperature[285]. The free hydrocarbon, although obviously extremely reactive, none-theless has a small but finite lifetime. However, spectroscopic or thermodynamic investigations on the free molecule (in its ground-state) have not yet been carried out.

Many qualitative arguments have been developed in the literature in order to explain the great instability of cyclobutadiene. The simple idea that this molecule should be unstable owing to considerable strain energy seems to be inconsistent with, for example, the recent synthesis of tetramethylenecyclobutane (II)[132] and of 3,4-dimethylenecyclo-butene[40, 145] (III); both compounds contain four sp_2-hybridised carbon atoms in a four-membered ring and should thus show

H_2C CH_2 H_2C CH_2 — II

CH_2 CH_2 — III

approximately the same strain-energy as cyclobutadiene. This energy has been estimated to be as large as 50–60 kcal. mole^{-1} [288]. However, tetramethylenecyclobutene is indefinitely stable in dilute solution at –78°; when warmed to room temperature, it undergoes dimerisation to a 1,5-cyclo-octadiene (IV). This behaviour is ascribed to the great reactivity of the exocyclic methylene groups of (II)[132]. Dimethylenecyclobutene on the other hand is a colourless liquid,

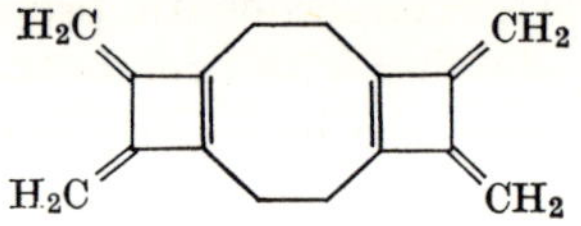

IV

b.p. 72°[145] and shows a marked tendency to polymerise. In the light of the modern thermodynamic definition of aromaticity, it should be emphasised that, according to Dewar[88], tetramethylenecyclobutane is a non-aromatic molecule ($E_R = 0$) and thus stabilising delocalisation energy cannot compensate for the presumed adverse effects of strain.

Particularly instructive are the several attempts which have been made to generate a cyclobutadiene by the Diels–Alder addition of tetracyanoethylene with a dimethylenecyclobutene[40, 41, 80, 132]. Tetramethylenecyclobutane (II) may be regarded formally as a potential precursor for substituted cyclobutadienes. Conceivably, advantage might be taken of Diels–Alder reactions in order to introduce endocyclic double bonds into the four-membered ring, and indeed, the reaction of (II) with tetracyanoethylene at 0° gave a colourless, relatively stable crystalline 1/1 adduct (V). However, although a large excess of dienophile was employed, no products resulting from the addition of two moles of tetracyanoethylene to (II) could be isolated[132]; the same is true with phenylmaleimide as dienophile at

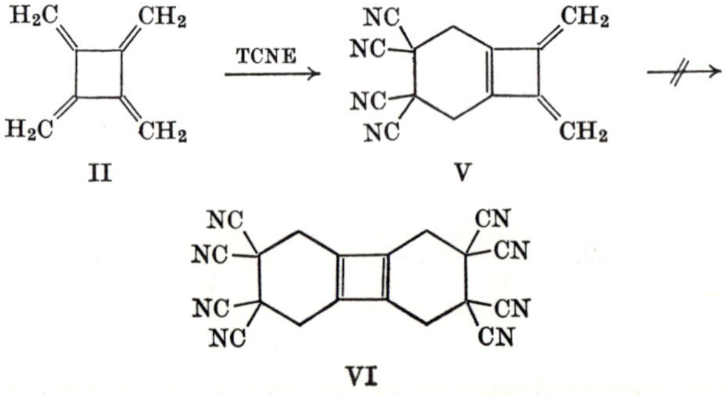

II TCNE V

VI

25°. An unusually large energy barrier seems to exist if the expected Diels–Alder adduct is a cyclobutadiene derivative. Thus, 1,2-dimethyl- (VII(a)) and 1,2-diphenyl-3,4-dimethylenecyclobutene (VI(b)) react with tetracyanoethylene to give the crystalline 1/1 cycloaddition product (IX) instead of the expected Diels–Alder

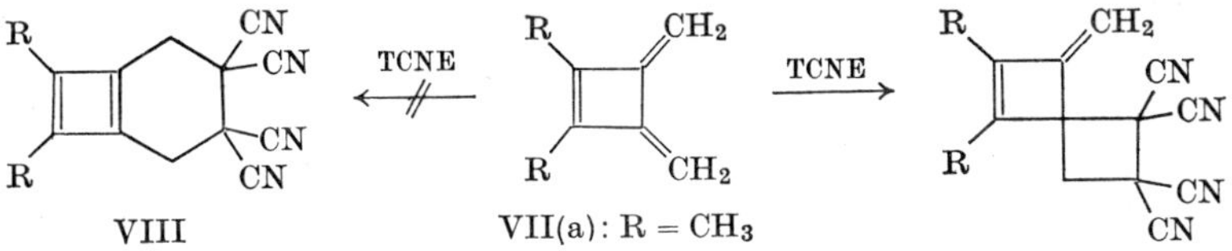

VIII

VII(a): R = CH_3

VII(b): R = C_6H_5

IX

adduct (VIII). The reason for this particular behaviour is almost certainly the high-energy electronic structure of the cyclobutadiene-like transition state of the reaction, which should closely resemble the final products.

Other evidence of this type includes the failure of 2,4-dichloro-3-phenylcyclobutene to undergo enolisation[222] and of a fluorocyclobutene anion to lose fluoride ion although the corresponding cyclobutane undergoes facile elimination[47].

The puzzling problem concerning the nature of the ground state of the cyclobutadiene molecule has been the subject of two recent experimental investigations. Skell and Petersen[240] presumably generated free tetramethylcyclobutadiene (X) in the gas phase by the reaction of XI with sodium-potassium vapour at 245–255° in

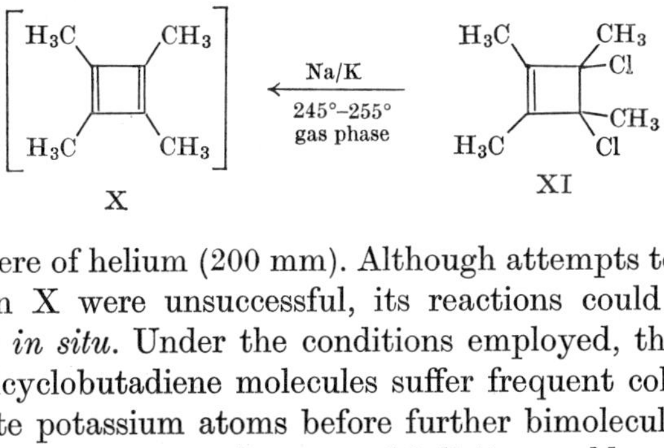

X

XI

an atmosphere of helium (200 mm). Although attempts to isolate the hydrocarbon X were unsuccessful, its reactions could be studied quite easily *in situ*. Under the conditions employed, the generated tetramethylcyclobutadiene molecules suffer frequent collisions with doublet state potassium atoms before further bimolecular reaction, so that relaxation to ground state multiplicity would occur. Besides other compounds, the *syn*-dimer XII and 1-methylene-2,3,4-tri-

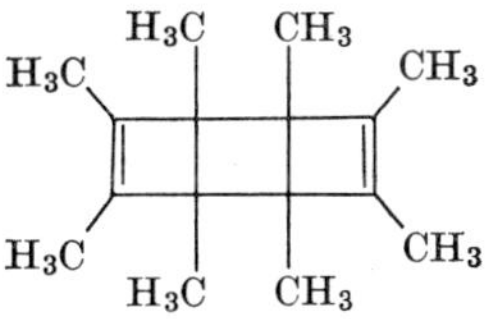

XII (syn)

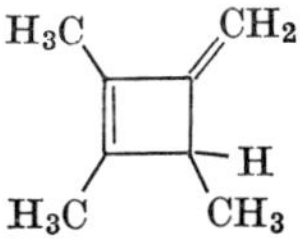

XIII

methylcyclobutene (XIII) were obtained from the reaction mixture in a ratio independent of the initial concentration of the dichloro compound IX in the gas phase. Since XII must be formed by a bimolecular process, the formation of XIII must also be bimolecular. The observed products are readily explained if a *triplet cyclobutadiene-structure* is postulated for the intermediate[240]: 'The analogous mono-radical reactions of coupling and disproportionation are well-established rapid processes. The disproportionation reaction applied to the interaction of two triplets

$$\uparrow\cdot C_8H_{12}\cdot\uparrow + \downarrow\cdot C_8H_{12}\cdot\downarrow$$

leads to two molecules of XIII, while coupling leads to XII'. When the reaction was carried out in the presence of triplet methylene $CH_2:\uparrow\uparrow$ generated *in situ* from CH_2Br_2, the coupling and disproportionation products XII and XIII were reduced drastically and the major compound was identified as 1,2-dimethylene-3,4-dimethylcyclobutene (XIV), which was not observed if CH_2Br_2 were not introduced

H_3C, CH_3, H_3C, CH_3 (triplet tetramethylcyclobutadiene) $+ \ \dot{\underset{\cdot}{C}}H_2 \downarrow\downarrow \longrightarrow$ H_3C, CH_2, H_3C, CH_2 $+ \cdots$

XIV

into the reaction medium. If, on the other hand, the reaction was carried out in the presence of methyl radicals (generated *in situ* from methyl bromide), no detectable amount of XIV was formed. Thus, tetramethylcyclobutadiene has a marked preference for loss of two hydrogen atoms in reaction with triplet methylene; according to Skell and Peterson, this can easily be visualised by a transition state involving simultaneous transfer of two hydrogen atoms from triplet tetramethylcyclobutadiene to triplet $CH_2:\uparrow\uparrow$.

However, as pointed out by Pettit and co-workers[286], Skell and Peterson's conclusion can be questioned on the grounds of the nature of the experimental conditions employed in the reaction: 'if free tetramethylcyclobutadiene is indeed involved, then reaction via thermally produced [low-lying] triplet excited species is readily conceivable'. On the other hand, cyclobutadiene, liberated *in situ* by the ceric ion oxidation of the stable cyclobutadiene tricarbonyl complex[285], reacts in a *stereospecific manner* both as a diene and as a dienophile. Reaction at 0° with dimethyl maleate produced *only* the endo, *cis*-dicarbomethoxybicyclohexene XV, while reaction with dimethyl fumarate gave *exclusively* the trans-derivative XVI. This is just what would be expected if cyclobutadiene acts as a *singlet*

XV

XVI

diene in the normal Diels–Alder reaction[230]; on the other hand, one would not expect stereospecific addition to occur if cyclobutadiene were a triplet[286]. Similarly, cyclobutadiene reacts in a stereospecific manner as a dienophile with cyclopentadiene. Its dimerisation seems, however, to be non-stereospecific; a mixture of the syn- and anti-isomers in the ratio 5/1 is produced, but even in this case, the syn-isomer is produced preferentially, as would be expected on the basis of the recent Woodward–Hoffmann rules[141]. According to Pettit and co-workers, 'since cyclobutadiene is both a very reactive diene and dienophile, it is not surprising that in the dimerisation, the stereochemical selectivity is diminished'[286].

Thus, although many qualitative experimental arguments can be advanced which seem to prove the theoretical conclusions regarding the anti-aromaticity and geometry of cyclobutadiene, no definite conclusion can be obtained from the available data. For example, it must be pointed out that neither Skell and Peterson's nor Pettit's arguments concerning its multiplicity (and hence geometry) are supported by completely reliable experimental or theoretical laws[286]. On the other hand, many unsuccessful attempts to synthesise cyclobutadiene may be due to the great reactivity which may be expected from this molecule[92]. It thus appears very likely that a definite answer could be provided only by spectroscopic investigations.

2. *Cyclo-octatetraene*

Cyclo-octatetraene was first prepared by Willstätter in 1911 from the alkaloid ψ-pelletierine[290], and later by W. Reppe *et al.* by catalytic tetramerisation of acetylene[218]. It is a yellow liquid, b.p.$_{760} = 141°$ and F. $-4{\cdot}3$ to $-3{\cdot}5°$, unstable to light and oxygen[233]. Electron diffraction measurements indicate a non-planar tub-form with D_{2d} symmetry for this molecule (Fig. 13), with alternating single and double bonds of length 1·462 and 1·334 Å respectively[30]. As a result, the double bonds may be considered as essentially localised, overlap of the π-orbitals of the different ethylene units being severely reduced. Cyclo-octatetraene (COT) should thus be considered as a slightly strained, weakly conjugated cyclic tetra-olefin. This conclusion seems

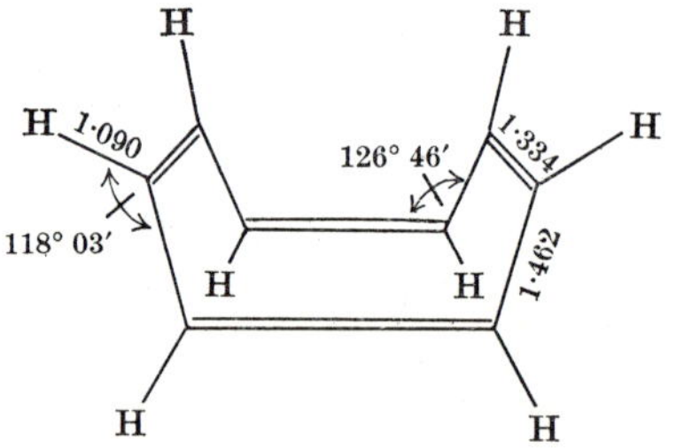

FIG. 13. Structure of cyclo-octatetraene according to Bastiansen *et al.*[30]

to be in good agreement with the experimental heat of formation from constituent atoms:

$$-\Delta H_f^0 = 1720{\cdot}8 \text{ kcal. mole}^{-1}$$

from which a resonance energy of only 4·8 kcal. mole^{-1} can be calculated[214, 257]. Using Dewar's bond energy values[88], a *negative* resonance energy (thus destabilising) of $-2{\cdot}3$ kcal. mole^{-1} is obtained. However, the strain energy in the tub form has been estimated[5, 7] to be 3·5 to 8·2 kcal. mole^{-1}; for the planar form, a value of 25·5 to 43·6 kcal. mole^{-1} has been advanced. Thus, although COT has been shown[270] to be less stable than isomeric styrene by 36·3 kcal. mole^{-1} it seems that thermochemical data are unable to verify the predicted anti-aromaticity of planar cyclo-octatetraene. Applying the Pariser and Parr LCAO method to the calculation of the electronic structure of the ground state of COT in the tub form as found by Bastiansen, it was also shown theoretically that this molecule should have very little resonance energy[232]. These results are however in definite disagreement with a recent study of the magneto-optical behaviour of COT, where, on the basis of an exceptionally large Faraday effect, considerable tridimensional conjugation is postulated in this molecule [160, 161].

The molar susceptibility of COT has been measured and a value of $\chi_M = -53{\cdot}9 \,.\, 10^{-6}$ cgs units has been obtained[84, 238]. Application of Pascal's rules gives, however, a value of $-54{\cdot}8 \,.\, 10^{-6}$ cgs units and accordingly, a diamagnetic susceptibility exaltation of $+0{\cdot}9$ can be

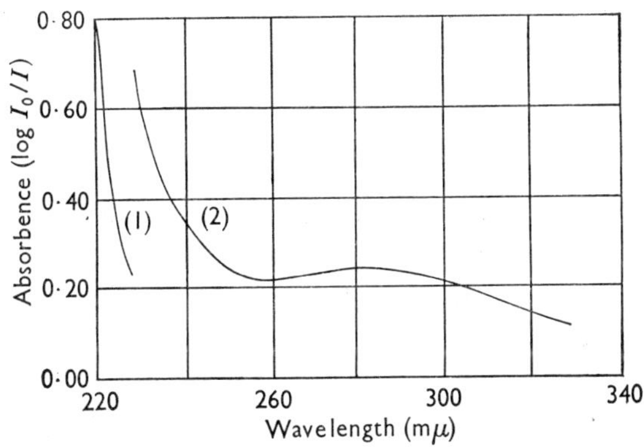

FIG. 14. Ultra-violet absorption spectrum of cyclo-octatetraene at 25° in solution in iso-octane; (1) and (2) represent concentrations of 29·8 and 99·3 mg/l respectively (1·00 cm cell). Reprinted from the *Journal of the American Chemical Society* (1950), **72**, 3866; reprinted by permission of the copyright owner.

deduced, classifying COT as an essentially non-aromatic substance[84]. The same conclusion is obtained in a more qualitative way by the observation that COT reacts as a characteristic cyclic tetra-olefin[233]. It should also be mentioned that, in contrast to the behaviour of dimethylenecyclobutene derivatives (see p. 292), the exocyclic double bonds of 7,8-dimethylene-octa-1,3,5-triene[12, 97] react readily with a variety of dienophiles to form substituted dihydro- or tetrahydro-benzocyclo-octatetraenes[97].

The ultra-violet (u.v.) spectrum of COT is shown in Fig. 14; its correlation with the known molecular structure presents an interesting problem. The non-planarity of the molecule and the properties we have already discussed suggest a π-electron system composed of four weakly interacting ethylene-units. On the other hand, the fact that it has an absorption maximum at 2800 Å in its electronic spectrum [95, 192] leads to the presumption that there is appreciable conjugation between the four double bonds. It has been shown by McEwen and Longuet-Higgins[176] that this is entirely due to electronic interactions between the *excited* configurations of the various π-electron subsystems. If we let θ_1, ω_1, χ_1 and ψ_1 represent the bonding molecular orbitals of the four ethylenic subsystems and θ_2, ω_2, χ_2 and ψ_2 their anti-bonding MO's, then the singlet excited states of the molecule can be expressed in terms of two types of excited configurations: (a) locally excited configurations such as $\theta_1^{-1}\theta_2$, and (b) electron transfer

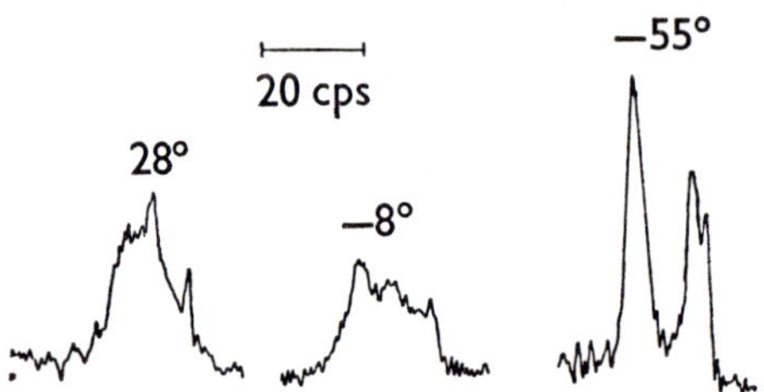

Fig. 15. High-field ^{13}C satellite of cyclo-octatetraene in carbon disulphide solution (66 per cent) at 60 Mc/s; the magnetic field increases from left to right. Reprinted from the *Journal of the American Chemical Society* (1962), **84**, 672; reprinted by permission of the copyright owner.

configurations such as e.g. $\theta_1^{-1}\psi_2$. McEwen and Longuet-Higgins were able to show, by use of the variation principle, that the wave-function of the first excited state of the molecule Ψ_α contained an important amount of configurations describing electron transfer between adjacent ethylene units.

Similarly, the interpretation[3, 4] of the unusual values of the adiabatic ionisation potential (IP = 8·04 eV) and of the half-wave reduction potential in anhydrous medium (DMF, $\epsilon_{1/2}$ v. SCE = − 1·62 eV) of COT should be undertaken with care, much attention being paid to the energy of the respective ions obtained. The proton n.m.r. spectrum of cyclo-octatetraene shows a singlet at 5·68 ppm with respect to TMS[221, 287], all protons being equivalent, in agreement with Bastiansen's tub form of the molecule. This frequency falls in the range of values encountered for olefinic protons, thus emphasising the absence of an important ring-current, in agreement with Dauben's susceptibility measurements[84].

Several authors[10, 11, 135, 201, 202, 221] have investigated the temperature-dependence of the n.m.r. spectrum of cyclo-octatetraene derivatives in order to get further information about the suspected 'mobile' structure of these molecules at room temperature (see sections B.2 and D). First came experimental evidence of the predicted phenomenon of *bond-shift* by observing the ^{13}C satellites of the proton signal of COT itself over a temperature range from $-55°$ to $28°$[10]. Figure 15 shows the high-field ^{13}C satellite of cyclo-octatetraene in carbon disulphide solution at various temperatures. At low temperatures, the rate of interconversion XVII(a) $\rightleftharpoons$ XVII(b) is very small; proton H_1 in structure XVII(a) would be expected to be coupled to

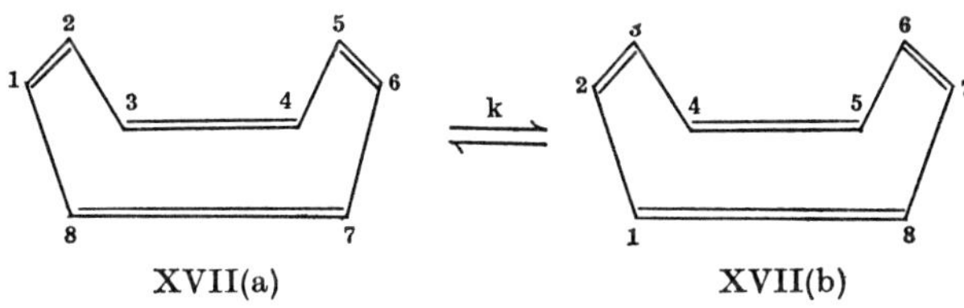

XVII(a) XVII(b)

H_2 by about 10 cps (the normal *cis*-vicinal coupling in alkenes[103]), and only slightly to H_8 (as a consequence of Karplus' dihedral angle rule[155]). Therefore, each proton attached to a ^{13}C atom in XVII(a) or XVII(b) should give rise to a doublet separated by about 10 cps, with additional but small splittings. The ^{13}C-satellites of the COT-protons at $-55°$ are indeed slightly split doublets with a separation of 11·8 cps[10]. When the temperature is increased, only the average spectrum of XVII(a) and XVII(b) will be observed: the ^{13}C satellites would be expected, in a very crude approximation, to be a 1/2/1 triplet with a spacing of about 6 cps. If additional coupling constants are introduced in the calculations, then the spectrum is expected to be a broad unresolved band, with only a vague reminiscence of a triplet structure. The shape of the ^{13}C satellite of the COT-protons at $28°$ fits very well with these expectations. From the coalescence of the doublet ($-10°$), a value of about 26 sec^{-1} is obtained for k, which gives a free energy of activation of about 13·7 kcal. $mole^{-1}$ at $-10°$ by the Eyring rate equation.

In a subsequent study of substituted cyclo-octatetraene derivatives, Anet and co-workers[11] were able to show that the eight-membered

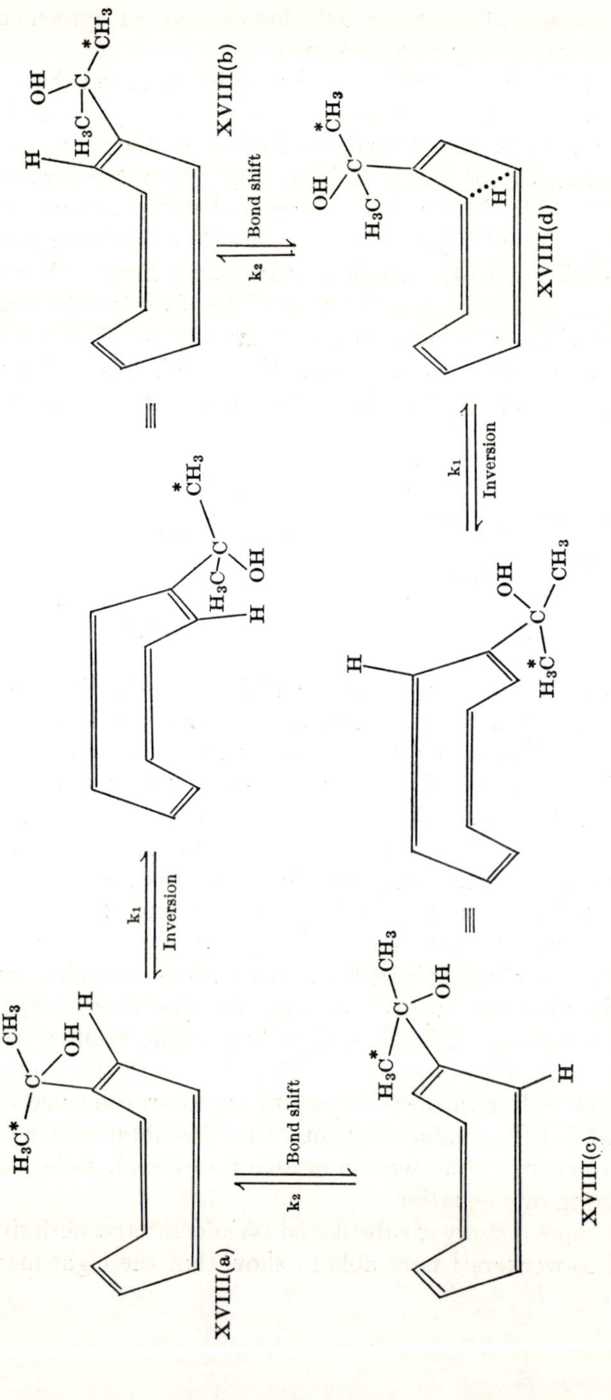

FIG. 16. Bond-shift and ring-inversion processes of cyclo-octatetraenyl-2,3,4,5,6,7-d_6-dimethylcarbinol.

ring underwent not only bond shift but also *ring-inversion* at room temperature. The n.m.r. spectrum of cyclo-octatetraenyl-2,3,4,5,6,7-d_6-dimethylcarbinol (XVIII) may be expected to be very simple. At low temperatures (–35°, CS_2 solution) and with double irradiation at the deuteron frequency, it gave rise to two sharp lines of equal intensity at 5·76 and 5·80 ppm; only the high-field band remained sharp in the absence of deuterium decoupling and thus this frequency was assigned to the proton in XVIII(a) and XVIII(b), while the low-field band must then be attributed to the proton in XVIII(c) and XVIII(d). The methyl protons in these structures are chemically non-equivalent and gave also two bands at 8·84 and 8·79 ppm. At –2°, the methyl doublet coalesced and finally became a sharp signal as the temperature was raised; the doublet arising from the ring-protons showed a similar behaviour but coalesced only at +41°. Kinetic analysis of the reaction paths shown in Fig. 16 showed that $k_1 \gg k_2$. The free energy values of activation $\Delta G^{\ddagger}$ for the two processes (ring inversion and bond-shift) calculated[134] from the observed spectra are given in Table 10.

TABLE 10

Free energy values of activation for bond shift and ring inversion in the cyclo-octatetraene derivative XVIII at –2°

Process	k (sec^{-1})	$\Delta G^{\ddagger}$
Bond shift	0·1*	17·1*
Ring inversion	7·8	14·8

* Calculated from $\Delta H^{\ddagger}$ and $\Delta S^{\ddagger}$ values determined by measurements of the rate constant for bond shift from 26 to 66°.

These values are of considerable theoretical interest. The most probable transition state structure for the ring-inversion process should be a *strained planar* state, with alternating 'single' and 'double' bonds (XIX(a)), i.e. exactly the one which was calculated by the LCAO theory to be the most stable planar COT structure

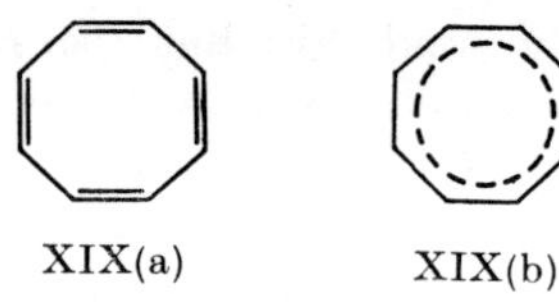

(see section B.2). The transition state for the bond-shift process on the other hand, should be the planar *symmetrical* structure XIX(b); this means that their energy difference should be exactly the energy barrier predicted by the Jahn–Teller theory. If the entropy of activation values ($\Delta S^{\ddagger}$) for the two processes are assumed to be roughly the same, then structure XIX(a) appears to be more stable than structure XIX(b) by about 2·4 kcal. mole^{-1} in rather good agreement with the predicted theoretical value[111, 241].

Bond shift[11, 135, 201] and bond shift-ring inversion[202] studies were also carried out on monosubstituted cyclo-octatetraenes where the substituent is directly conjugated to the eight-membered ring. The most extensive studies have been made on isopropoxy-cyclo-octatetraene by Oth[202]; the results are shown in Table 11. Note that the $\Delta S^{\ddagger}$ values for the ring-inversion and bond-shift processes are approximately the same; a Jahn–Teller effect of about 3·7 kcal. mole^{-1} can be obtained from these experimental values.

TABLE 11

Kinetic data for the bond-shift and ring inversion processes of isopropoxy-cyclo-octatetraene

Process	Temperature range	$\Delta G^{\ddagger}$ at 0°	$\Delta H^{\ddagger}$	$\Delta S^{\ddagger}$
Bond shift	0° to + 100°	15·6 ± 0·1	15·7 ± 0·4	+0·5 ± 1·2
Ring inversion	−60° to + 10°	12·7 ± 0·1	12·0 ± 0·4	−2·6 ± 1·8

Dr. J. F. M. Oth (Union Carbide, European Research Associates Brussels) kindly allowed us to describe these unpublished results.

It should be emphasised that, in the case of monosubstituted cyclo-octatetraenes, the two compounds obtained by bond-shift should be enantiomers. Attempts to resolve such compounds have been unsuccessful[72, 74] at room temperature; it appears now however that this could be achieved at a slightly lower temperature (about −40°).

Recently, a careful kinetic examination of the rather unusual reaction of COT derivatives with dienophiles, e.g.[231]:

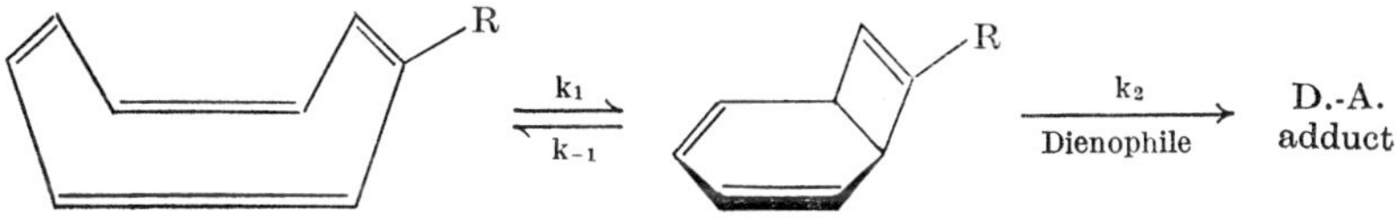

R $\underset{k_{-1}}{\overset{k_1}{\rightleftharpoons}}$ R $\xrightarrow[\text{Dienophile}]{k_2}$ D.-A. adduct

XXII

allowed Huisgen *et al.*[146] to discover another important feature characterising the structure of cyclo-octatetraene, namely the existence of *the bicyclic valence-tautomer* XXII. The resulting Diels–Alder adducts such as XXI can be derived formally from the bicyclo-[4,2,0]octa-2,4,7-triene structure XXII, presenting a nearly planar 1,3-diene system and thus having a favourable geometry for Diels–Alder reactions; this geometry should however be excluded in the

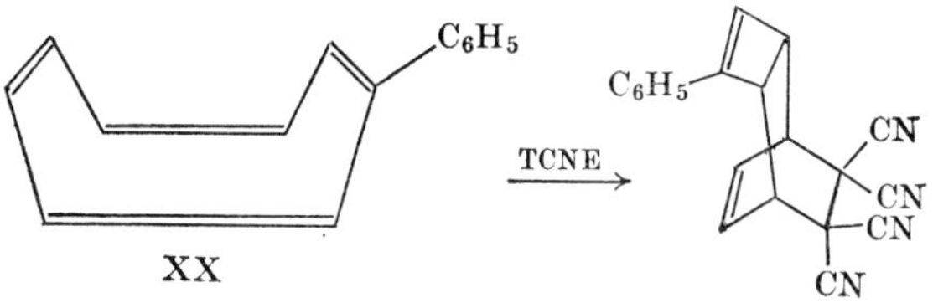

C_6H_5 $\xrightarrow{\text{TCNE}}$ C_6H_5 CN CN CN CN

XX XXI

tub form eight-membered ring structure. There was, however, the possibility that the tautomerism proceeded in an activated collision complex between the COT derivative and the dienophile, equilibrium $K_1 = k_1/k_{-1}$ existing then only within the reaction path. Huisgen showed that the rate of reaction of phenylcyclo-octatetraene with dicyano-maleimide in tetrahydrofuran was independent of the concentration of the dienophile; this implies that the first step (k_1) is the determining one, and thus that equilibrium K_1 should exist even in the absence of any reactant. The obtained kinetic data for the reaction process according to k_1 are: $\Delta H^{\ddagger} = 25{\cdot}0$ kcal. mole^{-1}, $\Delta S^{\ddagger} = -3{\cdot}0$ u.e. ($k_1 = 2{\cdot}28 . 10^{-4}$ sec^{-1} at 70°). Similarly, an equilibrium constant K_1 of about 0·01 per cent (dioxan, 100°) could be estimated for the valence-isomerisation of unsubstituted cyclo-octatetraene[146]. Recently, the valence-isomer XXII (R = H) was synthesised by Vogel and co-workers[279] in 95 per cent purity at −75°; isomerisation to cyclo-octatetraene took place readily when it was warmed (half-life time of 14 minutes at 0°).

3. *Higher Annulenes*

Almost all of the work on the synthesis of higher annulenes has been done by Sondheimer and co-workers [242]. The general scheme consists of oxidative coupling of suitably chosen α,ω-unconjugated diynes, followed by (prototropic) rearrangement of the cyclic products obtained to completely conjugated dehydro-annulenes, which can then be hydrogenated catalytically to the corresponding annulenes. In this

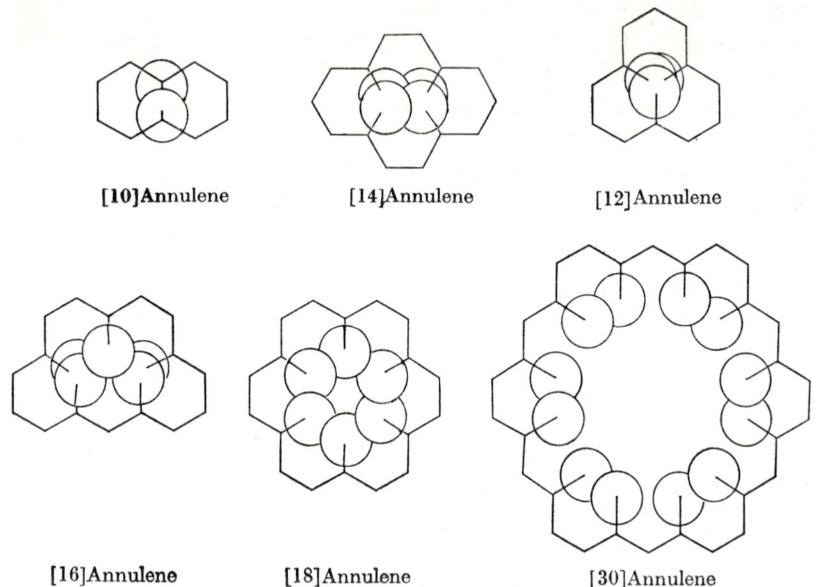

FIG. 17. Scale drawings of some annulenes (bond lengths: $R_{CC} = 1{\cdot}40$ Å all equal; $R_{CH} = 1{\cdot}10$ Å; all valence angles 120°; hydrogen radii $= 1{\cdot}00$ Å)[188].

way were synthesised [14] [43, 244, 248], [16] [245], [18] [252, 253, 296], [20] [246, 247], [24] [250, 253] and [30]annulenes [247, 253, 254, 255]. Recently, Schröder and Oth [234] reported a convenient photochemical synthesis of [16]-annulene from a cyclo-octatetraene-dimer. Substances have been reported in the literature for which the [12]annulene structure was tentatively suggested [218, 289], but no confirmation of this assignment has been provided subsequently; on the other hand, several analogues incorporating fused benzene-rings are known (see section D) and a fully conjugated trisdehydro- and a bisdehydro-annulene have been synthesised [242]. In the ten-membered ring size, some benzo-derivatives are known (see section D), but [10]annulene itself has only been postulated as a transient molecular species[274, 275]. All the annulenes so far isolated are highly coloured crystalline compounds, except [20]-annulene which was obtained as a yellow oil and could not be induced to crystallise. They are rather unstable substances sensitive both to air and light. It was pointed out very early on by Mislow[188] that all the annulenes from $C_{10}H_{10}$ to $C_{28}H_{28}$ presumably cannot be planar in view of the steric interactions of the internal hydrogen atoms in the planar molecules, as is shown in the rough scale drawings of Fig. 17. Experimental verification of this conclusion has been obtained at least for some of the known lower members of the annulene series.

The annulene for which the most extensive experimental data have been obtained in the last few years is without doubt [18]annulene. A careful tridimensional X-ray analysis[44] at 80° K shows the molecule to be centrosymmetric (Fig. 18(a)), and nearly planar, possessing six 'long' outer carbon–carbon bonds (mean value 1·419 Å) and twelve 'short' inner CC bonds (mean value 1·382 Å). It is virtually certain that, in the absence of any intermolecular forces, the molecule would adopt the structure shown in Fig. 18(b)[143]. This result obviously rules out the possibility of an alternating bond-length structure. The observed geometry (b) seems indeed to be derived more easily from an original structure with all C—C bonds equal in length, the observed inequalities as well as the non-planarity of the molecule being due to intramolecular hydrogen repulsions. Hirshfeld and Rabinovich[143] pointed out however that the observed molecular deformations and especially the observed variations in bond lengths cannot be explained successfully by valence-field calculations and suggested that σ-orbital bending, due to relatively large deviation of the bond angles from 120°, must be of some importance[143]. This hypothesis has been shown to be correct by Murrell and Hinchliffe[196] who calculated the $(\sigma + \pi)$ electronic structure of the molecule by the semi-empirical method of Pople and Santry[210].

The electronic spectrum of [18]annulene has been taken in an EPA* glass at 77°K[126]. Analysis of the vibrational structure established that the long-wavelength band consisted principally of the ^1L_a-transition (Clar's *p*-band), hiding partially the less intense ^1L_b-transition (Clar's α-band). Gouterman and Wagnière[126] suggested that the bond-alternation problem could be resolved by looking at the

* EPA is a mixed solvent containing ethyl ether, 2-methyl butane and ethyl alcohol in a volume ratio of 5:5:2 respectively.

distribution of the transition intensity among the vibrational components. Theoretical arguments exclude bond-alternation in the excited states of the molecule; accordingly, two different cases can be expected with regard to the molecular geometry: (a) absence of

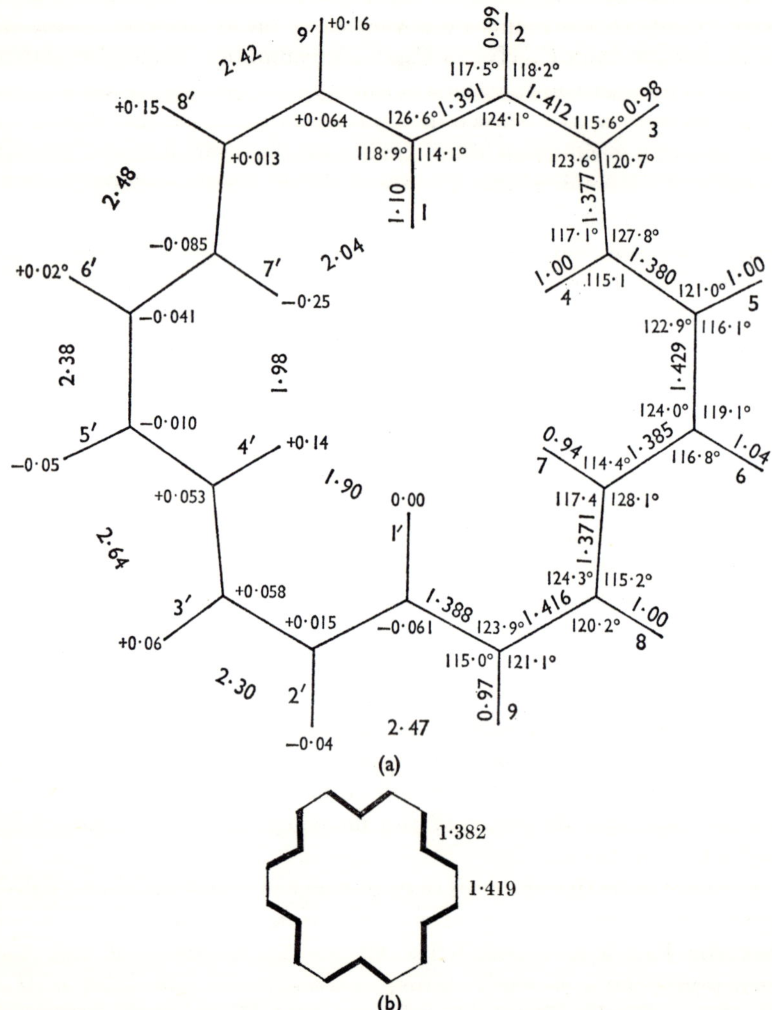

FIG. 18. Crystallographic structure of [18]annulene (interatomic distances (Å) and bond angles; near the atomic positions are given the displacements (Å) from the mean molecular plane). Reprinted by permission from *Acta Crystallographica* (1965), **19**, 234 [44].

bond-alternation in both fundamental and excited state, or (b) bond-alternation in the ground-state but a symmetrical excited state (Fig. 19). In case (a), according to the Franck–Condon principle, most of the intensity will be confined in the 0—0 band; in case (b) some vibrational component of the progression will be stronger than the 0—0 band. Yet the electronic spectrum at low temperature indicates no vibrational progression as depicted in case (b). Thus the vibrational structure argues in favour of similar potential curves in

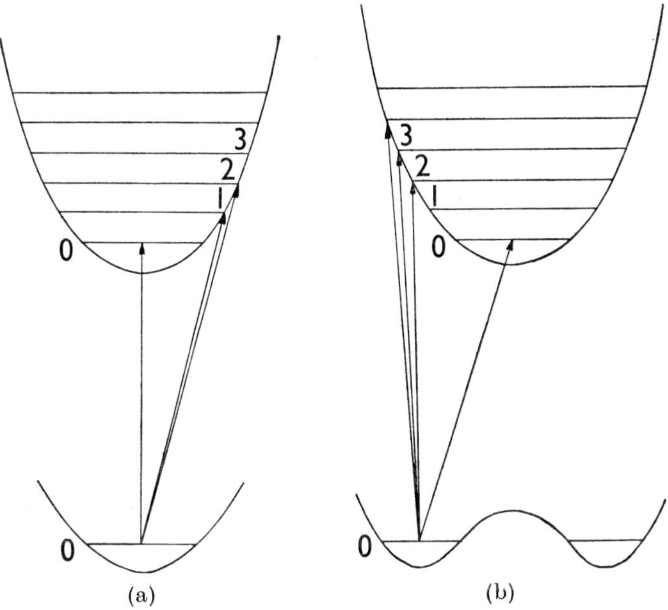

FIG. 19. Possible electronic energy curves for the ground and excited states of $C_{18}H_{18}$.

both ground and excited states, thus ruling out bond alternation. This result appears to be entirely consistent with X-ray data.

The u.v. spectrum of [18]annulene (see Fig. 20) has been discussed by several authors[48, 76, 86, 113, 126, 136, 143, 173]. It is well known that a plot of the frequency of the p-bands of benzenoid aromatic hydrocarbons against the HMO energy difference between the highest occupied and lowest unoccupied MO's (ΔE in β-units) shows an excellent linear correlation. According to Streitwieser[265], the p-band position can easily be calculated by the relation:

$$\nu_p\,(\text{cm}^{-1}) = 19{,}020\,.\,\Delta E + 10{,}520.$$

A scattering of the experimental values of about 1000 cm^{-1} seems however to occur commonly. Using this equation we may calculate for symmetrical $C_{18}H_{18}$ $\nu_p = 23{,}700$ cm^{-1} or 422 mμ, and for symmetrical $C_{30}H_{30}$ $\nu_p = 18{,}450$ cm^{-1} or 542 mμ, in excellent agreement with the experimental values (22,300 cm^{-1} or 448 mμ and ~18,520 cm^{-1} or 540 mμ respectively). The attributions of the various elec-

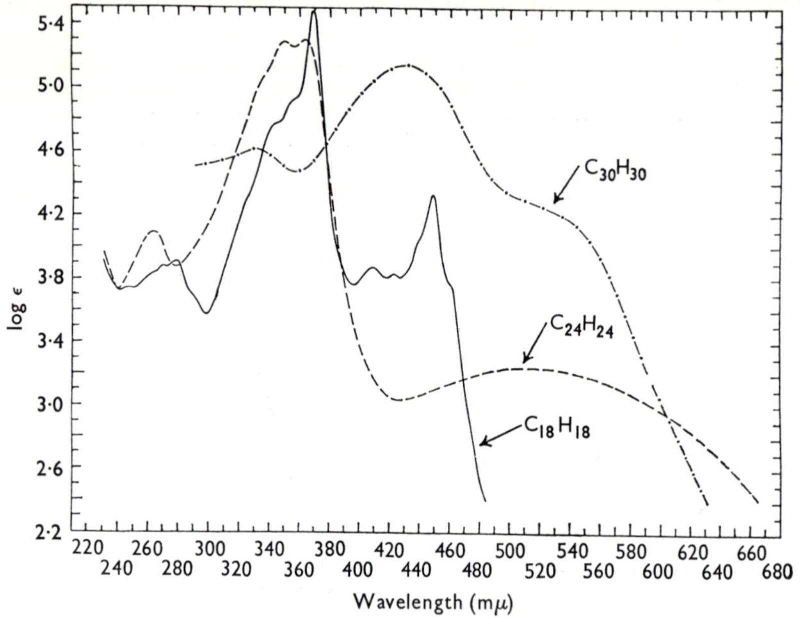

FIG. 20. Ultra-violet absorption spectra of [18]annulene ($C_{18}H_{18}$) in iso-octane, [24]annulene ($C_{24}H_{24}$) in iso-octane, and [30]annulene ($C_{30}H_{30}$) in benzene. Reprinted from the *Journal of the American Chemical Society* (1962), **84**, 274; reprinted by permission of the copyright owner.

tronic transitions which appear in the [24]annulene-spectrum have not yet been settled sufficiently to permit instructive comparison with theoretical computations.

The results seem to be in favour of a symmetrical structure for the [18] and [30]annulenes, all bonds having the same length. It should however be pointed out that the u.v. spectra of these molecules are rather insensitive to bond alternation[113]. Further, it appears that molecular distortions as revealed by the X-ray experiment on [18]-

annulene may cause a red shift of the *p*-band[76, 143]; an inverse hypsochromic shift may be expected if bond alternation is present[136, 173].

The observed heat of combustion $-\Delta H_c^0$ for [18]annulene in the crystalline form is 2346·8 ± 4·0 kcal. mole^{-1} [32]. The latent heat of sublimation for this compound has not yet been measured, but a value of 28 ± 2 kcal. mole^{-1} can be estimated, since for aromatic compounds of about the same molecular weight, the latent heats of sublimation lie in the range 26–30 kcal. mole^{-1} [181]. From these values, a heat of formation from constituent atoms of $-\Delta H_f^0 = 3946{\cdot}8$ kcal. mole^{-1} may be calculated. Using Dewar's bond energy values[88], an experimental resonance energy of about 70 kcal. mole^{-1} is obtained, in severe disagreement with the predicted theoretical value of 3·2 kcal. mole^{-1} for the unstrained bond-alternating molecule[88]. Although it may surely be concluded that bond-alternation sets in too early in the SCF scheme, the error appears too large to be accounted for simply by some defect in the π-electron theory alone. It should be mentioned that simple Hückel theory provides a delocalisation energy of a better order of magnitude[86, 113]. Thus, using, for example, Klages' bond-contributions to the heats of combustion of aromatic compounds[264], an empirical resonance energy of about 100 kcal. mole^{-1} may be estimated for [18]annulene; on the other hand, Streitwieser has shown that the resonance energies so obtained correspond well to the theoretical delocalisation energies calculated by the simple Hückel method, provided a β-value of -16 kcal. mole^{-1} is adopted[264]. For [18]annulene this theory[113] predicts RE $= 80$ kcal. mole^{-1}. The discrepancy lies however in the opposite direction from what would be expected if non bonded intramolecular interactions were important[76]. It seems likely therefore that a definite answer to this puzzling problem will only be obtained by a complete *a priori* calculation of the $(\sigma + \pi)$ energy content of the molecule.

The chemical properties of [18]annulene resemble in many aspects those of a normal conjugated polyene. Thus it is readily hydrogenated catalytically, undergoes facile addition of bromine and forms a Diels–Alder type of adduct with maleic anhydride in boiling benzene[253]. It has recently been found, however, that under suitably mild conditions, [18]annulene can be converted into a mononitro- and a monoacetyl-derivative, and it therefore to some extent shows 'aromatic' chemical behaviour[53]. These findings point once again to the inherent danger in relating aromaticity to some kind of reactivity.

The n.m.r. spectrum of [18]annulene and its derivatives have been studied extensively by Sondheimer *et al.*[52, 53, 54, 122, 148, 242, 243]. This seemed the most convenient method for determining whether or not the molecules were aromatic. If important bond-alternation may be excluded, the presence of an induced diamagnetic ring-current will shift the *outer* protons to low field relative to the 'olefinic' resonance

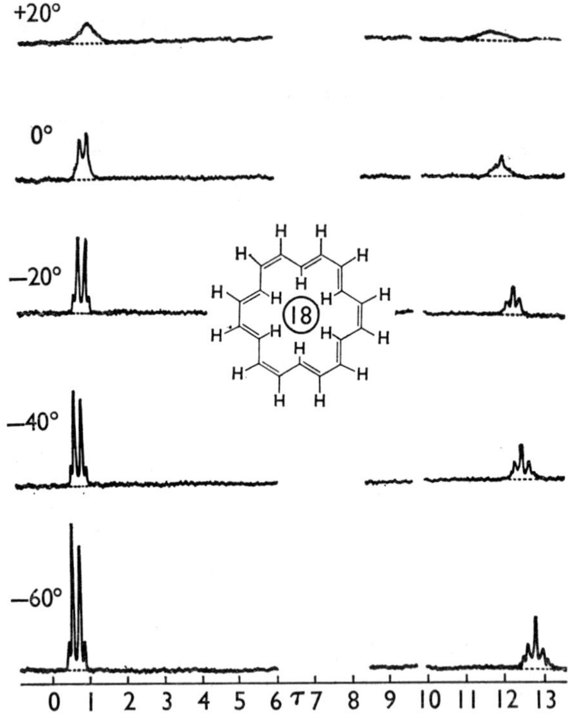

Fig. 21. N.m.r. spectra (60 Mc/s) of [18]annulene in perdeuteriotetrahydrofuran from +20 to −60°. Reprinted by permission from F. Sondheimer *et al.*, in *Aromaticity*, Special Publication No. 21 of The Chemical Society, London (1967), p. 90.

frequency and the *inner* protons to an unusually high field (Table 8). The spectrum of [18]annulene at room temperature (+20°; see Fig. 21) in perdeuteriotetrahydrofuran as a solvent consisted effectively of two broad unresolved bands, one at 8·94 ppm (outer protons) and the other at −2·0 ppm (inner protons). When the solution was cooled, the bands became progressively sharper and exhibited fine structure, the separation between the bands being increased. Finally, at −60°

the outer protons appeared at 9·28 ppm and the inner at −2·99 ppm. The relative area of the two bands was almost exactly 2/1, indicating twelve protons outside and six inside the ring. This spectrum clearly shows [18]annulene to be aromatic.

Comparison of the observed chemical shift with the theoretical values listed in Table 8 (page 288) for [18]annulene seems to indicate some bond-alternation in this molecule. Interpolation between these

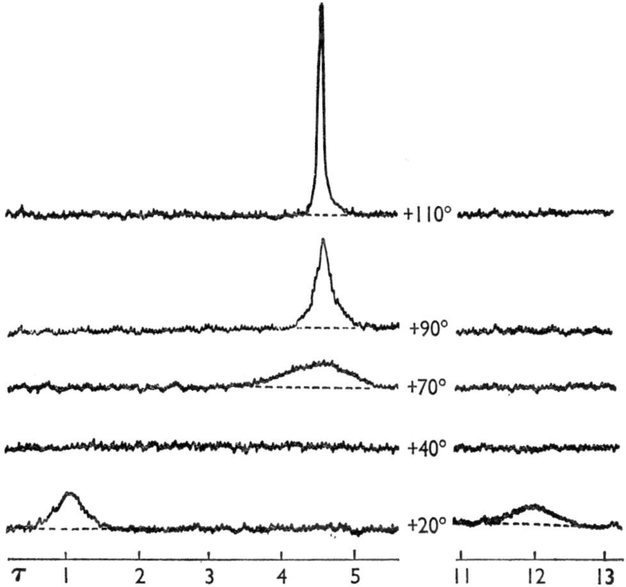

FIG. 22. N.m.r. spectra (60 Mc/s) of [18]annulene in perdeuteriotetrahydrofuran from +20° to +111°. Reprinted by permission from F. Sondheimer *et al.*, in *Aromaticity*, Special Publication No. 21 of The Chemical Society, London (1967), p. 91.

values for the outer-proton chemical shift would give a bond alternation characterised by $\lambda = 0{\cdot}88$, i.e. alternating bond lengths of about 1·35 and 1·45 Å[113]. The observed chemical shift of the inner protons can hardly be discussed in a quantitative manner; indeed, both the mutual Van der Waals contact effect[108] due to the overcrowding of these hydrogens and additional anisotropic contributions arising from adjacent carbon–carbon bonds[83, 209] should cause an important deshielding effect, moving these protons to lower field than expected. However, in a structure such as shown in Fig. 18(b), the mobility of the π-electrons around the aromatic ring will be reduced relative to a perfectly symmetrically delocalised structure: 'the π-electrons will tend to be localised in the six pairs of shorter inner bonds as a result of the smaller resonance integral of the outer bonds, which acts as a slight barrier between neighbouring inner-bond pairs'[143]. This perturbation should influence the ring-current intensity J_{μ} in the same manner as a slight bond-alternation, and although we have not attempted a quantitative estimate of this effect, we think it likely that the great dependence of J_{μ} on electron delocalisation may account for the observed chemical shift of the outer protons.

Figure 22 shows that, when the temperature is raised to 40°, the bands become so diffuse that they can no longer be detected. At still higher temperatures, a new broad band appears at about 5·4 ppm;

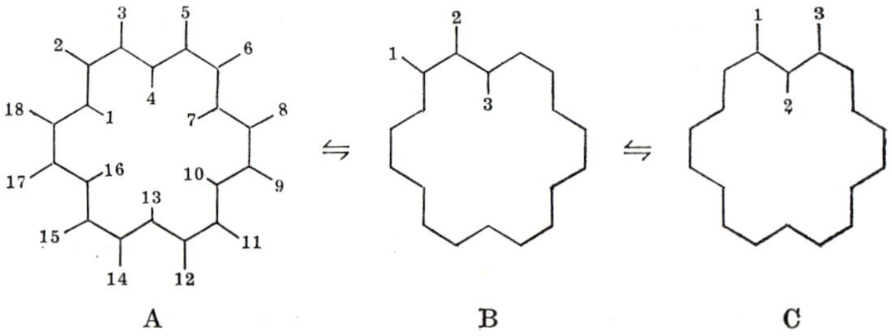

FIG. 23. The three possible conformers of [18]annulene.
$\delta_{mean} = \frac{2}{3} \times 9{\cdot}28 + \frac{1}{3} \times (-2{\cdot}99) = 5{\cdot}19$ ppm, close to the position of the high-temperature singlet (5·45 ppm).

finally at 111° there is a sharp signal at 5·45 ppm. The most likely explanation for this behaviour is that at the higher temperatures, the protons change position at such a rate that an average value results for the band location. If in [18]annulene, the protons are identified by H(1) to H(18), then the symmetry of the system allows three possible equivalent conformers (Fig. 23). If we suppose that these conformers are in rapid dynamic equilibrium, it is seen that each proton spends one-third of its time inside the ring and two-thirds outside the ring, occupying successively the three magnetically different positions. The mean value of the chemical shift should thus be identical for each proton and may easily be calculated (Fig. 23).

If H(1) is replaced by a substituent which, for steric reasons, cannot occupy an internal position, then conformer A is excluded. As a result, the five protons H(4,7,10,13,16) can never occupy an internal position, the twelve others spending half their time inside and half their time outside the ring. The n.m.r. spectrum of monosubstituted [18]-annulenes should thus show a temperature-independent 5H band at very low field attributable to these outer protons. The twelve remaining protons should appear at low temperature as two bands of equal intensity, one due to the six external protons at low field and one corresponding to the six internal protons at high field. At higher temperatures, these last two bands should coalesce and finally give rise to an averaged signal at a position which is half-way between the positions of the low-temperature bands. This is exactly the behaviour shown by mononitro-[18]annulene[53], thus proving once again the conformational mobility of the parent hydrocarbon. It should therefore be emphasised that the existence of a single n.m.r. band in the olefinic region should not be considered as evidence for lack of aromaticity or anti-aromaticity, unless an extensive temperature-dependence study has been made.

Of the two other $[4n + 2]$annulenes known, only [14]annulene has been the subject of physico-chemical studies. A preliminary X-ray crystallographic analysis[43] confirmed that this molecule indeed exists in the configuration shown in Fig. 17. However, thin-layer chromatography of the compound on Kieselgel coated with silver nitrate produced two unequal neighbouring spots. Immediate rechromatography of an ethereal solution of either spot produced in each case only one spot essentially unchanged from the original one; after 30 minutes standing at room temperature however, the ethereal solution of each spot produced again the original two spots. A solution freshly prepared from tho crystalline substance produced only one spot, but the equilibrium mixture was again formed after the solution had been kept for 30 minutes. The n.m.r. spectrum of [14]annulene at room temperature consists of only two sharp singlets at 5·58 and 6·07 ppm in an intensity ratio of about 6/1[122, 124]. A solution freshly prepared from crystalline [14]annulene showed only the 5·58 ppm band; when the solution was kept, the 6·07 ppm band appeared progressively.

These experiments were interpreted by Sondheimer[124] as being due to the existence of two *conformational isomers*. Owing to the over-crowding of the internal hydrogen atoms, two conformers may indeed be possible, differing in the relative positions of these protons (Fig. 24). The X-ray analysis of the crystalline substance indicates that the molecules are centrosymmetric, unless the molecular packing is disordered[43], and thus the unique conformer present in the crystal should be conformer A (centre of symmetry).

The change in the n.m.r. spectrum of [14]annulene with temperature[122, 242b] is similar to that observed for [18]annulene. When the temperature is lowered, the 5·58 ppm band corresponding to isomer A broadens and disappears (coalescence), and at –60° the spectrum consists of two broad bands at 7·6 (10H) and 0·0 ppm (4H) due to the outer and inner protons of conformer A, as well as the B conformer peak at 6·07 ppm which has broadened slightly.

FIG. 24. Possible conformers of [14]annulene.

It should however be pointed out that all the experiments described above can also be interpreted on the basis of a *configurational isomerism*

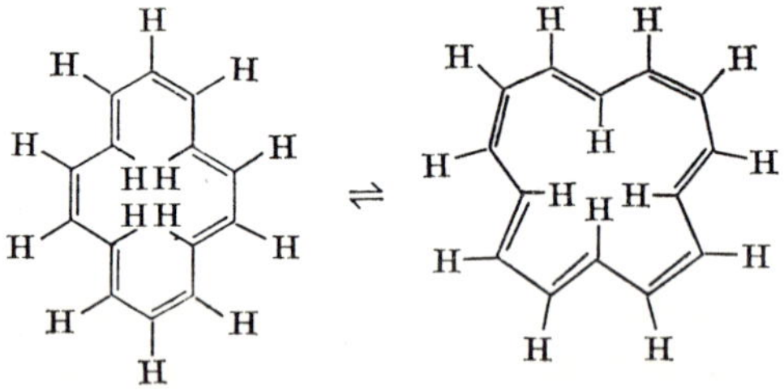

XXIII ⇌ XXIV

of the type XXIII ⇌ XXIV; configuration XXIV is indeed of the same type as the one obtained for [16]annulene[234] and molecular models suggest even slightly less strain energy for XXIV than for the [16]annulene structure. Obviously, the problem is not yet settled and additional experimental work seems necessary.

N.m.r. experiments similar to those described for the $[4n + 2]$-annulenes were also carried out by Schröder and Oth[234] and by Calder and Sondheimer[55] on [16] and [24]annulenes respectively, and the results obtained were similar (e.g. Fig. 25). At room temperature, the n.m.r. spectrum of [24]annulene evidently again represents an average, due to rotation of the carbon–carbon bonds (conformational isomerism). But in this case, it has been postulated that in addition bond-shift should be involved[55, 212, 234]. At −80° the spectrum exhibits bands centred at about 12·0 ppm and 4·6 ppm with an

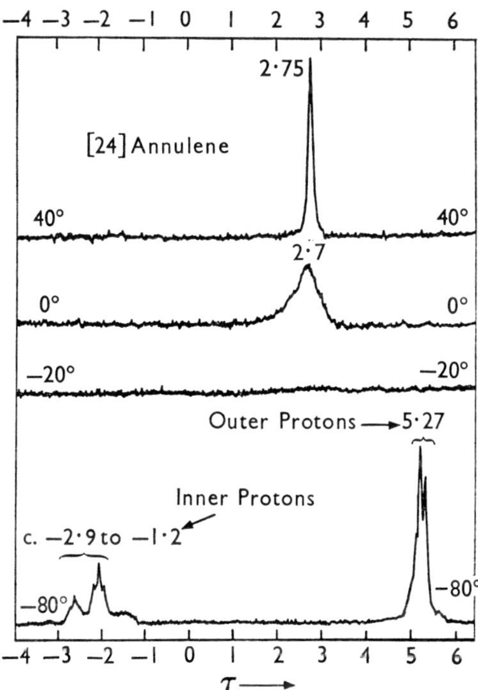

Fig. 25. N.m.r. spectra of [24]annulene in perdeuteriotetrahydrofuran solution (100 Mc/s, tetramethylsilane used as internal standard), from −80° to +40°. Reprinted by permission from I. C. Calder and F. Sondheimer, *Chemical Comm.* (1966) 904.

intensity ratio of c. 35/65. This is consistent only with either structure XXV (inner/outer proton ratio = 37·5/62·5) or XXVI (ratio = 33·3/66·7); the degree of accuracy of the integration is however insufficient to allow a definitive attribution. The important thing however is that the *inner protons appear at low field and the outer protons at high field*. This of course constitutes experimental evidence for an *induced paramagnetic ring-current*, as predicted by the theory.

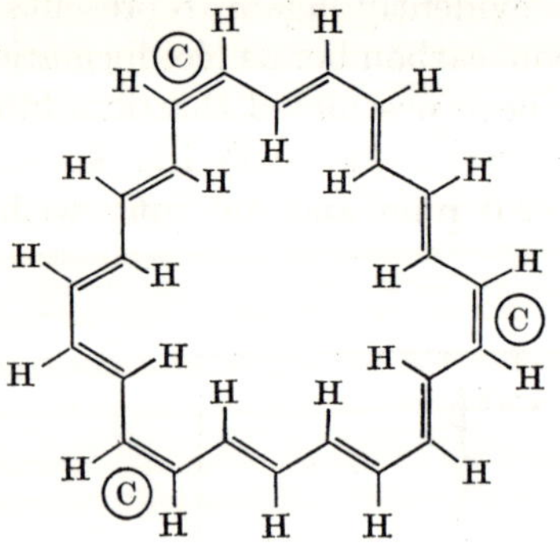

XXV

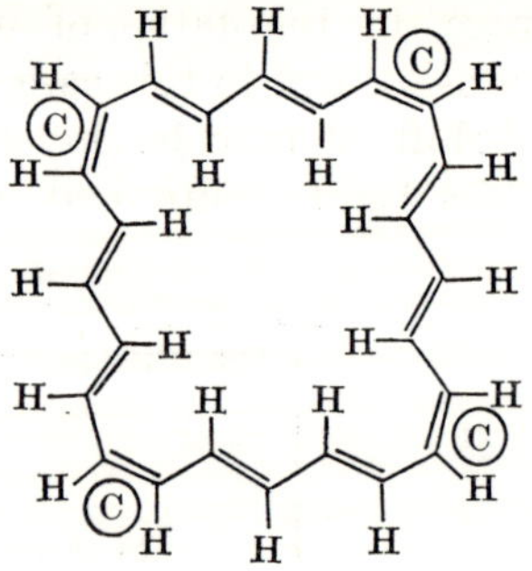

XXVI

In the case of [16]annulene, the n.m.r. spectrum at $-110°$ showed the inner protons appearing at 10·43 ppm, the outer at 5·40 ppm; the intensity ratio of the two bands is 4/12, in agreement with the proposed

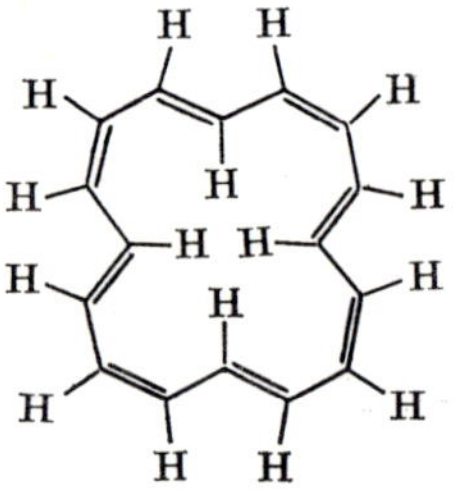

XXVII

conformational structure XXVII. Thus [16] and [24]annulenes appear to be *anti-aromatic compounds* on the basis of their magnetic properties. Very recently, this has been confirmed by diamagnetic susceptibility exaltation measurements on [16]annulene; an appreciable positive exaltation of $\Lambda = (4{\cdot}6 \pm 2) \times 10^{-6}$ cgs units has been observed [84].

In conclusion, it appears that the available experimental data with regard to the electronic structure of the annulenes are in fair agreement with the theoretical predictions. It should however be realised that most of the molecules investigated show out-of-plane distortions and/or bond-length modifications as a result of strong intramolecular steric repulsions in the planar, unstrained configurations. These interactions and deformations will perturb not only the π-electronic structure and π-energy of the molecule, but will alter to an even greater extent its σ-structure and σ-energy. Accordingly, it appears that, except for the very large annulenes ($N \gtrsim 30$), the thermodynamic criterion of aromaticity will hardly be applicable to this type of molecule. It appears also that, with increasing N, the resonance energy (negative or positive) becomes smaller and smaller, and thus the conclusions which may be obtained from the experimental values will become more and more dependent on small steric interactions and on bond bending. On the other hand, the magnetic criterion of aromaticity seems to be more adapted to the investigation of the effects of the π-electron delocalisation in the annulene family. Indeed, if the rings are nonplanar or buckled, only the *magnitude* of the ring-currents will be reduced, principally because of less effective overlap of $2p\pi$ atomic orbitals, and at the worst, an erronous statement of *non-aromaticity* can be reached. It appears thus that an *inverse* conclusion should never be reached by this method if the experimental results are properly analysed.

It should be emphasised that, at present, proton n.m.r. spectroscopy, especially at low temperature, seems to be the most convenient and most sensitive method for determining whether or not a compound of the type under discussion is aromatic.

D. Dehydro-annulenes and Benzo-annulenes

The aim of this section is to show that valuable information about the aromaticity of the annulenes may be obtained by careful analysis of some physico-chemical properties of their dehydro- and benzo-derivatives. It is not intended as a review of all the recent chemistry and spectroscopy of these compounds. It should be seen rather as a compilation of selected studies which have appeared in the literature, giving complementary and useful details about the electronic structure of ring-systems sometimes accessible otherwise only with difficulty.

1. *Polycyclic Cyclobutadienes*

Like the parent [4]annulene cyclobutadiene, benzocyclobutadiene is an extremely reactive molecule which has not yet been isolated*,

* The synthesis of a very stable benzocyclobutadieniron tricarbonyl complex has been reported recently[101].

although its existence as a transient reaction-intermediate has been postulated in many cases [63]. Recently, however, Cava and co-workers reported the synthesis of the two stable substituted naphthocyclobutadienes [59, 60] XXVIII and XXIX and of a stable 1,2-diphenylanthro[*b*]-cyclobutadiene (XXX) [58], which are of considerable theor-

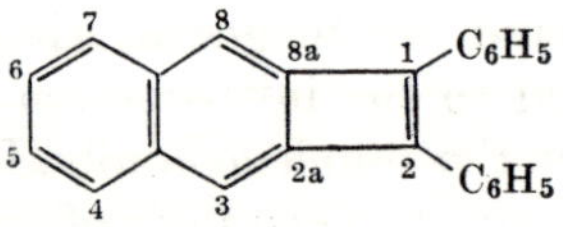

XXVIII 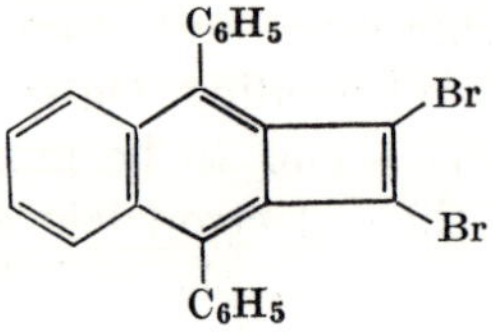 XXIX

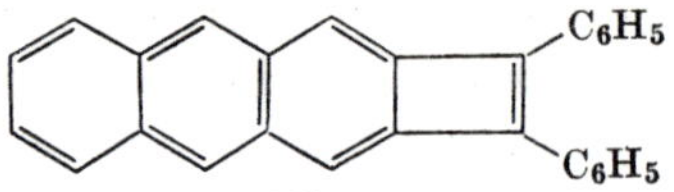

XXX

etical interest, for they represent the only known examples of molecules where the cyclobutadiene ring is fused to a benzene nucleus on one side only. The stability of this kind of compound has prompted many theoretical discussions [8, 77, 93, 164], from which it is generally concluded that little conjugation exists between the 'isolated' double bond and the rest of the molecule. The difference in resonance energy ΔE_R calculated by Pople's method [93] between the various cyclobutadiene derivatives and the pair of π-systems obtained by breaking the two 'single' cyclobutadiene bonds are listed in Table 12. A positive value of ΔE_R should imply that the corresponding cyclobutadiene ring is *aromatic*; a negative value on the other hand indicates that it is *anti-aromatic*. It is evident from these calculations that even in polycyclic derivatives, the cyclobutadiene ring remains anti-aromatic according to the thermodynamic criterion.

The ΔE_R values gathered in Table 12 seem to account well for the known stabilities of these compounds; compounds with large negative ΔE_R-values should indeed be chemically unstable, tending to undergo reactions which remove the unfavourable effect of the destabilising conjugation through the four-membered ring. This explains why the only compounds of this type which have as yet been synthesised are anthro[*b*]- and naphtho[*b*]-derivatives, benzo- [63] and phenanthro[*l*]-cyclobutadiene [61] being known only as unstable reaction inter-

TABLE 12

Anti-aromaticities of the four-membered rings in some cyclobutadiene derivatives[93]

Molecule	ΔE_R (eV)
benzo-CBD	–0·488
naphtho[*a*]-CBD	–0·615
naphtho[*b*]-CBD	–0·314
anthro[*a*]-CBD	–0·656
anthro[*b*]-CBD	–0·243
phenanthro[*l*]-CBD	–0·716

CBD: cyclobutadiene. In these calculations, all CC bond lengths were set equal to 1·40 Å[93].

mediates*. The chemical behaviour of this last substance led Cava *et al.* to postulate that the molecule might exist as a triplet in the ground state[61]; this assumption recently received support from a semi-empirical SCF-LCAO calculation[142].

The n.m.r. spectrum of XXVIII is of considerable interest[60]. In addition to the signals due to fourteen protons appearing in the usual aromatic region, the spectrum shows a sharp singlet attributed to H(3) and H(8) at 6·50 ppm. This was interpreted by Cava *et al.* as indicating an unusually large amount of double bond character in the 2*a*, 3- and 8,8*a*-bonds, shifting the resonance of H(3) and H(8) to the 'olefinic' position. However, it seems now more plausible (see, e.g., the spectrum of biphenylene, p. 333), to attribute this high-field frequency to the influence of an induced paramagnetic ring-current in the four-membered ring of XXVIII, thus providing experimental evidence regarding the anti-aromaticity of this ring with respect to the magnetic criterion.

Biphenylene may formally be regarded as a dibenzo-[4]annulene (or dibenzocyclobutadiene) and was first synthesised by Lothrop in 1941[174]. Since then, several other syntheses have been described and

* Recently, Bergmann *et al.*[34], on the basis of u.v. and n.m.r. arguments, tentatively assigned the phenanthro[*l*]-cyclobutadiene structure to a stable red hydrocarbon $C_{28}H_{18}$, m.p. 158°.

several reviews concerning the synthesis and chemistry of this most challenging molecule have appeared[23, 64].

X-ray analysis[105] has established that the molecule is completely planar, with a four-membered ring which is closer structurally to a tetramethylenecyclobutane than to a cyclobutadiene (Fig. 26), in agreement with LCAO-MO calculations[9, 14, 49, 93, 106, 116, 163, 213, 223, 283]. Important double bond character of the 2–3 bond has been shown both by chemical[24, 25, 38, 42, 177] and spectroscopic methods[116, 156]. Thus 2-hydroxy-biphenylene couples readily with diazonium salts at the 3-position to give 3-azo-2-hydroxy derivatives[38], as predicted by LCAO theory[170]. Similarly, elucidation of the A_2B_2 proton n.m.r. spectrum of unsubstituted biphenylene gave $J_{1,2} = 7{\cdot}1$ cps and $J_{2,3} = 8{\cdot}1$ cps[156], and since the magnitude of the ortho-coupling constants always follows the π-bond orders of the corresponding CC-bonds[153], it may be concluded that even in the unsubstituted

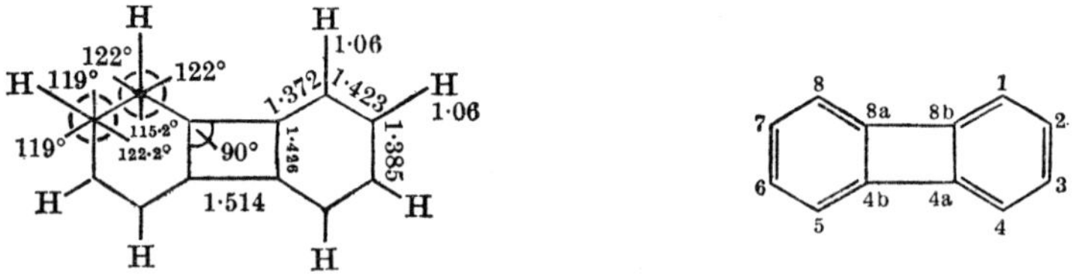

FIG. 26. X-ray structure of biphenylene according to Fawcett and Trotter[105].

hydrocarbon, partial double bond fixation as shown by the X-ray analysis is present. Similar conclusions may be obtained by a comparison of the n.m.r. spectra of 2-substituted biphenylenes and naphthalenes[116, 182], the largest ortho-effect of the substituent here being through the CC-bond with most double-bond character.

Although the two benzene-rings in the biphenylene molecule are linked together by almost 'single' bonds of length 1·514 Å (see Fig. 26), there is ample evidence to show that the two six-membered rings are not entirely independent and that substantial delocalisation is present over the entire molecular π-electron system. Thus, biphenylene absorbs u.v.-light at much longer wavelengths than biphenyl[23] (Fig. 27) and the long wavelength band at 350 mμ has been shown to be polarised along the long molecular axis[50, 140, 147]. Diacetylation and dinitration of the molecule give only 2,6-disubstituted products[21] thus proving deactivation of the other positions *through* the four-membered ring*. It has also been established that biphenylene forms

* A very small amount of 2,7-diacetylbiphenylene has also been identified[37].

a considerably more stable charge-transfer complex with tetracyanoethylene (TCNE) than does fluorene[104]. A much better argument, independent of the electronic structure of electronically excited states or transition states of reaction, is provided by the infra-red characteristics of the $\nu_{C \equiv N}$ vibration of cyano-biphenylenes and benzobiphenylenes (see Table 13). In each case, the frequency is lower and the intensity higher than in the corresponding benzenoid 'residues'

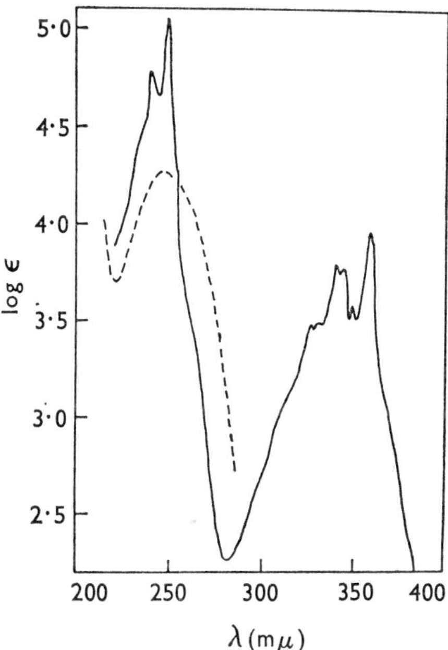

Fig. 27. Ultra-violet absorption spectra: —— biphenylene, ---- biphenyl (in ethanol). Reprinted by permission from W. Baker and J. F. W. McOmie, in D. Ginsburg, *Non-benzenoid Aromatic Compounds* (1959) Interscience, New York, p. 75.

obtained by removing the pseudo-single 4*a*,4*b*- and 8*a*,8*b*-bonds or their equivalent ones, indicating a more extended unsaturated system, and thus proving substantial π-delocalisation across the four-membered ring[118].

The heat of combustion of biphenylene has been determined recently[31]: $-\Delta H_c^0 = 1483{\cdot}3$ kcal. mole$^{-1} \pm 0{\cdot}7$ kcal. mole^{-1}. Taking the latent heat of sublimation[57] equal to 30·8 kcal. mole^{-1}, a heat of formation from constituent atoms of $-\Delta H_f^0 = 2352{\cdot}3$ kcal. mole^{-1}

TABLE 13

Characteristics of the $\nu_{C\equiv N}$ vibration in biphenylene- and benzo[a]biphenylene-derivatives[118]

Molecule	$\nu_{C\equiv N}$ (cm^{-1})	A_{CN} in units of 10^4 mole^{-1} l. cm^{-2}
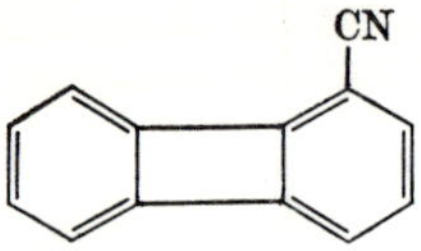 CN	2230·3	0·28
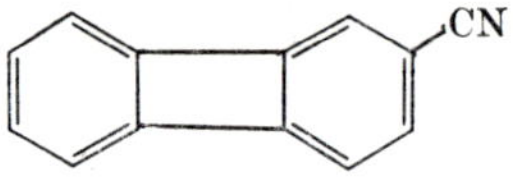 CN	2222·5	0·53
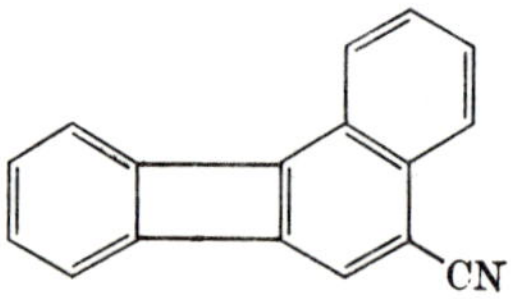 CN	2214·6	0·49
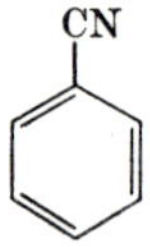 CN	2232	0·21
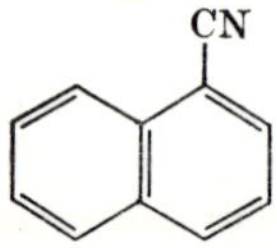 CN	2226	0·23

may be calculated, from which a resonance energy of 17·1 kcal. mole^{-1} is obtained[31]. Using Dewar's bond energy values, we obtain the rather astonishing value $E_R = -31{\cdot}56$ kcal. mole^{-1}. In either case, these are exceedingly low values for a substance with two benzene rings in the molecule (E_R of benzene is equal to 24·63 kcal. mole^{-1} with Dewar's parameters[88]). This may imply a high degree of strain associated with the four-membered ring system or a large anti-aromaticity of the central ring, or both. Strain energy has been estimated to be as large as 50–60 kcal. mole^{-1} for the cyclobutadiene molecule[288], and hence, if a similar value is assumed for the central ring of the biphenylene molecule, no definite conclusion concerning the aromaticity or anti-aromaticity of the four-membered ring can be obtained by thermochemical data.

The molar susceptibility of biphenylene has also been measured[15] and a value of $\chi_M = -41{\cdot}2 \times 10^{-6}$ cgs units has been obtained. This is less than the value for only *one* benzene ring (e.g. for benzene, $\chi_M = -54{\cdot}8 . 10^{-6}$ cm^3/mole[84]). In addition, the observed diamagnetic anisotropy is virtually zero: $\Delta\chi_M = (0{\cdot}0 \pm 0{\cdot}1) \times 10^{-6}$ cm^3/mole, as predicted by MO theory ($\Delta\chi_M \simeq -7{\cdot}0$ cm^3/mole, from $\chi_\pi = 0{\cdot}169 \times 10^{-17} \times \beta^{-1}$ and Fig. 10[36, 215]; compare with the values gathered in Table 5). Such a low experimental value for $\Delta\chi_M$ seems to indicate the presence of an induced paramagnetic ring-current in the four-membered ring, 'quenching' almost exactly the diamagnetic anisotropy of the two benzene nuclei. Full confirmation of this hypothesis has been obtained by Figeys, McOmie and Martin[117] in their theoretical and experimental studies of the proton n.m.r. spectra of biphenylene and some benzobiphenylenes.

Like biphenylene, the benzobiphenylenes are all formal derivatives of cyclobutadiene; all the possible members of this series are shown in Fig. 28. The two monobenzo-biphenylenes XXXI[28, 65, 66, 69, 70] and XXXII[22, 28, 151] are known. The first dibenzobiphenylene tc be synthesised was XXXVII by Curtis and Viswanath[81]; this was soon followed by the synthesis of the isomer XXXVI[28, 68, 69]. Since that time, the linear isomer XXXVII has been described by several research groups[28, 82, 205, 280, 281, 282]. The three other dibenzobiphenylenes XXXIII[29], XXXIV[27, 28] and XXXV[28] were all synthesised recently by Barton *et al.* No tribenzo- or tetrabenzo-biphenylenes have so far been prepared.

Physico-chemical studies on these compounds are few. Their u.v. spectra resemble that of biphenylene, except that the bands are shifted to longer wavelengths (see, e.g., Fig. 29); the available data are gathered in Table 14. Crawford's LCAO calculations showed that the long-wavelength transition in dibenzo[*b*,*h*]biphenylene (XXXVIII) is polarised along the long axis of the molecule[129] and should thus be an α-band[162]. The polarographic half-wave reduction potential of biphenylene and benzo[*b*]biphenylene in anhydrous medium

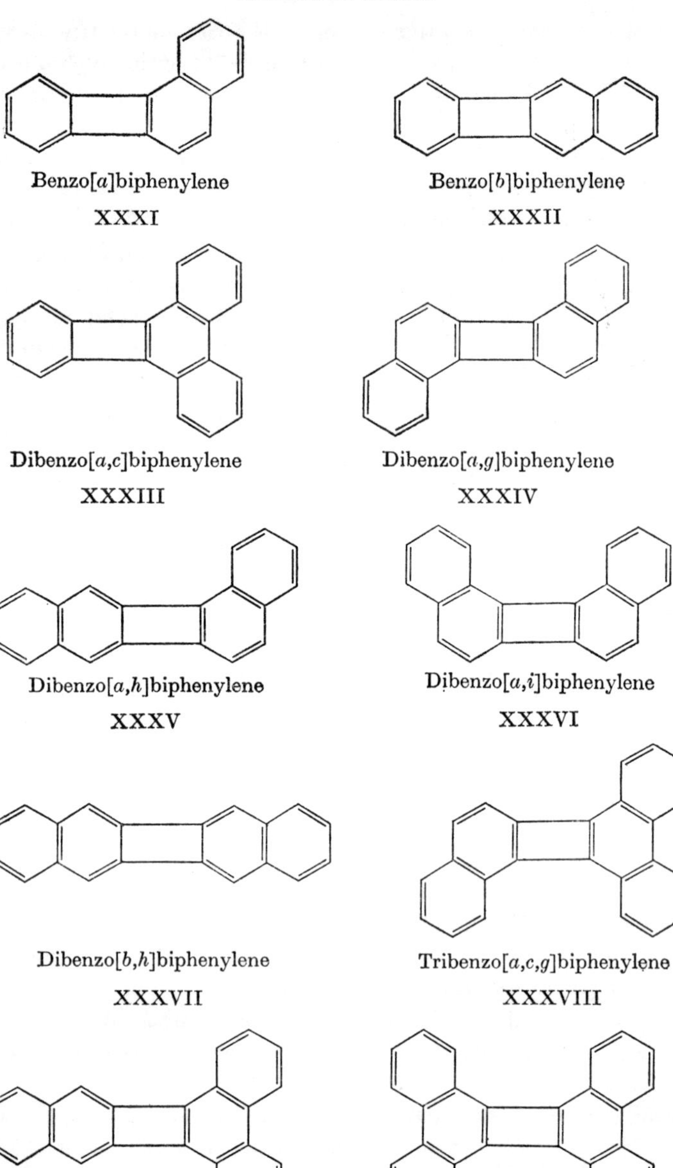

FIG. 28. Possible benzobiphenylenes.

(DMF) has been correlated with the energy of the lowest vacant antibonding orbital in these molecules [266]; it seems that this technique would provide an interesting method for measuring the extent of the conjugation between the two 'benzenoid' residues across the four-membered ring.

Qualitatively, striking differences exist between the properties of these benzobiphenylenes. Thus dibenzo[*a*,*i*]biphenylene decomposes readily at 160°, whereas dibenzo[*b*,*h*]biphenylene is an exceedingly stable substance which sublimes unchanged at 350°. From this and

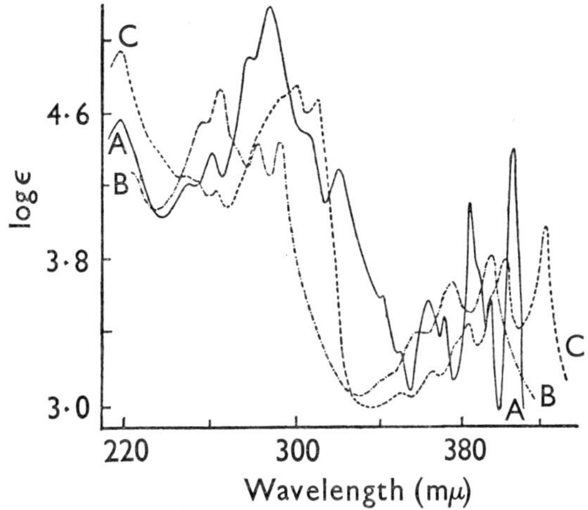

FIG. 29. Ultra-violet spectra of some benzobiphenylenes in ethanol solution. (A) Dibenzo[*b*,*h*]biphenylene; (B) benzo[*a*]biphenylene; (C) dibenzo[*a*,*i*]biphenylene. Reprinted by permission from R. F. Curtis and G. Visvanath, *J. Chem. Soc.* (1959), 1670.

similar observations, it appears that linear annellated benzobiphenylenes are more stable than biphenylene itself, while angular annellation reduces the stability. These conclusions have prompted several theoretical investigations [9, 13, 14, 26, 93, 117, 163] and it has been shown that the observed differences in stability between isomers closely parallels the differences in π-electron delocalisation energy. According to Coulson *et al.*[14], the stability of these compounds should be related to the electron distribution in the region of the central cyclobutadiene-like ring, and particularly to the ease with which the pseudo-single bonds joining the benzenoid portions of the molecule

TABLE 14

U.v.-absorption maxima of benzobiphenylenes

Reference	Biphenylene derivative and solvent	β-band λ(mμ)	β-band $\log_{10} \epsilon$	p-band λ(mμ)	p-band $\log_{10} \epsilon$	α-band λ(mμ)	α-band $\log_{10} \epsilon$
69	benzo[*a*]- (95% ethanol)	254	4·58	279	4·47	359	3·41
		262	4·77	291	4·48	375	3·71
						393	3·87
						331	3·68
22	benzo[*b*]- (95% ethanol)	255	4·87	285	4·40	348	3·71
		264	4·89	296	4·55	367	3·83
						386	3·78
29	dibenzo[*a,c*]- (ethanol)	251·5	4·33	292·5	3·57	359	2·97
		263	4·25	304	4·75	400	3·34
		267·5	4·22	316	4·73	420	3·40
27	dibenzo[*a,g*]- (ethanol)	271.5	4·50	296	4·02	389	3·19
		278	4·65	307·5	4·14	407	3·52
				320	4·12	428	3·59
28	dibenzo[*a,h*]- (ethanol)	288·5	4·87	316·5	4·00	367	3·64
		299	4·96	330	4·01	382	3·83
						405	3·85
69	dibenzo[*a,i*]- (95% ethanol)	247	4·26	298	4·79	350	3·08
		257	4·17	308	4·72	365	3·20
		261	4·20			382	3·47
						400	3·86
						420	4·05
82	dibenzo[*b,h*]- (ethanol)	250	4·23	278	4·93	341	3·61
		257	4·41	288	5·23	349	3·56
				301	4·47	362	3·63
				322	4·31	370	3·57
						380	4·15
						394	3·63
						406	4·43

can be broken. The energy required should be related, according to simple perturbation theory, to the bond-order of these bonds and this argument can then be used to predict the stability of any molecule of the series shown in Fig. 28. The results agree with both the conclusions obtained by considering delocalisation energies for isomers and with qualitative experimental observations; the tetrabenzo-derivative XL has been predicted to be less stable than all the known monobenzo- and dibenzo-biphenylenes[13].

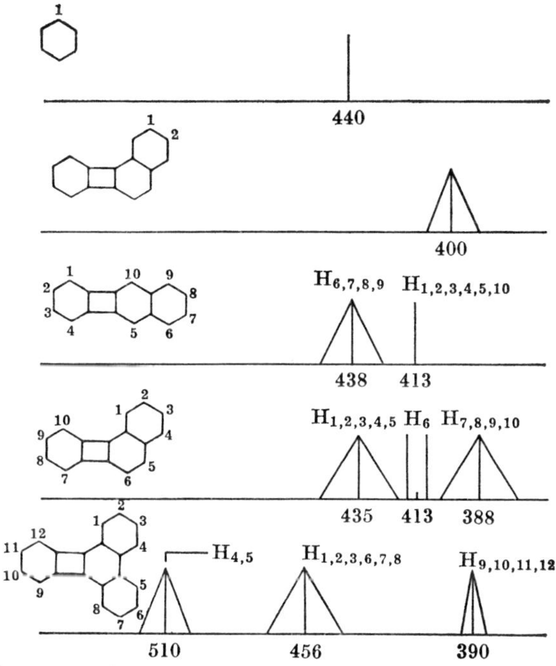

FIG. 30. Diagrammatic representation of the n.m.r. spectra of biphenylene and some of its benzo-derivatives relative to benzene.

The possible anti-aromaticity of the four-membered ring in biphenylene and benzobiphenylenes has been investigated by Dewar and Gleicher[93] from the thermodynamic point of view, and by Figeys, McOmie and Martin[117] according to the criterion based on the magnetic properties and especially the proton n.m.r. spectra. In contrast to the results obtained in the poly-condensed benzenoid hydrocarbons (see Table 6), the n.m.r. signals of the protons of biphenylene appear at higher field than the benzene resonance[119, 156, 182]; the same is true for certain protons of the benzobiphenylenes[117,

[119]. This is shown diagrammatically in Fig. 30 and can *a priori* be due to two different causes: either the occurrence of *very low deshielding ring-currents* in the various rings, or the presence of an *induced paramagnetic shielding ring-current* in the four-membered ring of this series of compounds, as has been observed in the monocyclic (4n) π-electron annulenes (see section C.3).

Experimental evidence for this second hypothesis was first obtained by Figeys[109] by analysing the spectrum of benzo[*b*]biphenyl-

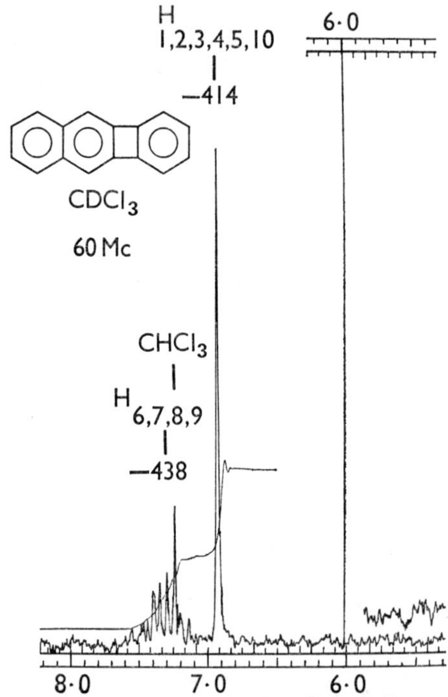

FIG. 31. N.m.r. spectrum of benzo[*b*]biphenylene ($CDCl_3$ solution, 60 Mc/s and TMS as internal standard).

ene which is shown in Fig. 31. This spectrum contains a broad singlet at 6·90 ppm which is attributed, by comparison with the spectra of several substituted derivatives, to protons 1, 2, 3, 4, 5 and 10[182] (see Fig. 32). Protons 6 to 9 give an A_2B_2 multiplet, centred at about 7·30 ppm. However, according to the hypothesis of deshielding ring-currents, $\delta H(1)$ should appear at a much lower field than $\delta H(2)$, whatever the relative values of the individual ring-currents, owing to the closer

TABLE 15

Ring-currents, calculated and observed proton chemical shifts in biphenylenes[117]

Molecule	Ring	Ring-current intensity†	Proton	Chemical shift* Observed	Chemical shift* Calculated
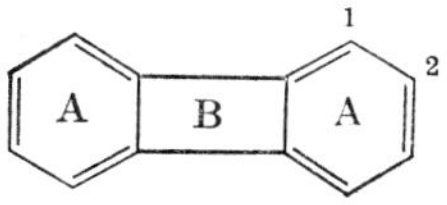	A	+0·563	1	6·598	6·662
	B	−1·028	2	6·702	6·756
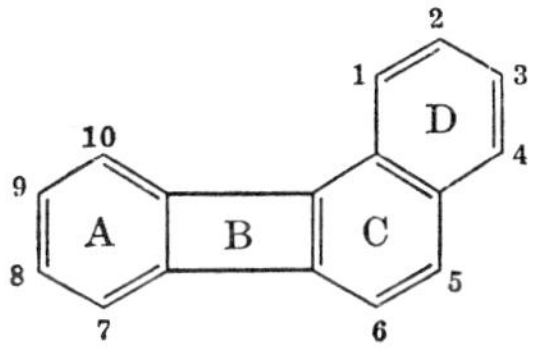	A	+0·374	1	centred at 7·250 (1–5)	7·223
	B	−1·474	2		7·169
	C	+0·705	3		7·177
	D	+0·875	4		7·273
			5		7·090
			6	6·883	6·806
			7	centred at 6·467 (7–10)	6·406
			8		6·526
			9		6·542
			10		6·436
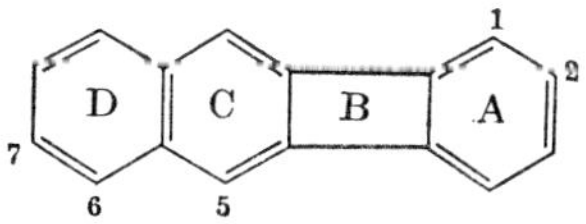	A	+0·578	1	6·907	6·716
	B	−0·949	2	6·907	6·788
	C	+0·577	3	6·907	6·899
	D	+0·929	4	7·427	7·314
			5	7·226	7·241
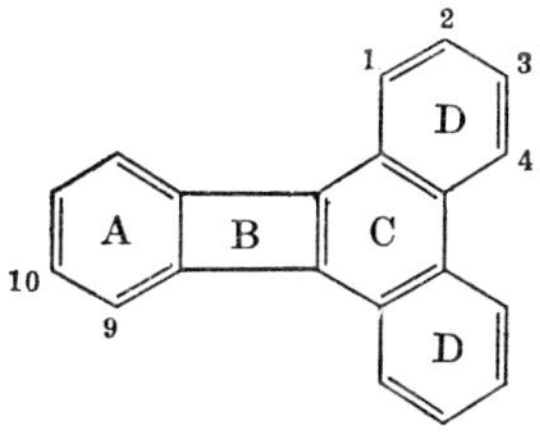	A	+0·285	1	centred at 7·600 (1–3)	7·382
	B	−1·668	2		7·329
	C	+0·720	3		7·352
	D	+0·995	4	8·500	7·580
			9	centred at 6·500 (9–10)	6·337
			10		6·440

* In ppm, TMS as internal standard.

† Ring-current intensity of benzene taken as unity.

proximity of rings II, III and IV. The only way to explain this experimental observation is to postulate a shielding ring-current in the four-membered ring, whose difference of shielding effect on protons 1 and 2 is nearly exactly compensated by the difference in deshielding effect of rings III and IV.

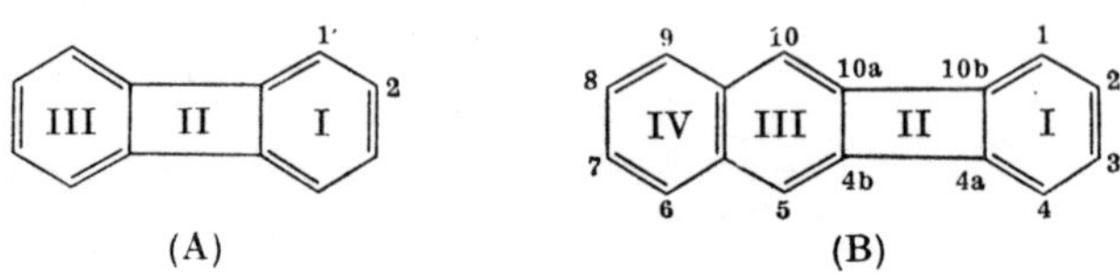

FIG. 32. Numbering of biphenylene and benzo[*b*]biphenylene.

The LCAO calculations of the ring-current effects in biphenylene, benzo[a]-, benzo[b]- and dibenzo[a,c]-biphenylene were made by McWeeny's perturbation theory[179]; the results are shown in Table 15[117]. A striking result is the occurrence of an induced paramagnetic ring-current in the four-membered ring of all these compounds, thus confirming Figeys' hypothesis[109]. As can be seen from Table 15, agreement between theory and experiment is rather good and

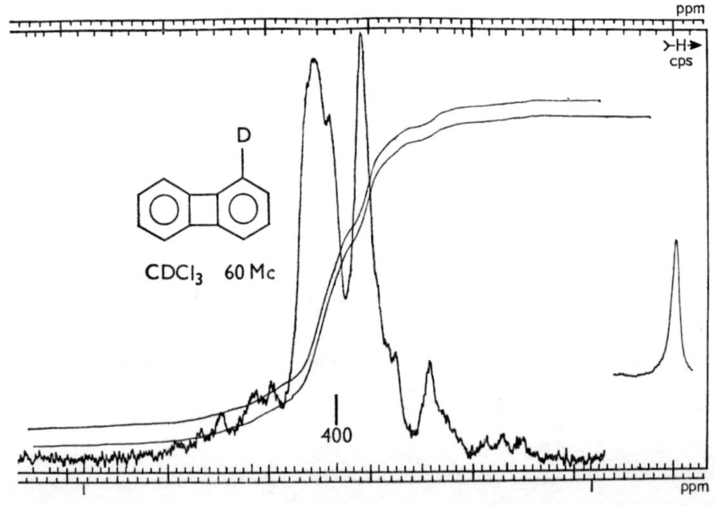

FIG. 33 (a).

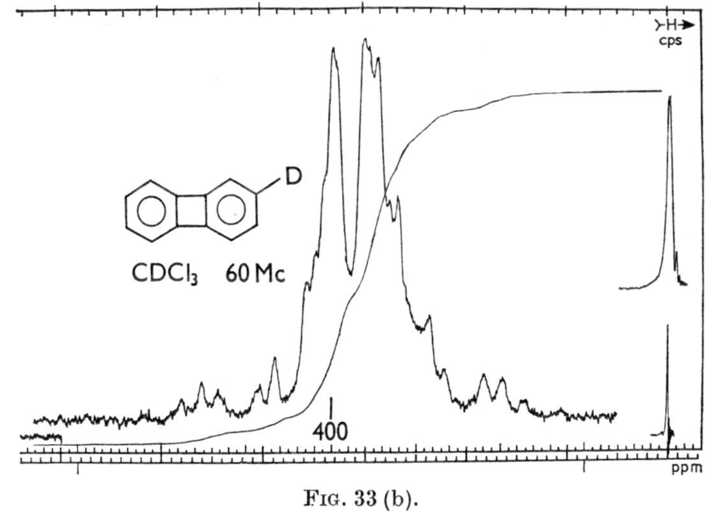

FIG. 33 (b).

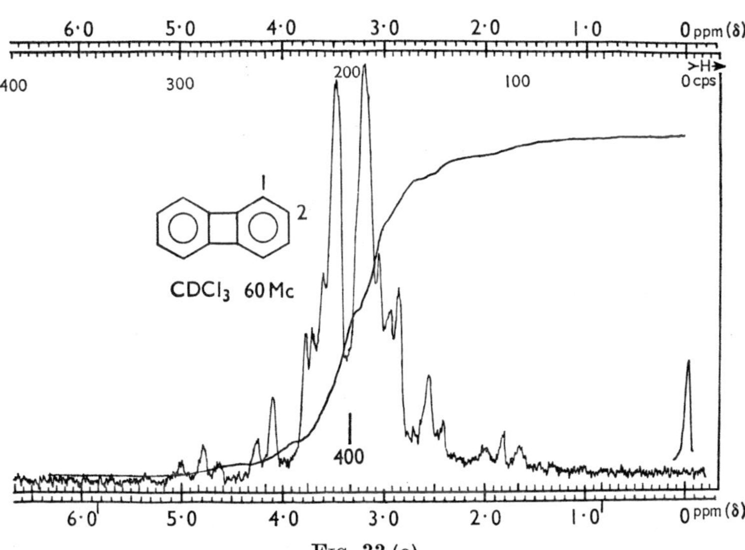

FIG. 33 (c).

FIG. 33. N.m.r. spectra of 1-deuterio-, 2-deuterio- and unsubstituted biphenylene in $CDCl_3$ solution at 60 Mc/s.

previous attributions, e.g. in benzo[*b*]biphenylene[182], are confirmed. It may thus be concluded that even when included in polycyclic compounds, (4*n*) π-electron rings may be regarded as anti-aromatic.

A rather complete theoretical and experimental investigation of the n.m.r. spectrum of biphenylene itself provided further evidence for the presence of an induced paramagnetic ring-current in the central ring of this molecule[117]. As shown in Fig. 33 and 34, the n.m.r. spectrum of biphenylene is a very compact A_2B_2 system from which it is impossible to tell, *a priori*, which of the two sets of protons, α or β, appears at lower field. Katritzky and Reavill[156] tried to assign the individual chemical shifts by selective deuteration with deuterio-trifluoro-acetic acid in CCl_4 at 34° during the determination of the n.m.r. spectrum, but owing to the non-specific character of this reaction[39, 157, 175, 267, 268], no definite conclusion can be obtained by this method. However, from the spectra of biphenylene, 1-deuterio-biphenylene and 2-deuterio-biphenylene shown in Fig. 33, it is clear that deuteration in the 1-position lowers the intensity at the high field side of the spectrum, while the inverse is observed for deuteration in the 2-position. Moreover, in Fig. 34 the n.m.r. spectra of biphenylene and of 2,3,6,7-tetradeuterio-biphenylene (XLI) are compared; in the case of the latter molecule, the signal of the remaining α-protons appears as a broad singlet at the *high-field side* of the A_2B_2 multiplet

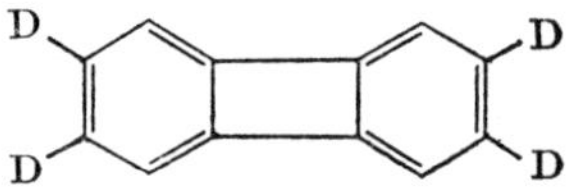

D D D D

XLI

of biphenylene taken under identical conditions. Thus the α-protons are less deshielded than the β-protons, in agreement with the theoretical predictions; the contributions δ_{RC} of the different rings[109] are given in Table 16.

The relation between the electronic structure and the anti-aromaticity of the four-membered ring in biphenylene and benzo-biphenylenes has been investigated by Figeys, McOmie and Martin[117] and by Dewar and Gleicher[93]. The first authors drew attention to the fact that a relation exists between the localisation of the π-electrons *in and around* the four-membered ring (as given by the corresponding bond orders) and the magnitude of the ring-current (Table 17); the results indicate that the shielding ring-current of the four-membered ring increases when its electron distribution goes from the 'tetra-methylenecyclobutane-like' to the 'cyclobutadiene-like' structure. For tetramethylenecyclobutane itself, a ring-current intensity of only

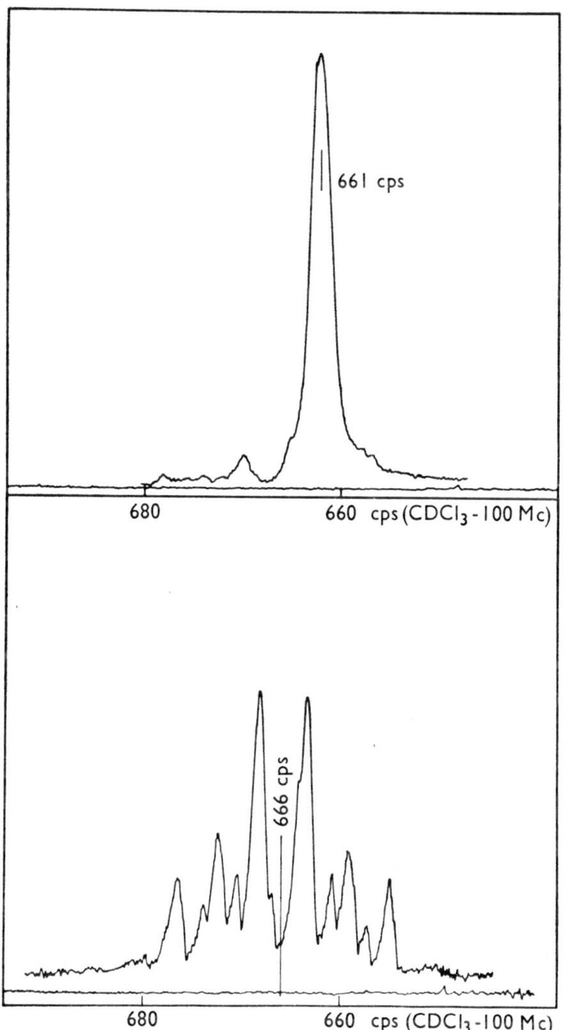

FIG. 34. N.m.r. spectra of 2,3,6,7-d_4-biphenylene (upper) and unsubstituted biphenylene (lower) in $CDCl_3$ solution at 100 Mc/s.

—0·094 benzene units has been calculated, whereas for cyclobutadiene a value of —2·976 was obtained. This conclusion seems to be in excellent agreement with Dewar's stability calculations, showing that tetramethylenecyclobutane is essentially a non-aromatic molecule[91]. Thus it may be concluded that in all the benzo-derivatives so far

TABLE 16

Contributions of the different rings to the calculated chemical shifts of protons 1 and 2 in biphenylene (Fig. 32(a))

Proton	δ_{RC}(ppm) Ring I	Ring II	Ring III	Calculated chemical shift (ppm)
1	+0·648	−0·165	+0·050	6·662
2	+0·648	−0·040	+0·019	6·756

TABLE 17

Mean π-bond-orders and ring-current in the four-membered ring of some biphenylene derivatives

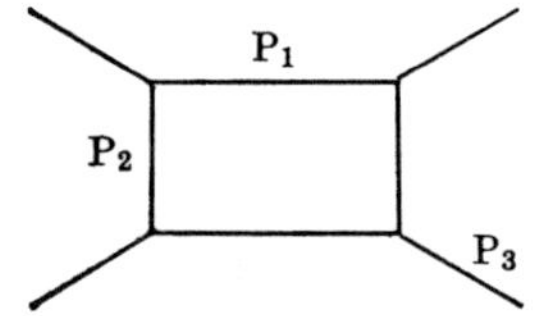

Molecule	Bond-orders P_1	P_2	P_3	Ring-current
Benzo[*b*]biphenylene	0·248	0·542	0·710	−0·949
Biphenylene	0·220	0·579	0·697	−1·028
Benzo[*a*]biphenylene	0·210	0·618	0·644	−1·474
Dibenzo[*a,c*]biphenylene	0·200	0·645	0·608	−1·668

studied experimentally, the cyclobutadiene ring remains anti-aromatic according to the magnetic criterion.

2. *Benzocyclo-octatetraenes and related compounds*

All the possible benzocyclo-octatetraenes are known, except dibenzo[*a,d*]cyclo-octatetraene (XLV) which has only been postulated

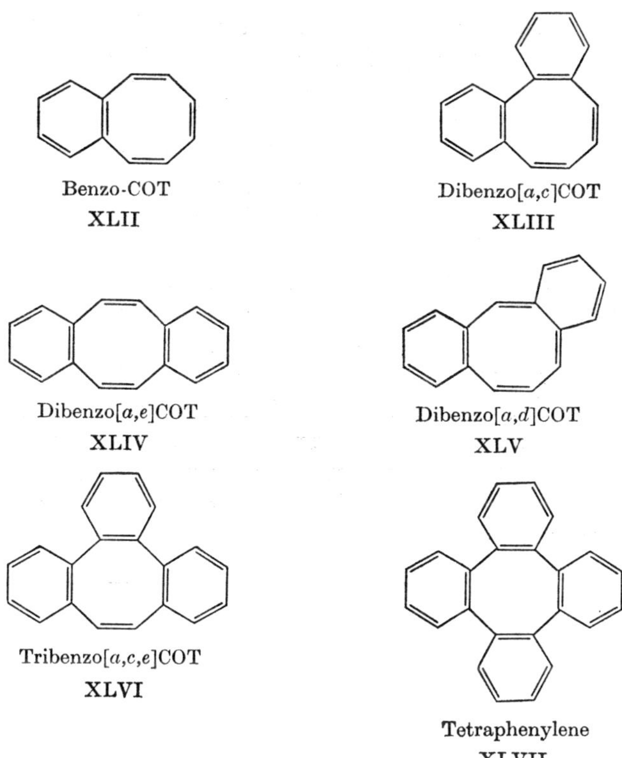

recently as a reaction intermediate[185]. Credit for the first synthesis of an authentic benzocyclo-octatetraene must go to Rapson, Shuttleworth and van Niekerk[217], who in 1943 prepared tetraphenylene (XLVII) in 16 per cent yield from 2,2′-dibromo-biphenyl; the other syntheses of this compound are all due to Wittig *et al.*[294]. Dibenzo-[*a*,e]cyclo-octatetraene (XLIV) was first prepared by Fieser and Pechet[107], but since that time, XLIV has been synthesised by several authors[17, 73, 131, 186, 198, 235, 291]. The structure and physico-chemical properties of these two cyclo-octatetraene derivatives have been studied fairly extensively, but there are few similar studies on the other derivatives XLII[98, 120, 185, 186, 187, 292], XLIII[278] and XLVI[239], although their syntheses were described some time ago.

The u.v. spectra of XLII, XLIV, XLVI and XLVII are shown in Fig. 35; from these spectra, it may be concluded that in the ground state, little interaction exists between the four separate π-electron subsystems (double bonds or benzene rings) and hence, that a non-planar structure for the eight-membered ring in these molecules is probable[176, 216]. This conclusion is in agreement with that derived from the n.m.r. spectra of XLII[99] and XLIV[158], where the resonance frequency of the protons of the double bonds appears in the olefinic region. Similarly, Karle and Brockway showed by electron diffraction study that the eight-membered ring of tetraphenylene (XLVII) has a non-planar 'tub' form with bonds alternating in length around the ring[154]. The shorter bonds are common to the six-membered rings and have a length of $1{\cdot}39 \pm 0{\cdot}02$ Å, the bonds joining the benzene rings have a length of $1{\cdot}52 \pm 0{\cdot}04$ Å and all the C–C–C angles are near 120°. The benzene rings are directed alternately above and below the average plane of the molecule[154]. This structure seems to be in excellent agreement with that predicted by quantum mechanical calculations[236]. A non-planar structure of the eight-membered ring in XLIV is also supported by the olefin-like reactions of the bridging double bonds of this molecule[16, 67].

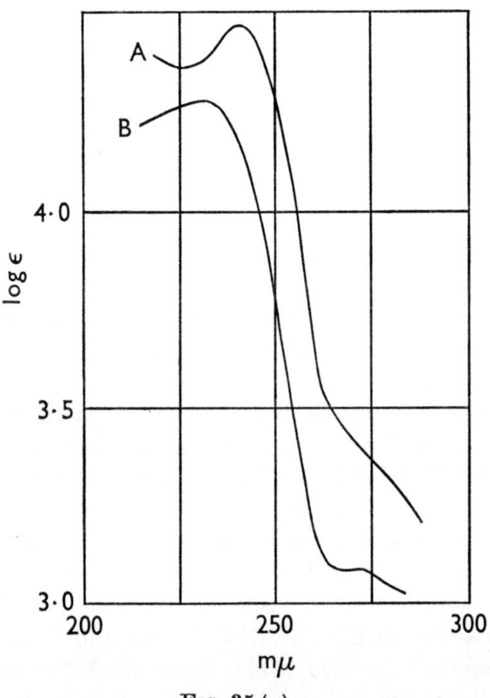

Fig. 35 (a).

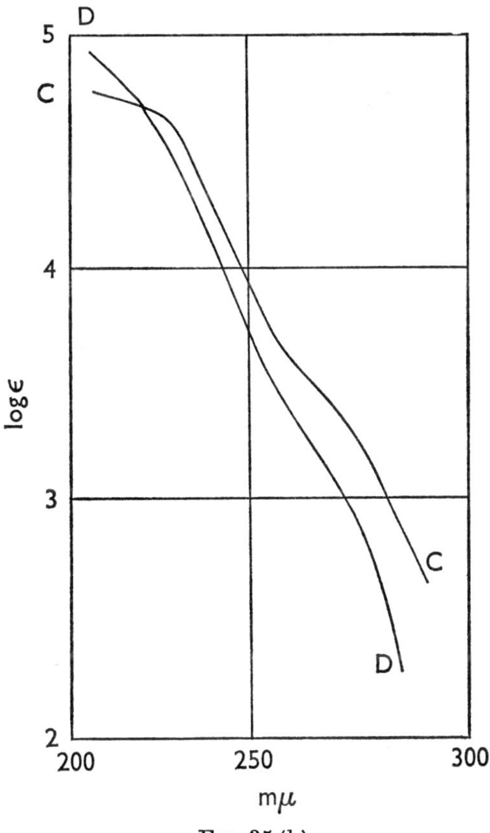

FIG. 35 (b).

FIG. 35. Ultra-violet absorption spectra of benzocyclo-octatetraenes; (A) benzocyclo-octatetraene (cyclohexane); (B) dibenzo[*a*,*e*]cyclo-octatetraene (cyclohexane); (C)tribenzo[*a*,*c*,*e*]cyclo-octatetraene (ethanol); (D) tetraphenylene (ethanol). Reprinted by permission from G. Wittig, H. Eggers and P. Duffner, *Ann.* (1958), **619**, 10 (curves A and B), and from W. S. Rapson, H. M. Schwartz and E. T. Stuart, *J. Chem. Soc.* (1944), 73 (curves C and D).

An attempt to estimate the energy difference between the 'tub' and planar forms of dibenzo[*a*,*e*]cyclo-octatetraene (XLIV) has been made by Mislow and Perlmutter[189, 206] by studying the racemisation of the dissymmetric derivative XLVIII. Resolution of this diacid by fractional crystallisation of its brucine salt afforded (+)- and (—)-2-bromodibenzo[*a*,*c*]cyclo-octatetraene-6,11-dicarboxylic acid, and the rate of racemisation by ring-inversion of the (+)-isomer was measured at several temperatures $t^\circ \lesssim 120^\circ$ in diethylene glycol diethyl ether; the estimated energy of activation[189] is 27 kcal. mole^{-1}. However,

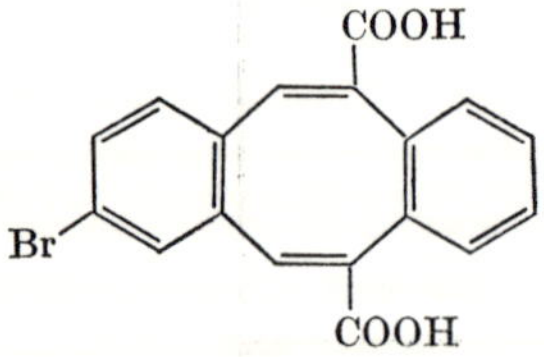

XLVIII

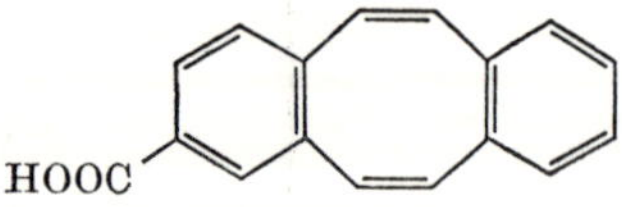

XLIX

inspection of models revealed that non-bonded interactions of the carboxyl groups and benzene hydrogens in the planar transition state are likely to contribute significantly to the destabilisation of this transition state and thus should raise the energy barrier for ring-inversion in this molecule. Attempts to resolve the acid XLIX by similar methods failed, establishing that the racemisation process involves a much lower energy barrier and that ring-inversion is rapid at room temperature[220].

Recently, the compounds L, LI and LII have been synthesised respectively by Breslow *et al.*[46], Sondheimer *et al.*[99] and Krebs[159]; from spectroscopic arguments (u.v., n.m.r.), it may be deduced that here again the eight-membered rings are non-aromatic; in particular

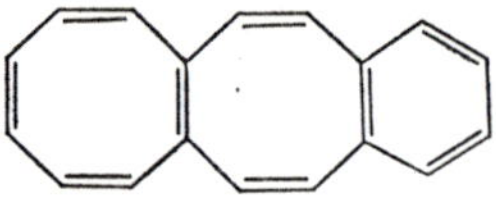

Benzo[*c*]octalene
L

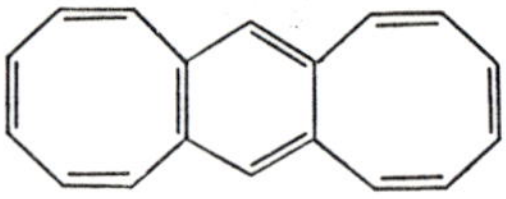

Benzo[1,2:4,5]dicyclo-octatetraene
LI

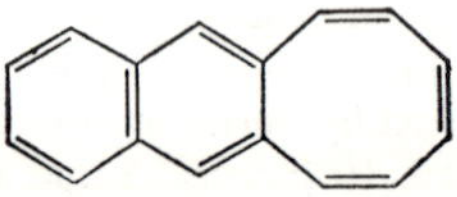

Naphtho[*b*]cyclo-octatetraene
LII

for L and LI, it should be pointed out that two 'wrongs' do not make a 'right' [46, 99]!

3. *Higher Benzo-annulenes and Dehydro-annulenes*

Although [10]annulene has been postulated as a reaction intermediate by several authors[274, 275], it is only in very recent years that the first compounds containing a fully conjugated unsaturated ten-membered ring were synthesised. Shortly after an unsuccessful attempt to obtain a monocyclic [10]annulene derivative by enolisation of cyclodeca-2,4,8-triene-1,6-dione (LIII)[195], Sondheimer *et al.* reported the synthesis of the tribenzo-derivative LIV[133] and of the dinaphtho-[10]annulene LV[190]. Their chemical behaviour and n.m.r. spectra showed that these molecules contained a non-planar and non-delocalised 10π-electron system, in agreement with Mislow's predictions[188]. It should however be pointed out that the bridged 10π-electron system LVI appears to be a $(4n + 2)$ aromatic molecule in all respects[276, 277], but the properties of this molecule will not be reviewed here. Another interesting molecule is 1,6-bisdehydro-[10]annulene* (LVII), a hypothetical planar aromatic compound[269], which however has not yet been prepared.

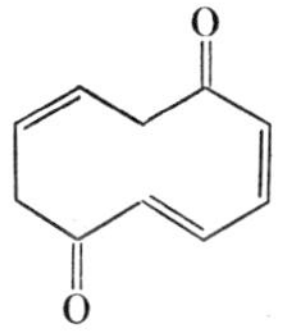

LIII

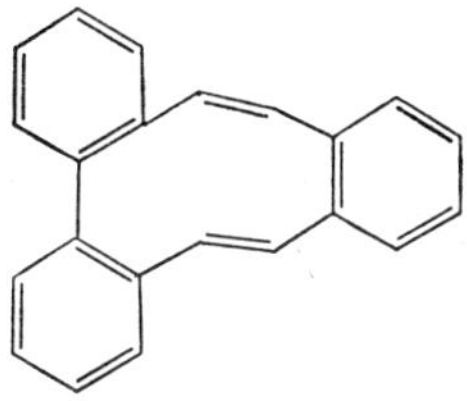

LIV

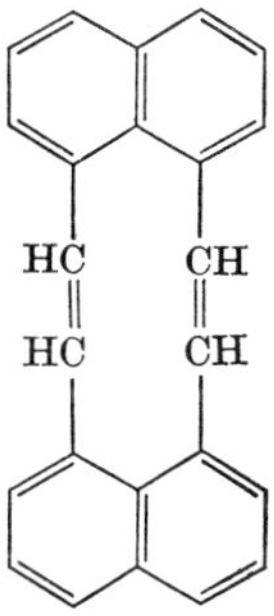

LV

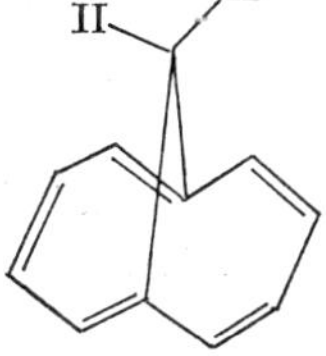

LVI

LVII

In the twelve-membered ring series, several benzo- and dehydro-annulenes are known. The first compound of this family to be prepared was the tetrabenzo-derivative LVIII, obtained by Wittig *et al.*[293]. Models show that this molecule cannot be planar but that it could exist in a strainless buckled *cis-cis* or *trans-trans* form, and indeed these two isomers were actually obtained. Since that time, several other benzo-derivatives have been prepared, namely tribenzo-[261],

* Throughout this article the common usage of bisdehydro, trisdehydro, tetrakisdehydro etc. to mean, respectively, two, three or four etc. triple bonds is employed. Strictly bis-, tris-, tetrakis-dehydro etc. are abbreviations for bis-, tris-, tetrakis-didehydro etc.—Editor.

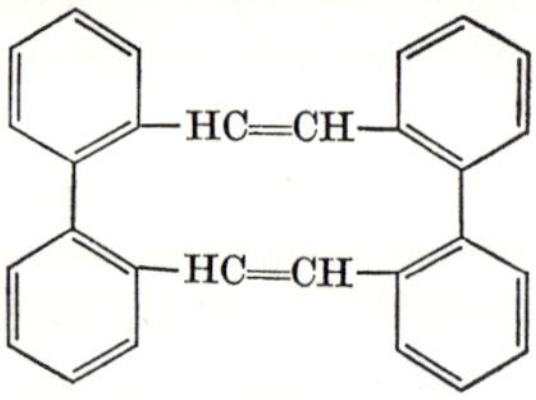

LVIII 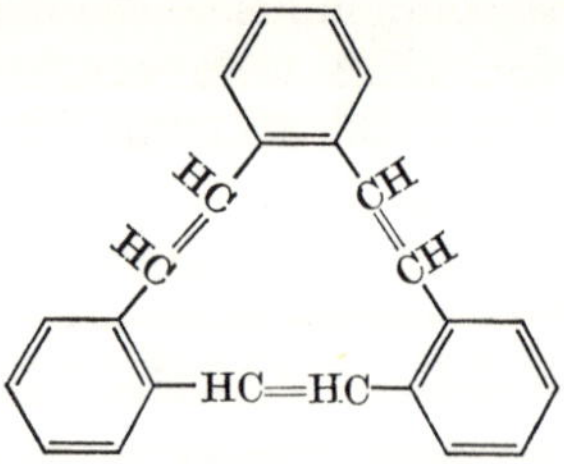 LIX

hexabenzo-[294(a), 294(d)], bisdehydro-tetrabenzo-[262], trisdehydro-tri-benzo-[56, 260] and tetrakisdehydro-dibenzo-[12]annulenes[33] (LIX–

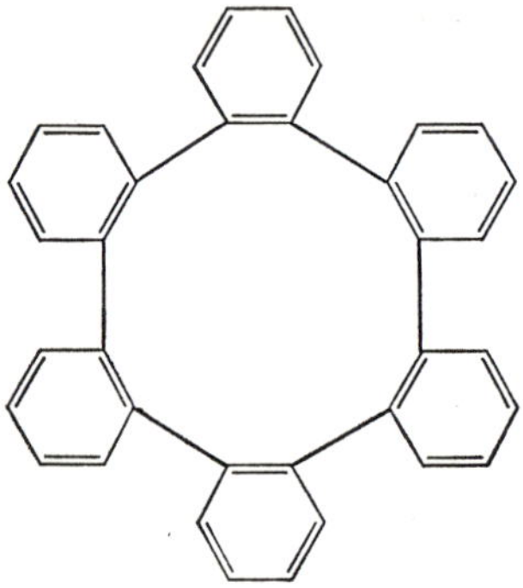

LX

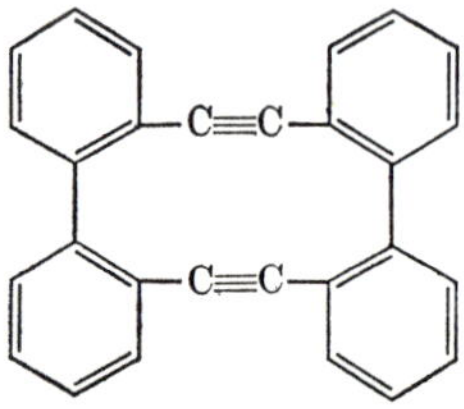

LXI

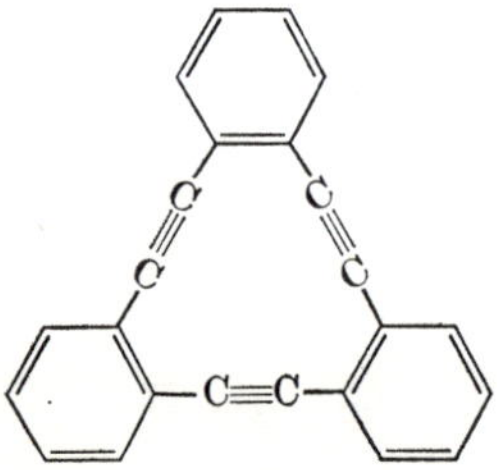

LXII

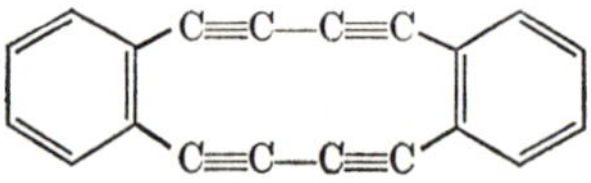

LXIII

LXIII). All these compounds have been shown to have a buckled unconjugated twelve-membered ring, except XLII which may be expected to be planar and perhaps LXIII which has bowed diacetyl-enic chains[128]. In all cases, no conclusions regarding the aromaticity of the 12π-electron ring can be drawn from the available experimental data.

In addition to these benzo-derivatives, a trisdehydro- (LXIV)[256, 271] and a bisdehydro-[12]annulene[297] (LXV) have been synthesised. The

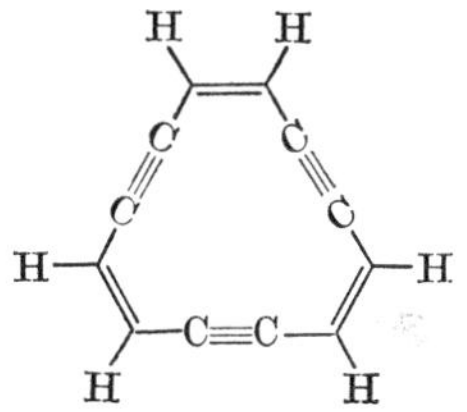

LXIV

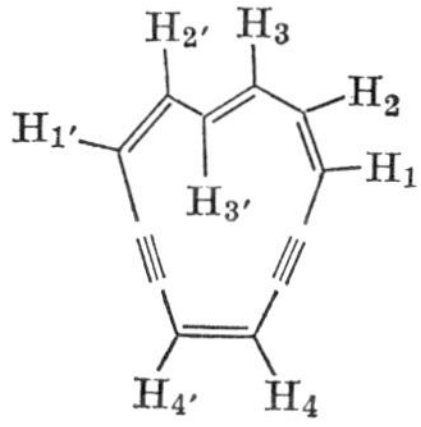

LXV

first of these is a strainless planar molecule whose n.m.r. spectrum consists of only one peak at 4·42 ppm (CCl_4). This comparatively high-field frequency was first discussed by Pople and Untch[212] and was postulated as experimental evidence for the existence of a paramagnetic ring-current in this molecule, although almost the same up-field shift could be due to the anisotropy of the triple bonds. A definite answer came from the comparison of the resonance frequency of the ethylenic protons in *cis*-hex-3-ene-1,5-diyne (LXVI), and in 1,3,7,9,13,15-hexakisdehydro[18]annulene (LXVII) which is a $(4n+2)$ ring system and should have a deshielding diamagnetic ring-current[200].

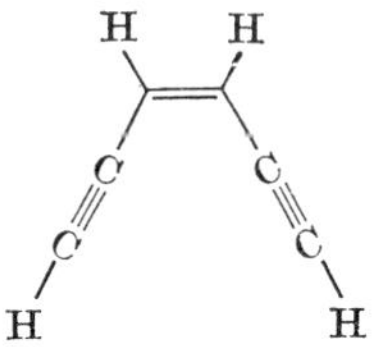

LXVI

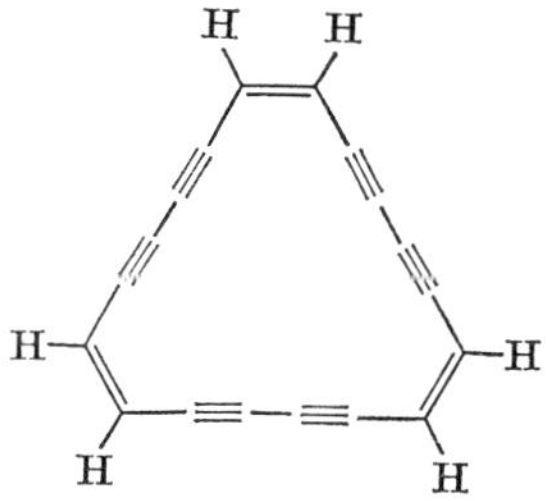

LXVII

The n.m.r. spectrum of LXVII consisted of a singlet at 7·02 ppm, which, compared with the *cis*-ene-diyne molecule XLVI as a model ($\delta_H = 5{\cdot}89$ ppm), constitutes a downfield shift of 1·13 ppm. On the other hand, the upfield shift in LXIV is 1·47 ppm, thus providing strong evidence for a paramagnetic ring-current in this molecule. It may therefore be assumed that this twelve-membered ring is an *anti-aromatic* ring. It should be noted that the ring-current effects in these molecules are rather small compared with the usual chemical shifts obtained for the outer protons of the annulenes themselves at low temperature. Introduction of triple bonds should cause strong bond alternation to occur and accordingly, a lowering of the ring-current intensity may be expected (see Fig. 11). This lowering should be more important for LXVII than for LXIV, in qualitative agreement with the experimental results.

Compound LXV, 1,5-bisdehydro-[12]annulene is a particularly interesting ($4n$) π-electron system, because of the unusually low-field resonance-frequency observed for the two protons of the *trans* double bond[212, 243, 272] at 11·18 ppm (perdeuterioacetone, +37°). The equivalence of these two protons can be achieved by rapid interconversion of two equivalent conformers via rotation of the *trans* double bond[212]. Cooling to −80° resulted in a downfield shift of about 0·4 ppm of the multiplet attributed to these hydrogens, with slight broadening and loss of fine structure[243]. Spectra at still lower temperatures, whereby the non-interconverting conformer might be observed, are not yet available. If proton H(3) is replaced by a bromine atom however, no such conformational mobility can occur, since the bromine atom is too large to move through the ring. Untch and Wysocki[272] examined carefully the n.m.r. spectrum of this compound, 5-bromo-1,9-bisdehydro-[12]annulene, and reported the resonance frequency of H(3′), at 16·4 ppm. This is the lowest resonance-value reported in the literature for a hydrogen bonded to carbon[272], indicating again an induced paramagnetic ring-current in this molecule and suggesting that LXV is planar or nearly so. However, in addition to the ring-current effect, the proximity of the π-electrons of the triple bonds in the same plane and the anisotropy of the *sp*-hybridised carbon atoms will certainly contribute significantly to the observed chemical shift[212].

In the fourteen-membered ring series, only one benzo-derivative LXVIII is known[191]; the product is however so unstable that it could be kept at room temperature in very dilute solution for only 1 hr and it was not possible to make accurate spectroscopic measurements.

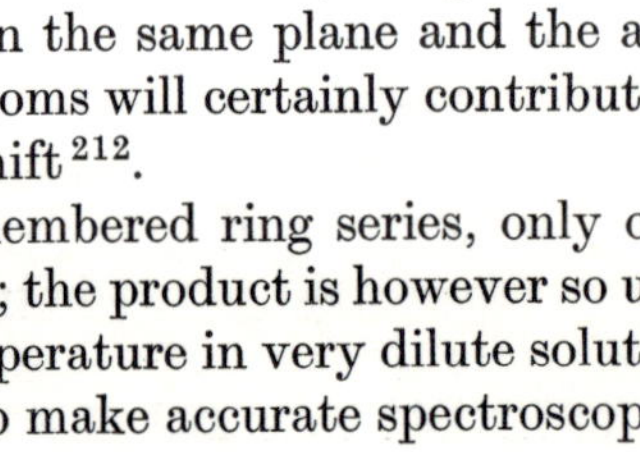

C≡C—C≡C

C≡C—C≡C

LXVIII

Five different dehydro[14]annulenes are known and n.m.r. spectroscopic studies on these molecules confirm the aromaticity of the fourteen-membered conjugated ring[242, 243]. Among them and of special interest is 1,8-bisdehydro-[14]annulene (LXIX), which has

LXIX

the peculiarity that no fully conjugated Kékulé-type formula can be written to represent its delocalised structure. It appears to belong to the same family of planar conjugated dehydro-annulenes as 1,6-bisdehydro-[10]annulene (LVII) and is the first example to be synthesised of this 'new' class of 'aromatic' compounds. Its structure and magnetic properties have been studied extensively by Mason *et al.*[19, 20, 249].

Three-dimensional X-ray analysis (Fig. 36) shows the molecule to be planar and centrosymmetric[20]. With the exception of the two bonds adjacent to the formal triple bonds, the C—C bonds are all of equal length with a mean value of $1{\cdot}398 \pm 0{\cdot}009$ Å, in agreement with simple Hückel calculations[20]. Also of interest is the very short 'non-bonded' distance of 1·85 Å between the internal hydrogen atoms (which corresponds to a Van der Waals radius for H of only $0{\cdot}93 \pm 0{\cdot}04$ Å). The small angular distortions of the molecule from simple diagonal and trigonal hybridisation may surprisingly enough be rationalised if one adopts the view that it is the repulsions between the internal hydrogens and the triple bonds which are important.

The n.m.r. spectrum of LXIX consists of three multiplets at $-5{\cdot}54$ (symmetrical triplet with $J = 13{\cdot}2$ cps), 8·43 (doublet with $J = 8{\cdot}0$ cps) and 10·45 ppm (double doublet), attributed respectively to protons of type H(3), H(1) and H(2)[249]. Proton H(3) has the largest *upfield* shift which is found in the literature for a hydrogen bonded to carbon[272]; obviously, LXIX is an *aromatic* molecule, which even undergoes substitution reactions[123, 242(b)]. This conclusion is strengthened by susceptibility-exaltation measurements[19].

The first large unsaturated ring-system to be prepared was the tetrabenzo-[16]annulene LXX[35, 130], synthesised by Bergman and Pelchowicz as early as 1953[35]. I.r. spectroscopy showed an all-*trans* configuration of the double bonds, the molecule having a buckled but

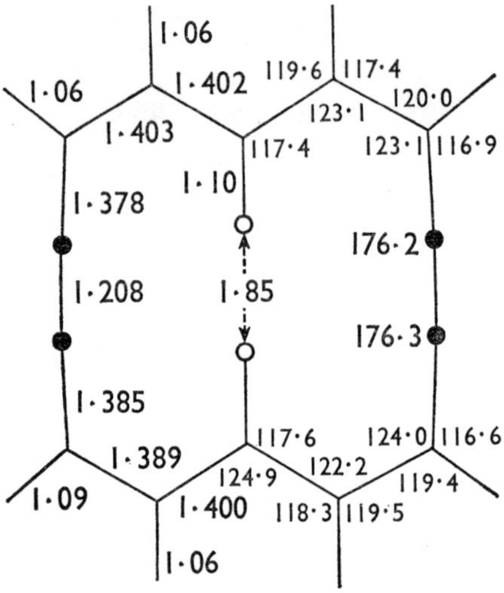

FIG. 36. X-ray structure of 1,8-bisdehydro[14]annulene: bond lengths (Å) and bond angles. Reprinted by permission from N. A. Bailey and R. Mason, *Proc. Roy. Soc.* (1966), **290A**, 94.

unstrained central ring. The corresponding tetrakisdehydro-derivative has also been reported[56] as well as octabenzo-[16]annulene or ortho-

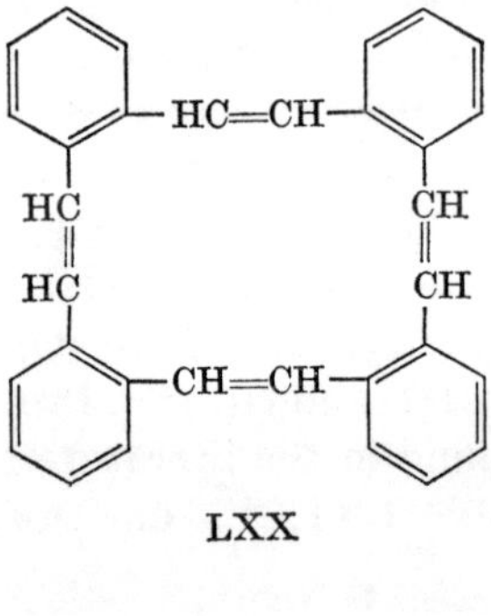

LXX

octaphenylene[294(a)–(c)]. Four dehydro-annulenes in the sixteen-membered ring series are known[242], but only the n.m.r. spectrum of 1,9-bisdehydro-[16]annulene has been investigated at low temperature, confirming the known anti-aromaticity of this $(4n)\pi$-electron system.

In the eighteen-membered ring series, only three benzo-derivatives are known, namely hexa-*m*-phenylene LXXI[258], 3,6′; 3′,6″; 3″,6-triphenanthrylene LXXII[259] and the dianthracenic derivative LXXIII[2]. N.m.r. spectroscopy has shown that the inner rings of LXXI and LXXII do not sustain appreciable ring-currents and this behaviour has been ascribed to strong bond-alternation in these molecules[96].

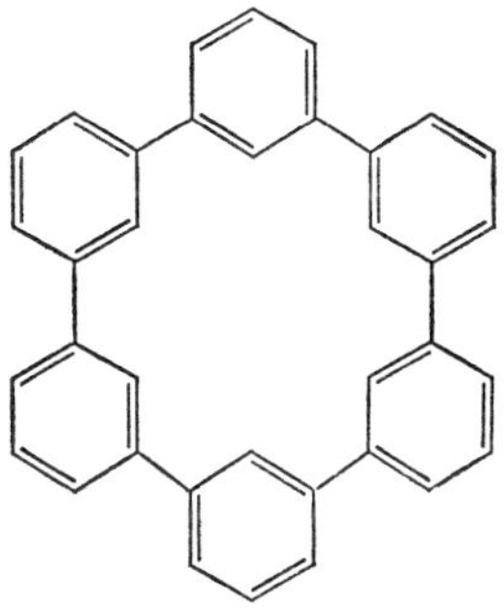

LXXI

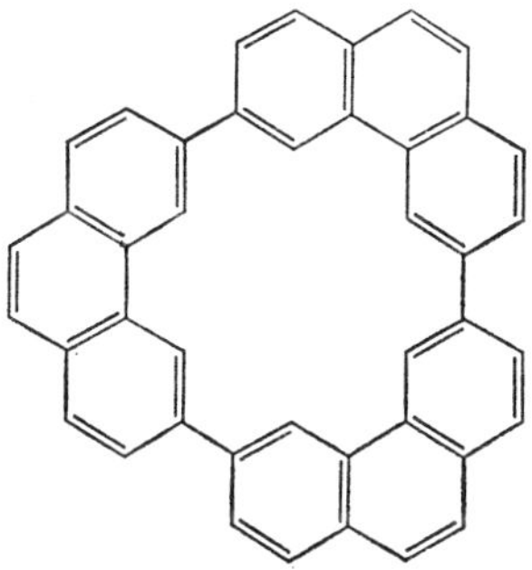

LXXII

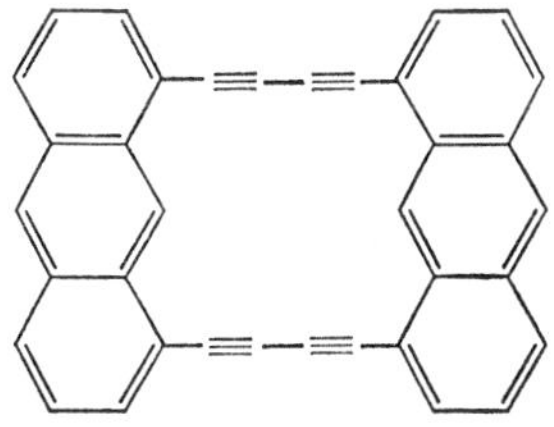

LXXIII

Five different dehydro-[18]annulenes have been synthesised; for all these compounds, n.m.r. spectroscopy has shown the presence of an induced diamagnetic ring-current in agreement with the presence of $(4n + 2)$ out-of-plane π-electrons[199, 200, 242, 296]. Models suggest that these compounds may be planar and their n.m.r. spectra are essentially temperature-independent. Particularly interesting is LXXIV[199], a higher homologue of 1,8-bisdehydro-[14]annulene (LXIX) in the series of 'non-Kekulé' aromatic compounds. In this molecule, the outer protons appear at 9·66 ppm and the inner protons at –5·23 ppm, thus suggesting a much more important ring-current than in LXVII.

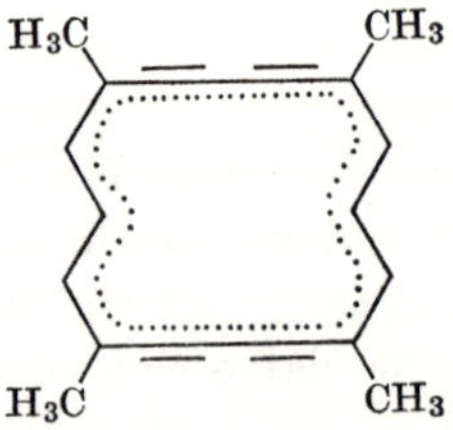

H_3C CH_3

H_3C CH_3

LXXIV

This has been shown to be due to a much more effective quenching of the π-diamagnetic anisotropy by bond-alternation in LXVII then in LXXIV[114].

Besides [20]annulene itself, only a mono-dehydro-[20]annulene[246] and 1,11-bisdehydro-[20]annulene[247] are known in the twenty-membered ring series. The n.m.r. spectrum of the latter compound has recently been investigated at low temperature[243] and the results are shown in Fig. 37. Obviously, 1,11-bisdehydro-[20]annulene is a molecule undergoing fast dynamic interconversion between different conformers at room temperature. Even at –80°, large bands without fine structure remain, indicating that there is still a high rate for the conformational mobility process. Nevertheless, the location of the inner protons H(3) and H(4) at chemical shift values larger than 10 ppm indicates the compounds to be *anti-aromatic*, as expected for a ($4n$) out-of-plane π-electron system.

Up to now, no twenty-two-membered fully conjugated ring system has been synthesised, so that no experimental verification is available for the expected aromaticity of this family. On the other hand, analysis of the n.m.r. spectrum of 1,7,13,19-tetrakisdehydro-[24]-annulene[243] confirms the anti-aromaticity of this ring-system. Experimental information about the physico-chemical properties of the larger rings is scarce. Recently, Leznoff and Sondheimer[166] reported the synthesis of a trisdehydro-[26]annulene, whose n.m.r. spectrum at room temperature exhibited a very broad multiplet going approximately from 5·5 to 8·0 ppm. Cooling to –60° caused essentially no change and no position could be attributed to discrete inner or outer protons. Similar spectra were obtained for a tris-dehydro-[247] and a pentakisdehydro[30]annulene[251], and all resemble closely the spectrum of a long open-chain conjugated polyene-polyyne. This behaviour has been interpreted[166] as indicating *the absence of a measurable ring-current* in the larger $[4n + 2]$annulenes, as a consequence of the bond-alternation process which may have set in (see section C.3). It should however be emphasised that these results

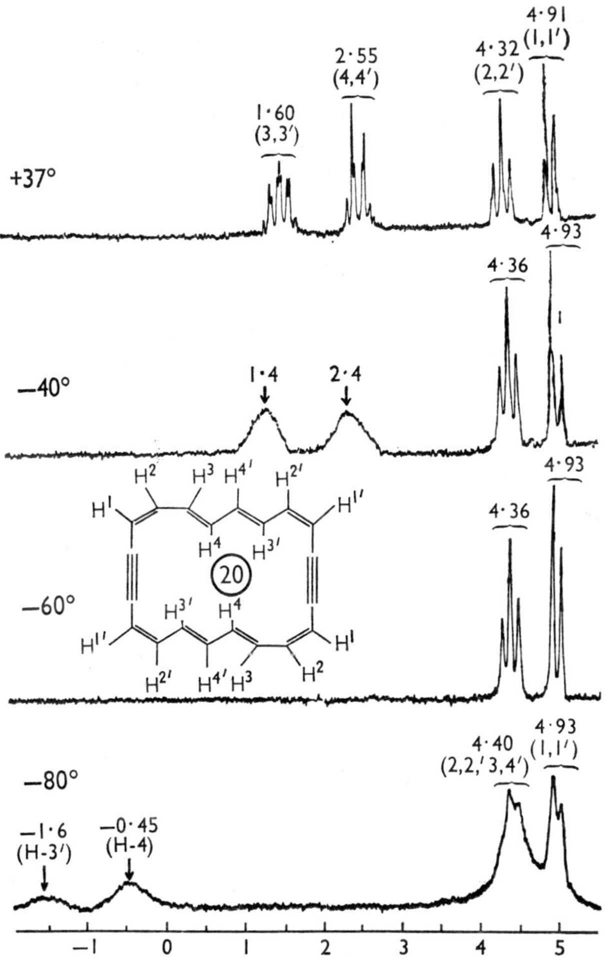

FIG. 37. N.m.r. spectra (100 Mc/s) of 1,11-bisdehydro[20]annulene in perdeuteriotetrahydrofuran. Reprinted by permission from F. Sondheimer *et al.*, in *Aromaticity* (1967), Special Publication No. 21 of The Chemical Society, London, p. 92.

cannot be used *a priori* to indicate the limiting ring-size above which bond-alternation would occur in the corresponding *annulenes*, since it has been shown, for example in the eighteen-membered ring series, that introduction of triple bonds in the ring enhances bond-alternation[114].

The largest benzo-annulene yet synthesised appears to be [2,2,2,2,2,2]*meta*cyclophane-hexaene LXXV, obtained by Jenny and Burri[150]; its n.m.r. spectrum, taken in hexadeuteriodimethylsulphoxide at a concentration of c. 0·1 per cent with a 'CAT' accessory, consists of an AB_2C-multiplet ascribed to the 'benzene' protons with δH_a = 7·39 ppm, δH_b = 7·54 ppm and δH_c = 7·92 ppm, and a sharp singlet at 7·40 ppm attributed to the vinyl protons. Evidently, the molecule has a 'mobile' structure at room temperature, so that no conclusions regarding the aromaticity of the central ring can be obtained from these results. Credit for the synthesis of the largest fully conjugated ring system must however go to Sondheimer and Wolovsky[251], who prepared in 1962 the hexakisdehydro-[36]annulene LXXVI (or an isomer), but no physico-chemical properties of this compound were determined.

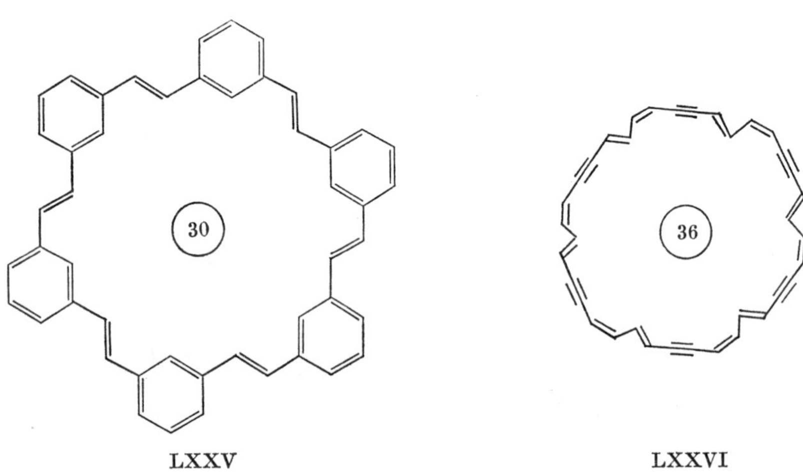

LXXV LXXVI

ACKNOWLEDGMENTS

My greatest debt is to Professor R. H. Martin (University of Brussels), to whom I owe my interest in this field, for several stimulating discussions and continuous encouragement.

I am also deeply grateful to Dr. J. F. W. McOmie (University of Bristol), and Professor F. Sondheimer (University College London) for reading the entire manuscript before publication and for making several critical suggestions.

My thanks are due to Professor M. J. S. Dewar (University of Texas, Austin) for his kind interest in our work.

Finally, I would like to thank the Editor, Douglas M. G. Lloyd (University of St. Andrews) who, in addition to making several pertinent scientific comments, corrected many faults in the English of the original manuscript.

Financial support of the work by a grant from the Fonds National Belge de la Recherche Scientifique (F.N.R.S.) is gratefully acknowledged.

References

1. Abraham, R. J. and Thomas, W. W. (1966) *J. Chem. Soc.* (*B*) 127.
2. Akiyama, S., Misumi, S. and Nakagawa, M. (1960) *Bull. Chem. Soc. Japan* **33**, 1293.
3. Al-Joboury, M. I. and Turner, D. W. (1964) *J. Chem. Soc.* 4434.
4. Allendoerfer, R. D. and Rieger, P. H. (1965) *J. Amer. Chem. Soc.* **87**, 2336.
5. Allinger, N. L. (1962) *J. Org. Chem.* **27**, 443.
6. Allinger, N. L., Gilardeau, C. and Chow, L. W. (1968) *Tetrahedron* **24**, 2401.
7. Allinger, N. L., Miller, M. A., Chow, L. W., Ford, R. A. and Graham, J. C. (1965) *J. Amer. Chem. Soc.* **87**, 3430.
8. Anastassiou, A. G. (1964) *Chem. & Eng. News* **42**, no. 3, 37.
9. Andrade e Silva, M. and Pullman, B. (1956) *Compt. Rend.* **242**, 1888.
10. Anet, F. A. L. (1962) *J. Amer. Chem. Soc.* **84**, 671.
11. Anet, F. A. L., Bourn, A. J. R. and Lin, Y. S. (1964) *J. Amer. Chem. Soc.* **86**, 3576.
12. Anet, F. A. L. and Gregorovitch, B. (1966) *Tetrahedron Letters*, 5961.
13. Asgar Ali, M., Carey, J. G., Cohen, D., Jones, A. J., Millar, I. T. and Wilson, K. V. (1964) *J. Chem. Soc.* 387.
14. Asgar Ali, M. and Coulson, C. A. (1960) *Tetrahedron* **10**, 41.
15. Atkinson, E. R. and Simon, L., unpublished work cited in ref. 62, p. 290.
16. Avram, M., Dinulescu, I. G., Dinu, D., Mateescu, G. and Nenitzescu, C. D. (1963) *Tetrahedron* **19**, 309.
17. Avram, M., Dinu, D. and Nenitzescu, C. D. (1959) *Chemy. Ind.* 257; Avram, M., Dinu, D., Mateescu, G. and Nenitzescu, C. D. (1960) *Chem. Ber.* **93**, 1789.
18. Baer, F., Kuhn, H. and Regel, W. (1967) *Z. Naturforsch.* **22a**, 103.
19. Bailey, N. A., Gerloch, M. and Mason, R. (1966) *Mol. Phys.* **10**, 327.

20. Bailey, N. A. and Mason, R. (1966) *Proc. Roy. Soc. (London)* **290A**, 94.
21. Baker, W., Barton, J. W. and McOmie, J. F. W. (1958) *J. Chem. Soc.* 2666.
22. Baker, W., Barton, J. W., McOmie, J. F. W. and Searle, R. J. G. (1962) *J. Chem. Soc.* 2633.
23. Baker, W. and McOmie, J. F. W. (1959) in D. Ginsburg, *Non-Benzenoid Aromatic Compounds*, Interscience, New York, chap. 2.
24. Baker, W., McOmie, J. F. W., Preston, D. R. and Rogers, V. (1960) *J. Chem. Soc.* 414.
25. Baker, W., McOmie, J. F. W. and Rogers, V. (1958) *Chemy. Ind.* 1236.
26. Barron, T. H. K., Barton, J. W. and Johnson, J. D. (1966) *Tetrahedron* **22**, 2609.
27. Barton, J. W. (1964) *J. Chem. Soc.* 5161.
28. Barton, J. W. and Jones, S. A. (1967) *J. Chem. Soc. (C)* 1276.
29. Barton, J. W., Rogers, A. M. and Barney, M. E. (1965) *J. Chem. Soc.* 5537.
30. Bastiansen, O., Hedberg, L. and Hedberg, K. (1957) *J. Chem. Phys.* **27**, 1311.
31. Bedford, A. F., Carey, J. G., Millar, I. T., Mortimer, C. T. and Springall, H. D. (1962) *J. Chem. Soc.* 3895.
32. Beezer, A. E., Mortimer, C. T., Springall, D. H., Sondheimer, F. and Wolovsky, R. (1965) *J. Chem. Soc.* 216.
33. Behr, O. M., Eglinton, G. and Raphaël, R. A. (1959) *Chemy. Ind.* 699; Behr, O. M., Eglinton, G., Galbraith, A. R. and Raphaël, R. A. (1960) *J. Chem. Soc.* 3614; Behr, O. M., Eglinton, G., Lardy, I. A. and Raphaël, R. A. (1964) *J. Chem. Soc.* 1151.
34. Bergmann, E. D. and Agranat, I. (1966) *Israel J. Chem.* **3**, 197.
35. Bergmann, E. D. and Pelchowicz, Z. (1953) *J. Amer. Chem. Soc.* **75**, 4281.
36. Berthier, G., Mayot, M. and Pullman, B. (1951) *J. Phys. Radium* **12**, 717.
37. Blatchly, J. M., Boulton, A. J. and McOmie, J. F. W. (1965) *J. Chem. Soc.* 4930.
38. Blatchly, J. M., McOmie, J. F. W. and Thatte, S. D. (1962) *J. Chem. Soc.* 5090.
39. Blatchly, J. M. and Taylor, R. (1964) *J. Chem. Soc.* 4641.
40. Blomquist, A. T. and Maitlis, P. M. (1961) *Proc. Chem. Soc.* 232.
41. Blomquist, A. T. and Meinwald, Y. C. (1959) *J. Amer. Chem. Soc.* **81**, 667.
42. Bosshard, H. H. and Zollinger, H. (1961) *Helv. Chim. Acta* **44**, 1985.
43. Bregman, J. (1962) *Nature, Lond.* **194**, 679.
44. Bregman, J., Hirshfeld, F. L., Rabinovich, D. and Schmidt, G. M. J. (1965) *Acta Cryst.* **19**, 227.
45. Breslow, R., Brown, J. and Gajewski, J. J. (1967) *J. Amer. Chem. Soc.* **89**, 4383.
46. Breslow, R., Horspool, W., Sugujama, H. and Vitale, W. (1966) *J. Amer. Chem. Soc.* **88**, 3677.
47. Breslow, R., Kivelevich, D., Mitchell, M., Fabian, W. and Wendel, K. (1965) *J. Amer. Chem. Soc.* **87**, 5132.
48. Briat, B., Schooley, D. A., Records, R., Bunnenberg, E. and Djerassi, C. (1967) *J. Amer. Chem. Soc.* **89**, 7062.
49. Brown, R. D. (1949) *Trans. Faraday Soc.* **45**, 296.
50. Brown, R. D. (1950) *Trans. Faraday Soc.* **46**, 146.
51. Buenker, R. J. and Peyerimhoff, S. D. (1968) *J. Chem. Phys.* **48**, 354.

52. Calder, I. C. and Garratt, P. J. (1967) *J. Chem. Soc.* (*B*) 660.
53. Calder, I. C., Garrat, P. J., Longuet-Higgins, H. C., Sondheimer, F. and Wolovsky, R. (1965) *J. Chem. Soc.* (*C*) 216.
54. Calder, I. C., Garratt, P. J. and Sondheimer, F. (1967) *Chem. Comm.* 41.
55. Calder, I. C. and Sondheimer, F. (1966) *Chem. Comm.* 904.
56. Campbell, I. D., Eglinton, G., Henderson, W. and Raphaël, R. A. (1966) *Chem. Comm.* 87.
57. Cass, R. C., Springall, H. D. and Quincey, P. G. (1955) *J. Chem. Soc.* 1188.
58. Cava, M. P. (1967) in *Aromaticity.* Special Publication no. 21 of the Chemical Society, p. 175.
59. Cava, M. P. and Hwang, B. (1965) *Tetrahedron Letters* 2297.
60. Cava, M. P., Hwang, B. and Van Meter, J. P. (1963) *J. Amer. Chem. Soc.* **85**, 4032.
61. Cava, M. P. and Mangold, D. (1964) *Tetrahedron Letters* 1751.
62. Cava, M. P. and Mitchell, M. J. (1967) *Cyclobutadiene and Related Compounds*, Academic Press, New York, chap. 1.
63. *Ibid.*, chapter 6.
64. *Ibid.*, chapter 10.
65. Cava, M. P. and Napier, D. R. (1956) *J. Amer. Chem. Soc.* **78**, 500.
66. Cava, M. P. and Napier, D. R. (1957) *J. Amer. Chem. Soc.* **79**, 1701.
67. Cava, M. P., Pohlke, R., Erikson, B. W., Rose, J. C. and Fraenkel, G. (1962) *Tetrahedron* **18**, 1005.
68. Cava, M. P. and Stucker, J. F. (1955) *Chemy. Ind.* 446.
69. Cava, M. P. and Stucker, J. F. (1955) *J. Amer. Chem. Soc.* **77**, 6022.
70. Cava, M. P. and Stucker, J. F. (1957) *J. Amer. Chem. Soc.* **79**, 1706.
71. Čížek, J. and Paldus, J. (1967) *J. Chem. Phys.* **47**, 3976.
72. Cope, A. C., Burg, M. and Fenton, S. W. (1952) *J. Amer. Chem. Soc.* **74**, 173.
73. Cope, A. C. and Fenton, S. W. (1951) *J. Amer. Chem. Soc.* **73**, 1668.
74. Cope, A. C. and Kinter, M. R. (1951) *J. Amer. Chem. Soc.* **73**, 3424.
75. Coulson, C. A. (1938) *Proc. Roy. Soc. A* **164**, 383.
76. Coulson, C. A. and Golebiewski, A. (1960) *Tetrahedron* **11**, 125.
77. Coulson, C. A. and Poole, M. D. (1964) *Tetrahedron* **20**, 1859.
77(a). Craig, D. P. (1950) *Proc. Roy. Soc.* **202A**, 498; *J. Chem. Soc.* (1951) 3175 (see ref. 127).
78. Craig, D. P. (1959) *Non-benzenoid Aromatic Compounds*, (ed. D. Ginsburg), Interscience, New York, chap. 1, p. 11.
79. *Ibid.*, p. 27.
79(a). Crawford, V. A. (1952) *Canad. J. Chem.* **30**, 47 (see ref. 129).
80. Criegee, R., Dekker, J., Engel, W., Ludwig, P. and Noll, K. (1963) *Chem. Ber.* **96**, 2362.
81. Curtis, R. F. and Viswanath, G. (1954) *Chemy. Ind.* 1174.
82. Curtis, R. F. and Viswanath, G. (1959) *J. Chem. Soc.* 1670.
83. Dailey, B. P. (1964) *J. Chem. Phys.* **41**, 2304.
84. Dauben, Jr., H. J., Wilson, J. D. and Laity, J. L. (1968) *J. Amer. Chem. Soc.* **90**, 811.
85. Daudel, R., Lefebvre, R. and Moser, C. (1959) *Quantum Chemistry: Methods and Applications*, Wiley, New York.
86. Davies, D. W. (1959) *Tetrahedron Letters* No. 8, 4.

87. Dewar, M. J. S. (1965) in I. Prigogine (ed.) *Advances in Chemical Physics*, Vol. 8. Interscience, London, p. 120.
88. Dewar, M. J. S. (1967) in *Aromaticity*, Special Publication No. 21 of the Chemical Society, London, p. 177.
89. *Ibid.*, p. 195.
90. *Ibid.*, p. 196.
91. Dewar, M. J. S. and Gleicher, G. J. (1965) *J. Amer. Chem. Soc.* **87**, 692.
92. Dewar, M. J. S. and Gleicher, G. J. (1965) *J. Amer. Chem. Soc.* **87**, 3255.
93. Dewar, M. J. S. and Gleicher, G. J. (1962) *Tetrahedron* **21**, 1817.
94. Dixon, W. T. (1962) *Tetrahedron* **18**, 875.
95. Eccleston, B. H., Colemann, H. J. and Adams, N. G. (1950) *J. Amer. Chem. Soc.* **72**, 3866.
96. Ege, G. and Fisher, H. (1967) *Tetrahedron* **23**, 149.
97/8. Elix, J. A., Sargent, M. V. and Sondheimer, F. (1966) *Chem. Comm.* 508.
99. Elix, J. A., Sargent, M. V. and Sondheimer, F. (1966) *Chem. Comm.* 509.
100/1. Emerson, G. F., Watts, L. and Pettit, R. (1965) *J. Amer. Chem. Soc.* **87**, 131.
102. Emsley, J. W., Feeney, J. and Sutcliffe, L. H. (1966) *High Resolution Nuclear Magnetic Resonance Spectroscopy*, Pergamon Press, Oxford, chap. 10, pp. 749–782.
103. *Ibid.*, p. 713.
104. Farnum, D. G., Atkinson, E. R. and Lothrop, W. C. (1961) *J. Org. Chem.* **26**, 3204.
105. Fawcett, J. K. and Trotter, J. (1966) *Acta Cryst.* **20**, 87.
106. Fernandez Alonso, J. I. and Domingo, R. (1955) *An. Real Soc. Espan., fis. y quim.* (*Madrid*) **51B**, 447.
107. Fieser, L. F. and Pechet, M. M. (1946) *J. Amer. Chem. Soc.* **68**, 2577.
108. Figeys, H. P. (1966) *Tetrahedron Letters* 625.
109. Figeys, H. P. (1967) *Chem. Comm.* 495.
110. Figeys, H. P., to be published.
111. Figeys, H. P., to be published.
112. Figeys, H. P., to be published.
113. Figeys, H. P., unpublished results.
114. Figeys, H. P., unpublished results.
115. Figeys, H. P. and Dedieu, P. (1967) *Theor. Chim. Acta* **9**, 82.
116. Figeys, H. P., McOmie, J. F. W. and Martin, R. H., to be published.
117. Figeys, H. P., McOmie, J. F. W. and Martin, R. H., to be published.
118. Figeys, H. P., De Smedt, R. and McOmie, J. F. W., to be published.
119. Fraenkel, G., Asahi, Y., Mitchell, M. J. and Cava, M. P. (1964) *Tetrahedron* **20**, 1179.
120. Friedman, L. (1967) *J. Amer. Chem. Soc.* **89**, 3071.
121. Fukui, K., Imamura, A., Yonezawa, T. and Nagata, C. (1960) *Bull. Chem. Soc. Japan* **33**, 1591.
122. Gaoni, Y., Melera, A., Sondheimer, F. and Wolovsky, R. (1964) *Proc. Chem. Soc.* 397.
123. Gaoni, Y. and Sondheimer, F. (1964) *J. Amer. Chem. Soc.* **86**, 521.
124. Gaoni, Y. and Sondheimer, F. (1964) *Proc. Chem. Soc.* 299.

125. Gompper, R. and Seybold, G. (1968) *Angew. Chem.* **80**, 804; *Angew. Chem. Internat. Ed.* **7**, 824 (1968).
125(a). Gouterman, M. and Wagnière, G. (1960) *Tetrahedron Letters* No. 11, 22 (see ref. 136).
126. Gouterman, M. and Wagnière, G. (1962) *J. Chem. Phys.* **36**, 1188.
127. Craig, D. P. (1950) *Proc. Roy. Soc.* **202A**, 498; *J. Chem. Soc.* (1951) 3175.
128. Grant, W. K. and Speakman, J. C. (1959) *Proc. Chem. Soc.* 231.
129. Crawford, V. A. (1952) *Canad. J. Chem.* **30**, 47.
130. Griffin, C. E., Martin, K. R. and Douglas, B. E. (1962) *J. Org. Chem.* **27**, 1627.
131. Griffin, C. E. and Peters, J. A. (1963) *J. Org. Chem.* **28**, 1715.
132. Griffin, G. W. and Peterson, L. I. (1963) *J. Amer. Chem. Soc.* **85**, 2268.
133. Grohman, K. and Sondheimer, F. (1967) *J. Amer. Chem. Soc.* **89**, 7119.
134. Gutowsky, H. S. and Holm, C. H. (1956) *J. Chem. Phys.* **25**, 1228.
135. Gwynn, D. E., Whitesides, G. M. and Roberts, J. D. (1965) *J. Amer. Chem. Soc.* **87**, 2862.
136. Gouterman, M. and Wagnière, G. (1960) *Tetrahedron Letters* No. 11, 22.
137. Haberditzl, W. (1966) *Angew. Chem.* (*Internat. Ed.*) **5**, 288.
138/9. Hobey, W. D. and McLachlan, A. D. (1960) *J. Chem. Phys.* **33**, 1695.
140. Hochstrasser, R. M. (1961) *Canad. J. Chem.* **39**, 765.
141. Hoffmann, R. and Woodward, R. B. (1965) *J. Amer. Chem. Soc.* **87**, 4388.
142. Hilpern, J. W. (1965) *Mol. Phys.* **9**, 295.
143. Hirshfeld, F. L. and Rabinovich, D. (1965) *Acta Cryst.* **19**, 235.
144. Hückel, E. (1931) *Z. Physik* **70**, 204; **72**, 310.
145. Huntsman, W. D. and Wristers, H. J. (1963) *J. Amer. Chem. Soc.* **85**, 3308.
146. Huisgen, R. and Mietzsch, F. (1964) *Angew. Chem.* **76**, 36.
147. Hush, N. S. and Rowlands, J. R. (1963) *Mol. Phys.* **6**, 317.
148. Jackman, L. M., Sondheimer, F., Amiel, Y., Ben-Efraim, D. A., Gaoni, Y., Wolovsky, R. and Bothner-By, A. A. (1962) *J. Amer. Chem. Soc.* **84**, 4307.
149. Jahn, H. A. and Teller, E. (1937) *Proc. Roy. Soc.* **161A**, 220.
150. Jenny, W. and Burri, K. (1967) *Chimia* **21**, 534.
151. Jensen, F. R. and Coleman, W. E. (1959) *Tetrahedron Letters* No. 20, 7.
152. Johnson, Jr., C. E. and Bovey, F. A. (1958) *J. Chem. Phys.* **29**, 1012.
153. Jonathan, N., Gordon, S. and Dailey, B. P. (1962) *J. Chem. Phys.* **36**, 2443.
154. Karle, I. L. and Brockway, L. O. (1944) *J. Amer. Chem. Soc.* **66**, 1974.
155. Karplus, M. (1959) *J. Chem. Phys.* **30**, 11.
156. Katritzky, A. R. and Reavill, R. E. (1964) *Rec. Trav. Chim.* **83**, 1230.
157. Katritzky, A. R. (1968) personal communication.
158. Katz, T. J., Yoshida, M. and Siew, L. C. (1965) *J. Amer. Chem. Soc.* **87**, 4516.
159. Krebs, A. (1965) *Angew. Chem.* **77**, 966; Krebs, A. and Byrd, D. (1967) *Ann.* **707**, 66.
160. Labarre, J. F. and Chalvet, O. (1967) *Tetrahedron Letters* 5053.
161. Labarre, J. F. and Crasnier, F. (1967) *J. Chim. Phys.* **64**,1664.
162. Layton, Jr., E. M. (1960) *J. Mol. Spectroscopy* **5**, 181.
163. Lee, H. S. (1962) *Chemistry* (*Taiwan*) No. 4, 137.
164. Lee, H. S. (1963) *Chemistry* (*Taiwan*) No. 1, 22.
165. Lennard-Jones, J. E. and Turkevich, J. (1937) *Proc. Roy. Soc.* **158A**, 297.

166. Leznoff, C. A. and Sondheimer, F. (1967) *J. Amer. Chem. Soc.* **89**, 4247.
167. Liehr, A. D. (1956) *Z. phys. Chem.* **9**, 338.
168. Lloyd, D. (1966) *Carbocyclic Non-benzenoid Aromatic Compounds*, Elsevier, Amsterdam, p. 6.
169. London, F. (1937) *J. Phys. Radium* **8**, 397.
170. Longuet-Higgins, H. C. (1957) *Proc. Chem. Soc.* 161.
171. Longuet-Higgins, H. C. (1967) in *Aromaticity*, Special Publication No. 21 of the Chemical Society, p. 109.
172. Longuet-Higgins, H. C. and Salem, L. (1959) *Proc. Roy. Soc.* **251A**, 172.
173. Longuet-Higgins, H. C. and Salem, L. (1960) *Proc. Roy. Soc.* **257A**, 445.
174. Lothrop, W. C. (1941) *J. Amer. Chem. Soc.* **63**, 1187.
175. Lunelli, B. and Pecile, C. (1968) *Canad. J. Chem.* **46**, 391.
176. McEwen, K. L. and Longuet-Higgins, H. C. (1956) *J. Chem. Phys.* **24**, 771.
177. McOmie, J. F. W. (1962) *Rev. Chim. Acad. Rep. Populaire Roumanie* **7**, 1071.
178. McWeeny, R. (1958) *Mol. Phys.* **1**, 311.
179. McWeeny, R. (1958) *Mol. Phys.* **1**, 311; Veillard, A. (1962) *J. Chim. Phys.* **59**, 1056; Memory, J. D. (1963) *J. Chem. Phys.* **38**, 1341.
180. McWeeny, R. (1954) *Proc. Roy. Soc.* **223A**, 63, 306; **227A**, 228 (1955).
181. Magnus, A., Hartmann, H. and Becker, F. (1951) *Z. Phys. Chem.* **197**, 75.
182. Martin, R. H., Van Trappen, J. P., Defay, N. and McOmie, J. F. W. (1964) *Tetrahedron* **20**, 2373.
183. Mataga, N. and Nishimoto, K. (1957) *Z. Phys. Chem.* (*Frankfurt*) **13**, 140.
184. Memory, J. D. (1963) *J. Chem. Phys.* **38**, 1341.
185. Merk, W. and Pettit, R. (1967) *J. Amer. Chem. Soc.* **89**, 4787.
186. Merk, W. and Pettit, R. (1967) *J. Amer. Chem. Soc.* **89**, 4788.
187. Miller, R. G. and Stiles, M. (1963) *J. Amer. Chem. Soc.* 1798.
188. Mislow, K. (1952) *J. Chem. Phys.* **20**, 1489.
189. Mislow, K. and Perlmutter, H. D. (1962) *J. Amer. Chem. Soc.* **84**, 3591.
190. Mitchell, R. H. and Sondheimer, F. (1968) *J. Amer. Chem. Soc.* **90**, 530.
191. Mitchell, R. H. and Sondheimer, F. (1968) *Tetrahedron* **24**, 1397.
192. Miyakawa, S., Tanaka, I. and Uemura, T. (1951) *Bull. Chem. Soc. Japan* **24**, 136.
193. Moffitt, W. and Scanlau, J. (1953) *Proc. Roy. Soc.* **220A**, 530.
194. Moffitt, W. and Thorson, W. R. (1957) *Phys. Rev.* **108**, 1251.
195. Mulligan, P. J. and Sondheimer, F. (1967) *J. Amer. Chem. Soc.* **89**, 7118.
196. Murrell, J. N. and Hinchliffe, A. (1966) *Trans. Faraday Soc.* **62**, 2011.
197. Nakajima, T. and Kohda, S. (1966) *Bull. Chem. Soc. Japan* **39**, 804.
198. Nozaki, N. and Noyori, R. (1966) *Tetrahedron* **22**, 2163.
198(a). Öpik, U. and Pryce, M. H. L. (1957) *Proc. Roy. Soc.* **238A**, 425 (see ref. 273).
199. Ojima, J., Natakami, T., Nakaminami, G. and Nakagawa, M. (1968) *Tetrahedron Letters* 1115.
200. Okamura, W. H. and Sondheimer, F. (1967) *J. Amer. Chem. Soc.* **89**, 5991.
201. Oth, J. F. M., Merényi, R., Martini, Th. and Schröder, G. (1966) *Tetrahedron Letters* 3087.
202. Oth, J. F. M. (1968) unpublished results.
203. Pauling, L. (1936) *J. Chem. Phys.* **4**, 673.

204. Pauncz, R., de Heer, J. and Löwdin, P.-O. (1962) *J. Chem. Phys.* **36**, 2247, 2257; de Heer, J. and Pauncz, R. (1960) *J. Mol. Spectr.* **5**, 326.
205. Pearson, B. D. (1960) *Chemy. Ind.* 899.
206. Perlmutter, H. D. (1965) *Diss. Abstr.* **26**, 1351.
207. Polansky, O. E. (1960) *Monatsh.* **91**, 916.
208. Pople, J. A. (1958) *Mol. Phys.* **1**, 175.
209. Pople, J. A. (1964) *J. Chem. Phys.* **41**, 2559.
210. Pople, J. A. and Santry, D. P. (1965) *Mol. Phys.* **9**, 302.
211. Pople, J. A., Schneider, W. G. and Bernstein, H. J. (1959) *High Resolution Nuclear Magnetic Resonance*, McGraw-Hill, New York, pp. 180–183 and 247–265.
212. Pople, J. A. and Untch, K. G. (1966) *J. Amer. Chem. Soc.* **88**, 4811.
213. Pritchard, H. O. and Sumner, F. H. (1954) *Proc. Roy. Soc.* **126A**, 128.
214. Prosen, E. J., Johnson, W. H. and Rossini, F. D. (1950) *J. Amer. Chem. Soc.* **72**, 626.
215. Pullman, B. and Pullman, A. (1952) *Les théories électroniques de la chimie organique*, Masson, Paris, Chap. IX.
216. Rapson, W. S., Schwartz, H. M. and Stewart, E. T. (1944) *J. Chem. Soc.* 73.
217. Rapson, W. S., Shuttleworth, R. G. and van Niekerk, J. N. (1943) *J. Chem. Soc.* 326.
218. Reppe, W., Shlichting, O., Klager, K. and Toepel, T. (1948) *Ann.* **560**, 1.
219. Reppe, W. *et al.* (1951) summarized by Cope, A. C. and Fenton, S. W. *J. Amer. Chem. Soc.* **73**, 1195.
220. Riccobono, R. X. (1965) *Diss. Abstr.* **26**, 97.
221. Roberts, J. D. (1963) *Angew. Chem.* **75**, 20.
222. Roberts, J. D., Kline, G. B. and Simmons, H. E. (1953) *J. Amer. Chem. Soc.* **75**, 4765.
223. Roberts, J. D., Streitwieser, Jr., A. and Regan, C. M. (1952) *J. Amer. Chem. Soc.* **74**, 4579.
224. Rosenblum, M. and Gatsonis, C. (1967) *J. Amer. Chem. Soc.* **89**, 5074.
225. Salem, L. (1961) *Proc. Cambridge Phil. Soc.* **57**, 353.
226. Salem, L. (1966) *The molecular orbital theory of conjugated systems*, Benjamin, New York, pp. 110–127.
227. *Ibid.*, p. 467.
228. *Ibid.*, p. 486.
229. *Ibid.*, chapter 4.
230. Sauer, J., Wiest, H. and Mielert, A. (1964) *Chem. Ber.* **97**, 3183.
231. Scheiner, P. and Vaughan, W. R. (1961) *J. Org. Chem.* **26**, 1923.
232. Schiess, P. and Pullman, A. (1956) *J. Chim. Phys.* **53**, 101.
233. Schröder, G. (1965) *Cyclooctatetraen*, Verlag Chemie.
234. Schröder, G. and Oth, J. F. M. (1966) *Tetrahedron Letters* 4083.
235. Schröder, G., Martin, W. and Oth, J. F. M. (1967) *Angew. Chem.* (*Internat. Ed.*) **6**, 870.
236. Senent Pérez, S. (1955) *Anales real. soc. espan. fis. y quim.* (*Madrid*) **51B**, 263.
237. Shida, S. (1954) *Bull. Chem. Soc. Japan* **27**, 243.
238. Shida, S. and Fujii, S. (1951) *Bull. Chem. Soc. Japan* **24**, 173.

239. Shuttleworth, R. G., Rapson, W. S. and Stewart, E. T. (1944) *J. Chem. Soc.* 71.
240. Skell, P. S. and Petersen, R. J. (1964) *J. Amer. Chem. Soc.* **86**, 2530.
241. Snyder, L. C. (1962) *J. Phys. Chem.* **66**, 2299.
242. (a) Sondheimer, F. (1963) *Pure Appl. Chem.* **7**, 363 (Review).
(b) Sondheimer, F. (1967) *Proc. Roy. Soc.* **A**, 173 (Review).
243. Sondheimer, F., Calder, I. C., Elix, J. A., Gaoni, Y., Garratt, P. J., Grohmann, K., di Maio, G., Mayer, J., Sargent, M. V. and Wolovsky, R. (1967) in *Aromaticity*, Special Publication No. 21 of the Chemical Society, London, p. 75.
244. Sondheimer, F. and Gaoni, Y. (1960) *J. Amer. Chem. Soc.* **82**, 5765.
245. Sondheimer, F. and Gaoni, Y. (1961) *J. Amer. Chem. Soc.* **83**, 4863.
246. Sondheimer, F. and Gaoni, Y. (1961) *J. Amer. Chem. Soc.* **83**, 1259.
247. Sondheimer, F. and Gaoni, Y. (1962) *J. Amer. Chem. Soc.* **84**, 3520.
248/9. Sondheimer, F., Gaoni, Y., Jackman, L. M., Bailey, N. A. and Mason, R. (1962) *J. Amer. Chem. Soc.* **84**, 4595.
250. Sondheimer, F. and Wolovsky, R. (1959) *J. Amer. Chem. Soc.* **81**, 4755.
251. Sondheimer, F. and Wolovsky, R. (1962) *J. Amer. Chem. Soc.* **84**, 260.
252. Sondheimer, F. and Wolovsky, R. (1959) *Tetrahedron Letters* No. 3, 3.
253. Sondheimer, F., Wolovsky, R. and Amiel, Y. (1962) *J. Amer. Chem. Soc.* **84**, 274.
254. Sondheimer, F., Wolovsky, R. and Ben-Efraim, D. A. (1961) *J. Amer. Chem. Soc.* **83**, 1686.
255. Sondheimer, F., Wolovsky, R. and Gaoni, Y. (1960) *J. Amer. Chem. Soc.* **82**, 755.
256. Sondheimer, F., Wolovsky, R., Garratt, P. J. and Calder, I. C. (1966) *J. Amer. Chem. Soc.* **88**, 2610.
257. Springhall, H. D., White, T. R. and Cass, R. C. (1954) *Trans. Faraday Soc.* **50**, 815.
258. Staab, H. A. and Binning, F. (1964) *Tetrahedron Letters* 319.
259. Staab, H. A. and Braünling, H. (1965) *Tetrahedron Letters* 45.
260. Staab, H. A. and Graf, F. (1966) *Tetrahedron Letters* 751.
261. Staab, H. A., Graf, F. and Junge, B. (1966) *Tetrahedron Letters* 743.
262. Staab, H. A. and Wehinger, E. (1968) *Angew. Chem.* (*Internat. Ed.*) **7**, 225.
263. Streitwieser, Jr., A. (1961) *Molecular Orbital Theory for Organic Chemists*, Wiley, New York, chapter 10.
264. *Ibid.*, chapter 9.
265. For example, Streitweiser, Jr., A. (1961) *Molecular Orbital Theory for Organic Chemists*, Wiley, New York, p. 220.
266. Streitwieser, Jr., A. and Schwager, I. (1962) *J. Phys. Chem.* **66**, 2316.
267. Streitwieser, Jr., A. and Schwager, I. (1963) *J. Amer. Chem. Soc.* **85**, 2855.
268. Streitwieser, Jr., A., Ziegler, G. R., Mowery, P. C., Lewis, A. and Lawler, R. G. (1968) *J. Amer. Chem. Soc.* **90**, 1357.
269. Sworski, T. J. (1948) *J. Chem. Phys.* **16**, 550.
270. Turner, R. B., Meadow, W. R., Doering, W. von E., Mayer, J. R. and Wiley, D. W. (1957) *J. Amer. Chem. Soc.* **79**, 4127.

271. Untch, K. G. and Wysocki, D. C. (1966) *J. Amer. Chem. Soc.* **88**, 2608.
272. Untch, K. G. and Wysocki, D. C. (1967) *J. Amer. Chem. Soc.* **89**, 6386.
273. Öpik, U. and Pryce, M. H. L. (1967) *Proc. Roy. Soc.* **238A**, 425.
274. Van Tamelen, E. E. and Burkoth, T. L. (1967) *J. Amer. Chem. Soc.* **89**, 151; Masamune, S., Chin, C. G., Hoyo, K. and Seidner, R. T. (1967) *J. Amer. Chem. Soc.* **89**, 4804; Jones, Jr., M. (1967) *J. Amer. Chem. Soc.* **89**, 4236; Jones, Jr., M. and Scott, L. T. (1967) *J. Amer. Chem. Soc.* **89**, 150; Doering, W. von E. and Rosenthal, J. W. (1966) *J. Amer. Chem. Soc.* **88**, 2078; Vedejs, E. (1968) *Tetrahedron Letters* 2633.
275. Van Tamelen, E. E. and Burkoth, T. L. (1967) *J. Amer. Chem. Soc.* **89**, 151.
276. Vogel, E. (1967) in *Aromaticity*, Special Publication No. 21 of the Chemical Society, London, p. 113.
277. Vogel, E. (1968) *Chimia* **22**, 21.
278. Vogel, E., Frass, W. and Wolpers, J. (1963) *Angew. Chem.* **75**, 979.
279. Vogel, E., Kiefer, H. and Roth, W. R. (1964) *Angew. Chem.* **76**, 432.
280. Ward, E. R. and Marriott, J. E. (1963) *J. Chem. Soc.* 4999.
281. Ward, E. R. and Pearson, B. D. (1959) *J. Chem. Soc.* 1676.
282. Ward, E. R. and Pearson, B. D. (1961) *J. Chem. Soc.* 515.
283. Waser, J. and Schomaker, V. (1943) *J. Amer. Chem. Soc.* **65**, 1451.
284. E.g. Watanabe, H., Ito, K. and Kubo, M. (1960) *J. Amer. Chem. Soc.* **82**, 3294; Matsunaga, Y. (1957) *Bull. Chem. Soc. Japan* **30**, 227; Mulay, L. and Fox, S. M. E. (1962) *J. Amer. Chem. Soc.* **84**, 1308; Bailey, N. A., Gerloch, M. and Mason, R. (1966) *Mol. Phys.* **5**, 327.
285. Watts, L., Fitzpatrick, J. D. and Pettit, R. (1965) *J. Amer. Chem. Soc.* **87**, 3253.
286. Watts, L., Fitzpatrick, J. D. and Pettit, R. (1966) *J. Amer. Chem. Soc.* **88**, 623.
287. Waugh, J. S. and Fessenden, R. W. (1957) *J. Amer. Chem. Soc.* **79**, 846.
288. Weltner, Jr., W. (1953) *J. Amer. Chem. Soc.* **75**, 4224.
289. Wilke, G. (1957) *Angew. Chem.* **69**, 397.
290. Wilstättor, R. and Waser, E. (1911) *Ber.* **44**, 3423.
291. Wittig, G. (1951) *Ang. Chem.* **63**, 15; Wittig, G., Tenhaef, H., Schock, W. and Koenig, G. (1951) *Ann.* **572**, 1.
292. Wittig, G., Eggers, H. and Duffner, P. (1958) *Ann.* **619**, 10.
293. Wittig, G., Koenig, G. and Clauss, K. (1955) *Ann.* **593**, 127.
294. (a) Wittig, G. and Lehmann, G. (1957) *Chem. Ber.* **90**, 875; (b) Wittig, G. and Bickelhaupt, F. (1958) *Chem. Ber.* **91**, 883; (c) Winkler, H. J. S. and Wittig, G. (1963) *J. Org. Chem.* **28**, 1733; (d) Wittig, G. and Klar, G. (1967) *Ann.* **704**, 91.
295. Wolfsberg, M. (1953) *J. Chem. Phys.* **21**, 943.
296. Wolovsky, R. (1965) *J. Amer. Chem. Soc.* **87**, 3638.
297. Wolovsky, R. and Sondheimer, F. (1965) *J. Amer. Chem. Soc.* **87**, 5720.

ADDENDUM

A new calculation on the nature of the ground state of cyclobutadiene[298] has confirmed the earlier results; the rectangular singlet was found to be more stable than the square configuration by 2·77 kcal. mole^{-1}. The lowest triplet should be square.

In all the cyclo-octatetraene derivatives studied so far by variable temperature n.m.r. spectroscopy, non-equivalent bond-shift isomers are not possible. The photochemical low-temperature (–30 to–50°) interconversion of 1,2-disubstituted cyclo-octatetraenes has now been reported, re-equilibration of the irradiated solution having been followed by n.m.r. spectroscopy. A value for $\Delta G^{\ddagger}$ of 18·8 kcal. mole^{-1} at –12° has thus been obtained for methyl 2-methylcyclo-octatetraenecarboxylate[299].

Evidence has been presented for the first time that [10]annulenes are formed in moderate yields by photolysis of *cis*-9,10-dihydronapththalene at –60°[305]. The n.m.r. spectra at –70° showed the presence of new signals, a temperature-dependent multiplet centred at 5·84 ppm and a sharp temperature-independent singlet at 5·66 ppm, attributed respectively to a mono-*trans*-[10]annulene (I) and an all-*cis*-[10]annulene (II). Hydrogenation of the solution at –70° with a rhodium catalyst produced cyclodecane in more than 80 per cent yield (based on the calculated amount of I and II present). The annulenes I and II isomerised stereospecifically to 9,10-dihydronaphthalenes. The [10]annulenes seem to exhibit no significant diamagnetic ring-current[305].

Important experimental results concerning [16]annulene have been obtained. A single-crystal X-ray structure analysis was carried out[304] and the earlier proposed structure (derived from n.m.r. spectra) has been confirmed. The molecule consists of alternating 'single' and 'double' bonds (mean lengths, 1·46 and 1·34 Å) and is non-planar, the maximum deviation for a carbon atom from the mean molecular plane being 0·57 Å. On the other hand, careful re-examination of the n.m.r. spectrum of [16]annulene at –115 to –150° revealed the presence of two configurational isomers which are in rapid interconversion at room temperature[307]; however, only one isomer exists in the solid state.

It has been shown[300] that the electronic spectrum of [18]annulene as recorded so far has been incomplete. In fact there is a long-wavelength transition of the $^1L_b \leftarrow {}^1A$ type (α-band) at 13,000 cm^{-1}

($\log \epsilon = 2$), in perfect agreement with theoretical predictions based on simple configuration interaction theory[300].

In the field of dehydro- and benzo-annulenes very little new work has appeared. The stability of benzobiphenylenes has been re-investigated theoretically[306], and evidence for the presence of a paramagnetic ring-current in the four-membered ring of biphenylene has been presented, based on ^{13}C n.m.r. spectroscopy[303]. Benzo- and dibenzo[a,e]-cyclo-octatetraenes have been prepared in good yields ($\sim$ 80 per cent) by photochemical isomerisation of the corresponding barrelenes[308]. New dehydro[16]annulenes have been synthesised[301]. The low-temperature (–75°) n.m.r. spectra of some of them (1,3,9-trisdehydro- and 1,3-bisdehydro-[16]annulenes) are due to frozen mixtures of non-equivalent configurational isomers obtained by rotation around *trans* double bonds; as in the case of the corresponding [16]annulene, rapid interconversion occurs at room temperature[302].

Additional References

298. Allinger, N. L. and Tai, T. C. (1968) *Theor. Chim. Acta* **12**, 29.
299. Anet, F. A. L. and Bock, L. A. (1968) *J. Amer. Chem. Soc.* **90**, 7130.
300. Blattmann, H. R., Heilbronner, E. and Wagnière, G. (1968) *J. Amer. Chem. Soc.* **90**, 4786.
301. Calder, I. C., Gaoni, Y. and Sondheimer, F. (1968) *J. Amer. Chem. Soc.* **90**, 4946.
302. Calder, I. C., Gaoni, Y., Garratt, P. J. and Sondheimer, F. (1968) *J. Amer. Chem. Soc.* **90**, 4954.
303. Jones, A. J. and Grant, D. M. (1968) *Chem. Comm.* 1670.
304. Johnson, S. M. and Paul, I. C. (1968) *J. Amer. Chem. Soc.* **90**, 6555.
305. Masamune, S. and Seidner, R. T. (1969) *Chem. Comm.* 542.
306. Meyer, A. Y. (1968) *Tetrahedron* **24**, 6215.
307. Oth, J. F. M. and Gilles, J. M. (1968) *Tetrahedron Letters* 6259.
308. Rabideau, P. W., Hamilton, J. B. and Friedman, L. (1968) *J. Amer. Chem. Soc.* **90**, 4465.

INDEX

24

UNIV. COLL. N. W.
BANGOR
LIBRARY